EXPLORER RACE
ORIGINS and the NEXT 50 YEARS

Zoosh
through Robert Shapiro

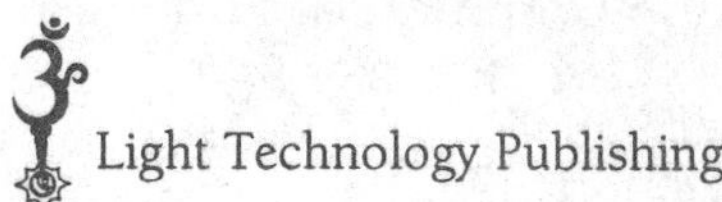

ISBN 0-929385-95-0

Michael Tyree, Illustrator
Margaret Pinyan, Editor

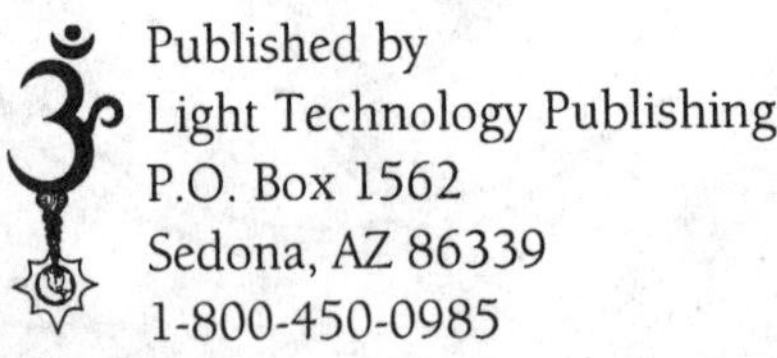

Published by
Light Technology Publishing
P.O. Box 1562
Sedona, AZ 86339
1-800-450-0985

Printed by
MISSION POSSIBLE
Commercial Printing
P.O. Box 1495
Sedona, AZ 86339

Introduction

All right, Zoosh speaking. Greetings.

As you read this book, look within yourself not only for the type of person society considers you to be, but also for the type of person you feel like. You might find that you identify with the qualities of one particular group of people more than the qualities of another group, and that group might or might not fit a description of you. The main thing is to understand your soul's calling, your personality's identity and your true purpose in life.

If a part of the description of one group touches on your purpose, fine. If it is from another group, fine. Do not feel that you must be exclusive to any one description, for ultimately it is intended that the human race blend, not only racially and nationally, but ultimately universally.

Someday there will be a being who will be known as a universal being. It will be made up of Earth people, of Saturn people, of Pleiadians, Orions, Sirians and planet names you've never even heard of on the other side of this universe. This is not to say that individual groups will not keep their identities, but the ultimate purpose, specifically on Earth, of having so many different races and nationalities and even subgroups within them, is that you all draw on each other's strengths and bring out the best in each other.

This book, which reminds you of how special you are, is dedicated to all of you who might find out more about yourselves and in so doing appreciate yourselves and others more than you have in the past.

Other Books by Robert Shapiro

The Explorer Race series:

The Explorer Race

ETs and the Explorer Race

Explorer Race: Origins and the Next 50 Years

Explorer Race: Creator and Friends of the Creator

Explorer Race: Particle Personalities

Explorer Race and Beyond

The Sedona Vortex Guidebook,
with other channels

Shining the Light series,
co-channeled with Arthur Fanning:

Shining the Light

Shining the Light II: The Battle Continues

Shining the Light III: Humanity Gets a Second Chance

Shining the Light IV: Humanity's Greatest Challenge

About Zoosh

April 11, 1991:

Have you been embodied on any other planet as an inhabitant?

No, but I have been embodied as a planet twice. That is part of the reason I tend to have a larger view. I assure you that Mother Earth also has a world view, so I tend to relate more to things on that level.

Can you tell us about your experience as a planet?

I was a planet on Orion that blew itself up because of what the people did on the planet. When I ended that form, the remnants were going to fuel a small, new star. But they were also going to act as a signpost to say, "This is not the way to do it; find another way." Ofttimes a big bang can be useful as a way of saying, "No, this won't work."

So I was that. I was also (in terms of time, you understand) a planet far, far away from here, in another time, that has a tone for its name. If I had a xylophone I could do it better, but it would be close to [he sounds a note, about A below middle C]. It's hard to reproduce that in print, but it was a planet that was associated with the origins of your souls, all the souls that are moving through Earth now and will move into other futures. It has do do with the origins of your souls.

Zoosh, were you masculine or feminine in your planet?

Well, even though there might be some objections, I was no particular sex, all right? If anything, I would perhaps have been the third sex.

Where did you obtain your wisdom?

The same place you all did. From existence. Every one of you is just as wise if not more of a wise guy than Zoosh. It is only that I am not in your world, so I do not have to be ignorant. On the other hand, my Christmas stocking is a little smaller than yours because I don't have the gift of ignorance. So while I might whine about that (like you before this life), wisdom is the natural state of being. Ignorance is the blessed state of being because you are allowed in those moments to have the joy of discovery. Wisdom is natural, wisdom is normal. It ain't no big deal. But because you think it is, you pursue it. And I say, "Have fun. Have a good time."

September 15, 1991:

Long before Robert incarnated in this life, he and I were like this [holds up two fingers entwined]. The arrangement was made for myself, as well as others, to speak through him about knowledge that in my case would help you understand the nature of your true history. This is why

I have referred to myself as the end-time historian.

December 11, 1994:

Without revealing too much of who I am, I will say this much: I have not only seen what you will do in the future, I'm proud of you for doing it.

May 11, 1995:

You once said that you had embodied two planets. I had only remembered one, which was the first one before Earth. What was the other one?

When I talk about having done something, it's outside the context of your time. Thus I could have done something in your future. I will probably be in seventh-dimensional Earth when you are there; I have been anchoring it in your now time. I will essentially walk in, hold the dream together and inhabit the molecules of your image of how you would like seventh-dimensional Earth to be. You will probably, as a race of souls, not inhabit seventh-dimensional Earth right now, but you will visit it. It will be almost like a park; perhaps you could think of it as a reward. As a pleasure, I will inhabit seventh-dimensional Earth as the spirit of the planet, and I will make it possible for any mortal personality who visits there to experience sequentially a full range of all they could be. One could come to the planet, say, for a vacation—but you'd have to be very focused spiritually—and say out loud or even think or pray, "What could I be that I have not yet been, that I would like to be?" and then become it, one after another, discovering expressions of your immortal personality in forms you may wish to accommodate in a future chain of lives. In this way many people will discover that many of the members of the so-called animal kingdom on your planet have essences in other planetary areas; they act differently on their home planets, and they represent true means of delightful creations of one's immortal personality.

That's wonderful. How did you choose that?

Well, I've always been interested in your potential. I've always wanted to support ways you could express it that you hadn't seriously considered. Then I heard that seventh-dimensional Earth was going to be a place where this could be experimented with, and where beings from all over the universe would come to see what they could be that would be compatible with them.

So would you form this planet similar to the Earth prototype?

I think it would be a little larger, maybe about one-and-a-half times this size, which will help to support individuality as I see it. I think there will be a lot of water on the surface, but no great oceans, creating in this way a lot of separate areas of land. There would be rivers that might be two or three miles wide in places, creating the effect of an island, so that when one arrives, either through projection or on a vehicle, one could land somewhere and have a space to oneself to become all these different things without having to run across anybody else in that place. As a result we could probably accommodate at any given moment several hundred thousand individual immortal personalities exploring the full range of their true potential.

Do we have to make a reservation?

Well, I think that would be appropriate!

1996:

Zoosh was ensouling the planet where the soul group of humanity started. This means that those 144,000 souls evolved on your planet, so you have an intimate connection with those souls and with us. Is that true?

That's right.

Can you say more about that?

Not too much because you'll deify me.

I promise not to.

Maybe the readers will.

But we want to know.

The material mass of your souls—and souls do have a material mass—are a portion of my own.

At what point, then, is there a coming together again, when we all go back to . . . ?

There's never a coming apart. That is only an illusion.

So you are with us until when?

I have always been with you, I will always be with you, I am with you now. Understand that when I say the material mass, I mean not only the material mass itself, but the *ability* to be material, the potential.

When will you be free to take your own path?

I *am* taking my own path. I have never been off my path, just like you have never been off your path. All this is your path, and more.

So as we become the Explorer Race and go out to other galaxies, you get the experience of going out to all of them because you're connected with us.

Yes, of course I have been there; I am there now, I will be there, and I will be there with you.

Can't we say the same thing of each of us in that larger sense?

That's right.

You're just not compartmentalized in consciousness like we are because we have this gift of ignorance?

Exactly. I am not compartmentalized. You are speaking from a fixed point of reference. I am speaking from all points of reference. That is why I am necessarily paradoxical. That's why I can generally answer questions about myself as yes *and* no all the time if I choose to.

But that's because there's no fixed truth. If there were a thousand viewpoints from the circular surface of the Creator, you can have all those viewpoints.

Yes, yes; also, there's no fixed truth, and there are many fixed truths. That is true, too. So I am necessarily paradoxical within the mental framework you now have.

Zoosh through Robert Shapiro

When he is not channeling, Robert is a shaman and spiritual teacher in his own right. He lives with his wife Nancy in Sedona, Arizona. He is available for personal appointments and can be reached at (520) 282-5883; P.O. Box 2840, Sedona, AZ 86339.

CONTENTS

THE ORIGINS OF EARTH RACES

THE NEXT 50 YEARS

The Origins of Earth Races

And There Shall Be Negative . . .

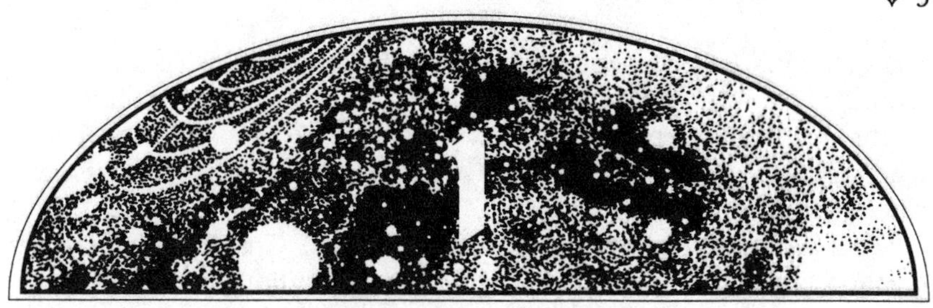

Our Creator and Its Creation

April 2, 1996

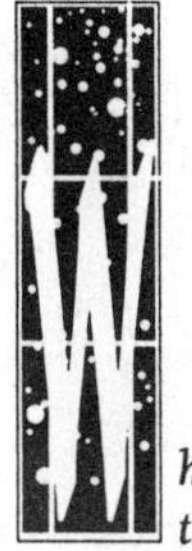

hat did the Creator learn from His tour of the other creations, and how did that affect the Explorer Race?

Well, the Creator was moving about – staring, as it were – before arriving at the glimmer of the creation you now occupy. Creator was involved in the opposite of what this universe is about. This universe at its core is about individuality, which is why the apparent variety and separation is a function here. Although separation is idealized by harmony between all beings, it is still a place of definition.

Before Creator arrived here, Creator was one of the many. And as one of the many, harmony and blending of all beings was an absolute, meaning that there was no "mine," only "ours." Creator is not old, in philosophical or self-expression terminology, but is young at heart. [Chuckles.] And so, not unlike most young people, Creator wanted to do something that She could step back from and say, "This is a good thing," not unlike the words in your Bible.

Harmony through Individuality

At the point of Creator's inspiration that harmony could be accomplished through individuality, Creator naturally needed to move away from the many, which He did. That's how I came to be with Creator – because of my previous experience of singularity, meaning (from Creator's point of view) that I had a working relationship with individuality. And since Creator was going to explore all facets of individuality during

Her creation, then Creator thought it might be advisable (if not at least amusing) to have somebody along in a companionable fashion who might be able to keep up the spirits [chuckles] of all concerned. This is when I joined Creator, and I have been hanging around with Her ever since.

This is probably not going to last indefinitely. Creator's plan for Her creation is going to last . . . well, I give time definitions in terms of what you have already done rather than years, because years are arbitrary. The point when you left Creator on your souls' journeys to become sufficiently individual as well as motivated to apply your curiosity to tasks (ergo, to take action) was really the beginning of your true conception as beings. From that time of leaving Creator to being here now is a length of time, all right? When I depart from your Creator and move on (perhaps with another creator who is being trained right now) will probably be about one-tenth the amount of time or experience that got you to this point.

So in terms of a long time [chuckles], it's one-tenth of a long time until your Creator and I mostly separate. A little bit of me will go on with your Creator, who will step up to Its next level. And you, the Explorer Race, who are being trained to take Creator's place, will get to have most of me with you, not only because we are companionable after so many experiences together, but because you might find it amusing to have me around.

Stimulating Change and Growth in This Universe

As the Explorer Race, you have been carefully prepared for your mission, which is to in time go out into space, initially within your own solar system, but then out into your galaxy and beyond. Actually, the primary function of your travels will be to stimulate change and growth in all you contact. Although to you your mission will seem to be to explore, to discover new worlds and so on, it will actually be, in terms of those you contact, to change their concepts, which are rather fixed, about who they are in relationship to what the universe is.

There's a tremendous level of spiritual evolution already in place in other places than the Earth, so you won't exactly be bringing in any new ideas that they can immediately integrate into their cultures. On the other hand, you will bring something that they do not now have, and that is [chuckles] *uncertainty!* Right now most extraterrestrial cultures are fairly (if not totally) certain that they possess the truth about life.

Now, you can't blame them for this apparent egocentric position, because their cultures have been in motion for at least millions of years and sometimes even more, in terms of your time. These concepts have proven themselves to be absolutely trustworthy, predictable and dependable, so you can see why they might have a sense of belief that their truth *is* truth. Certainly, if a thousand generations have been raised with the same truth and are none the worse for it, if not tremendously better, then they are going to believe in it. Your job, basically, is to spread doubt! [Chuckles.] I realize that sounds a bit extreme, but you have been raised without benefit of the exclusive application of the "truth" as most extraterrestrials know it. You have had that truth available to you here, but it has been a mixed bag, mixed with other truths that were believed to be so – some possibly or partly so, most not really so but apparently so for short periods of time.

You have not been raised in ways that would allow you to feel certain about your experience in life. You have largely been raised with doubts and uncertainty, and you have had to make your own way the best you can. You have arrived here as babies, and as babies you were quite telepathic and could see extraterrestrials, angels, spirits, fairies and so on around you at higher dimensions. Since *you* could see it, you naturally assumed that those larger versions of yourselves could see it, too. Imagine your shock when your telepathic messages to your parents and doctors and nurses and grandma and granddad went entirely unnoticed and you discovered that you had to revert to something basically (from a baby's point of view) on the outer bounds of sophistication – otherwise known as yelling – to get what you needed. Only later did you learn sound symbols, known to you now as language.

So you naturally forgot that you were telepathic because there was no means to apply it to the human world, although those of you who remembered might have applied it to the animal or plant world. Animals and plants have no need to forget who they are so that they can have the joy of discovery and a gift of ignorance. You have ignorance so that you will reinvent your reality. The animals don't need to do this, so they are totally telepathic now. This telepathy takes the form primarily of feeling and pictures, though words will sometimes work in terms of *their* feelings, *their* pictures interpreted by your words. But your speaking to them does not work unless it is accompanied by feelings and pictures. (Just a tip to those of you involved in telepathy exercises.)

Now, your function in your travels through space will largely be to disturb the continuity [chuckles]. You have been very carefully nurtured to do this, and it shows up unpleasantly between individuals from time to time here on Earth. You might ask (reasonably, of course) why we must disturb each other's continuity so much. But you know, when people grow up and live in a school to perform one certain function . . . if everybody, for instance, is raised to be a carpenter, there will obviously be a glut of carpenters and little chance to express those carpentry abilities. Therefore, the true gifts and abilities you all have achieved from your experience here on Earth are not particularly appreciated or needed here. It is only when you travel into space in a later life as a member of the Explorer Race that you will discover that what was a glut experience on Earth is something rare and precious in space and beyond, on other planets and in cultures where certainty is the rule and uncertainty is unknown.

So you will arrive elsewhere with the gift of uncertainty. You cannot give them the gift of ignorance because they believe they already know, and they can show you proof that what they know is true. The only thing you can give them is uncertainty, but you will not give it to them directly. You will demonstrate the *value* of uncertainty. Right now you might question that particular value, but when you go someplace, no matter how advanced the civilization, no matter how magnificent their spirituality or awesome their technology, people there will say to you, "Look at our civilization. Isn't it beautiful? It's been like this for millions" (or billions) "of years" – and it might well be. You will say, "That's magnificent, that's awesome, that's inspiring," and they will smile and nod – not exactly with conceit, but something bordering on it and with perhaps a little too much self-assurance.

The Gift of Doubt

You will be trained or at least prebriefed for your mission when you incarnate on Earth in the future to be astronauts or when you incarnate as an extraterrestrial in that civilization, not as a member of the extraterrestrial culture but as an undercover member of the Explorer Race. You will then, as an extraterrestrial, have an opportunity relatively unknown in these ET cultures: You will possess the gift of *doubt!* You will question your own knowledge, your own wisdom, your own values. You might give an opinion, and then ten minutes later completely contradict that opinion.

That experience is unknown in ET cultures. You will not find any ETs that you will come into direct contact with now or in the significant future who would say one thing and ten minutes later entirely contradict what they have just said, even if it is something mundane and unimportant. That does not happen, because what the rest of the universe is missing is *growth.* In order to have growth, it is *required* that you have doubt or uncertainty.

If you do not have doubt or uncertainty, there is virtually no need to change because even if you have a life lesson, your life will last roughly 800, 1200 years (maybe more, maybe a little less) because life is benign and has no strain, no stress. You might have a minor lesson to learn in that life, and you will take your whole life to learn it because there is no motivation, no constant annoyance or wracking anxiety.

The Invention of Hope

[Chuckles.] But you know, you here on Earth have done a lot. You've invented things that did not exist before. For example, *hope* is an emotion that has no place on other planets because there is no need for it. You have uncertainty here because you have doubt and many other much more extreme things. You do not need anything that involves the continuity that ET cultures have. You need instead to have emotions that will quickly change, adapt and structure themselves to a moving environment.

On extraterrestrial places they already have what they need and want. You invented hope because you are living in a world where you are sometimes in such overwhelming circumstances that without hope —even hope based upon nothing but the desire for an improvement —it would be difficult to go on. So you invented hope, which sprang out of negativity; without negativity there is no need to have hope. (I will let that one sink in for a moment.)

I am not trying to tell you that negativity is God's great gift to all souls. Negativity or discomfort existed long before your creation, and it existed uncomfortably. [Chuckles.] That is, it existed but there was no place for it to serve benevolently. It could serve, but not benevolently.

As you know, creation is not comfortable. It *cannot* be comfortable — certainly Creator cannot be comfortable —with some function of creation that does not work within a level of harmony. Well, negativity didn't, but it existed. True, you are experiencing too much negativity here, but that is temporary; you are moving beyond that. It is always the case that when something is going to be let go of, it comes on a little stronger at the end, like a rainstorm that rains a little bit harder just before it passes. Thus this experience of negativity has spawned several emotions that were either unknown before or applied only rarely, such as self-sacrifice. You might say that self-sacrifice is not good thing, but many times it is applied in such a way that it *can* be a good thing —such as courage that disregards one's own safety for the safety of others, sometimes many others. This was applied here. It was invented someplace else, but it was applied here on a wide range. There have been other creations, but I mention significant ones.

Impetus and Annoyance

You, then, are learning how to be less. [Chuckles.] In your larger concept of yourselves you are not unlike the extraterrestrials and their philosophies. Here you have to stretch from your larger conceptual selves "down" to this density in order to examine in its infinite steps the function, application and (perhaps most important) the responsibility or consequences of creation. To speed up the experience so that it does not take an infinite amount of time to accomplish, a certain amount of negativity was added to the mix so that you would have the gift of *impetus*. Negativity did get out hand, but it does that when it is not universally used in some advantageous way and when you are not paying attention to the subtle messages in your body.

For example, annoyance, however unpleasant, is actually a valuable means by which to motivate change. But if you do not pay attention to annoyances and make changes that will benefit at least yourself, then something more extreme than annoyance comes along eventually, getting your attention and causing you to make those changes.

So in the long run you as souls will, whether you incarnate as Earth people or in other cultures in this universe, take your experience here to other planets and cultures. You will help them to discover that there is more than they realized. You will bring annoyance there, not because you will be annoying, but because you will help to bring about 2% negativity to their cultures, which they *need.* It's not going to get beyond 2%. As a matter of fact, in some of the cultures it will immediately shrink to 1-1¼, maybe 1½% negativity. Only in cases of dire need for growth will it remain at 2%, but this much negativity will give them the gift of annoyance, which will allow them to make changes.

I am backtracking a bit, all right? Most of this is in *The Explorer Race.* I wanted to make sure you understand the basic model behind the Explorer Race concept.

The question was asked about Creator and what Creator was doing. I answered the question briefly to provide you with the infinite things that Creator was doing before Creator was inspired to produce this creation upon which you are now residing and all else in your universe. There was so much variety, and some of it was so far beyond the concepts that you are now dealing with that it would be difficult to explain short of simply acting out chaos. Chaos is the moment before creation, and while it may seem extremely difficult, it is basically the variety that in your now universe precedes the application of individuality or, specifically, definition.

Are there any other creations from the same source that are part of our reality now?

I don't think you understand. This is your Creator's *first* creation entirely on Its own.

Well, if He went around and looked at everything . . .

Looked, yes.

Did He learn anything or say anything like, "Oh, that's a good idea!"?

Creating Something New

No, because most of the other creations She looked at were involved in a function of universality, meaning that there was no individuality. What Creator primarily did on that journey was notice what *wasn't* in the other creations and then base His creation on what was missing – otherwise known as individuality in all of its expressions as well as the illusion of individuality – to achieve the conceptual application of something new. Creator's whole motivation was to create something *new.*

They say there is nothing new under the sun, but you have to consider that there *is* something new.

I can say outright that before Creator's creation, hope did not exist. It is hard for you to conceive of a world without hope, but if you can conceive of worlds where there is nothing unpleasant, there is no need for hope. And yet hope exists now. Creator can lay claim to His creations as having spawned hope, if nothing else. That was Creator's entire desire – to create something new, obviously something of use and value. Yet in order to do it, it was necessary to round up souls that would have depth.

Anybody born on Earth in the time of the genetic experiment on Earth (of which you, the Explorer Race, are the result) has had at least one life of spiritual mastery. So Creator, right out the door, said, "I am going to have a creation that will be the infinite variety of individuality. I am going to foster a portion of Myself that is rooted in curiosity and send it off to explore. Then I am going to take away its memory of all it has learned for a short time" (which is your lives here on Earth) "and give the gift of ignorance so this portion of Myself can re-create. It is My objective," says Creator to the Council of Creators before She is allowed to create this universe, "to create something new. I will not do it directly, but portions of Myself that you know as souls will do so because (a) they will have the depth from their previous spiritual mastery to call upon levels of themselves they are not consciously aware of; (b) they will be deprived of the knowledge of who they are by the gift of ignorance; and (c) because of these two factors as well as the factor of the growth function" (meaning motivation based upon need to change because of discomfort) "they will create something entirely new – perhaps many things entirely new – that will be a gift to My universe as I have created it and in time, I believe, a gift to all of your creations as well."

It is basically this statement that allows your Creator to have the go-ahead, basically a blank check, to create this universe, which your Creator has done, using negativity, a function of creation that had been outlawed before the creation of this universe for a time that is almost immeasurable (in terms of even my concepts). Negativity in application on any planet or on any culture or beings had been absolutely and totally outlawed because all previous applications of it had created terrible, painful experiences – which you know about, of course, because here you are living with it.

But your Creator said, "I will take this outlawed function, negativity, and apply it in only a small area of my universe – on Earth. I will allow some of my most experienced soul beings to live there so that they will, with the impetus of negativity, create new things and find a way for negativity to be expressed throughout all creation in a way that works and provides a gift."

The Council of Creators said, "This is allowed only if negativity proves itself beyond a doubt to have the capacity to stimulate something worthy."

Now, all this might all appear to be conceptual, at the very least erudite, to most of you, yet this was so long ago that many of the concepts (to say nothing of story lines that you all know) were still uninvented. Creator was doing something you all know about very well —bluffing. Even though the Council of Creators *knew* Creator was bluffing, they allowed your Creator to do it because they had faith that your Creator had the lifelines into inspiration that would allow Him to really do what He was bluffing that He could do.

Now, Creators don't tend to bluff each other very much, but because your Creator was young, youth will bluff. Creator's lifelines to inspiration went straight into the Council of Creators, and the Council knew this. Therefore they were willing to allow your Creator to go off and fulfill the bluff, because Creator was told that He/She could not return to sit on the Council or even communicate with it until the promise had been accomplished.

Trying Out Negativity

So your Creator took plenty of time and experience to achieve what has been achieved largely by *you*. Creator created many, many, many planets and cultures with a variety of different people and expressions of beings before He decided to try negativity here, there. You are only the third planet and culture where negativity has been tried in a reasonably successful manner. There have been two other successes, but they didn't last very long because with negativity, instability is a given. But you have achieved something with negativity here; you have achieved a level of stability.

Granted, some levels of negativity are beyond what is reasonable and some experiences are beyond what is all right. Yet even the littlest child two or three years old knows and understands annoyance. And though annoyance might be new to that child, the child will often have the means by which it can cure that annoyance by changing its immediate situation, even if it is to cry out for Mom or Dad (usually Mom) to come and help.

This is an application of the value of negativity in growth. This child does not have the capacity when it is seven or eight years old to call on any recollected past lives. When they are very young they can do that, but it doesn't really help them [chuckles], because mental or even emotional telepathy isn't received by most beings in your culture. When a child is older and asks Mom to help, it is applying that growth concept. Very often Mom helps by showing her child how to do it for itself, giving the child in the early years applications or solutions for its problems.

But most important, it teaches the child that *problems and solutions go together!* That is *very* important, and is almost unknown throughout the rest of creation that a child would possess such knowledge.

Generally speaking, throughout the rest of creation the experience – not just the theory, but the actual applied experience – that problems and solutions go together is applied only by very high and evolved spiritual teachers. But here on your planet in your culture a five-, six- or seven-year-old child knows that problems and solutions go together. That is very important.

Much of what you are doing here is revolutionary. That is why you are in training, in the long sense, to give Creator a coffee break. Creator, as you have known Creator to be, will go on and you will take over Creator's creation, ideally being more than She is, as Creator would like you to be. You *will* be, because you have invented things that were unknown elsewhere.

You have already proved your ability. It's important to understand these things so you can realize that Earth is not a prison, though it has been used that way in the past by some cultures. It is a school of applied creation. People here are not given the option of creation; they are *required* to learn creation. Because problems and solutions exist here, you *must* create. Sometimes they do create problems, but you will also create solutions. That is what you do here. That is why this is the Ph.D. program, not nursery school.

That is also why your lives are so short here, because obviously no one wants to spend a huge amount of time in school. You want to have recess and then go out and apply what you have learned. One hundred years here is a lot, but after a hundred years on another planet you would still be a kid.

In the book you said that the Explorer Race would take over this creation and you would go on with the Creator upward. Now you are saying you will stay with the Explorer Race?

I am not exactly saying that; I am saying that I will do both.

Are you going to split off?

I can't split off because I am not divisible, so I will do both – why not? This is something *you* can do. After all, right now every one of you has other lives going on elsewhere, and yet are you divided, are you cleaved? I do not think so. We can do more than one thing at once.

Is this a new idea, or you just couldn't talk about it before?

Let's just say Creator asked me to do it. So I have got a new assignment.

Keep an eye on the kids.

Keep an eye on the kids, that's right.

You also mentioned that Creator had a connection to a certain inspiration, which somehow connected to the Council of Creators.

Before Creator was jumped out of chaos, where He was —well, it was like a dream where Creator was in the Council of Creators (which I just described) and had that discussion. After that discussion took place, boom! Creator exited and started off on Its path.

Vertical Dreams and Vertical Thought

You know, dreams are not separate from life; they are fully integrated. Just because you do not interpret them easily and completely in your now life does not mean they cannot be interpreted. It is simply that you do not have all the tools that help you fully understand the concepts suggested in vertical dreaming (which you experience now, by the way). You do not dream in linear fashion; you remember only bits and pieces of your dreams in a linear way. But in fact you dream vertically, meaning that everything happens at once and several dreams happen concurrently.

Dreaming is essentially the total connection to not only all that is, but all you have ever been, all you are now, all you will be. It is essentially the visual version of thought. This is what Creator had while Creator was one of the many, but it is a *real* thing. When you have the tools here to learn how to analyze vertical dreaming, you will do so through the function of instinct, or what could also be called vertical thought. You are not so very far from having this tool. Some of you in this room are very close to applying that in its completion and certainly will achieve it in this life.

This generation being born now will certainly, even the densest of them, have almost all of that at the end of their natural cycle. The next generation that will be born in eight or nine years will achieve, at least by midlife, vertical thought applied in their life, which will allow them not only to understand their dreams completely but to utilize them as a means toward inspiration and to disperse excessive discomforts, which is perhaps the primary function of dreams at their vertical level.

If you should happen to have excessive discomfort about anything — even a disease —you can, by utilizing dream technology, spread this discomfort throughout a variety of expressions both visually and on the felt level, and through the use of these unlimited capacities (for example, a person could be in a wheelchair and yet dream that they were running on a beach) disperse this energy and cure the most lingering of diseases. In ten years the generation that will do that in midlife will be born. As I say, you here in this room are well on your way to accomplishing that.

This is the brave new world. The temporary aberration of computers is really a symbolic means by which you externalize your own minds. Your minds are more truly externalized through the dream world. It is likely that psychology is the field of study that will begin to apply vertical dreaming and vertical thought. It is not likely that schools of philosophy will do this, but psychology's future is inevitably tied to vertical thought and vertical dreams. So there is a future for psychology! Much of psychology, you know, has been rooted in looking behind, but vertical thought will allow a tremendous amount of understanding of the present and how choice is rooted in the *present,* not in the past or the future.

Is it safe to assume that all Creators of all creations have a mind of inspiration through the Council of Creators?

Yes. It is not possible to create as a Creator without being given the level of inspiration from those who have done something you have not done—at the very least achieved confidence though multilevel creations. All creators who are applying, who are creating, as it were, or who are active, have a connection to the Council of Creators.

And is there at the end someplace an ultimate creator or a Council of Creators with lines of inspiration beyond itself?

There is no end.

Did they have a mind that had a connection to some thing or some one or some council?

The Focus of Individuality

Eventually everything connects to everything else, but there is no [organizational] chart where something is at the top. If we had to create a chart, it would be a circle, and everything within would be involved with everything else. Creation, we must remember, does not begin with the small; it begins with the large, which is then experienced (at least in a universe of individuality) as an individual personality such as your own. But you *begin* with everything, and you *focus* in one basic thing.

You can hold this focus of individuality for only a short time. That is why you need sleep. Your body has the capacity to function without sleep, you know, but you need sleep because holding the function of individuality is a strain for any soul, and souls need to function without the encumbrance of the physical self within the many. Of course, you do not remember this experience while you are awake and functioning, but that is so you can have the gift of ignorance and thus re-create from a position of not knowing what you have already created.

Since this Creator came up with a new idea of focusing on individuality, are there other creations that work which are set up like this one?

No, because you are still at the experimental level. Also, this Creator did *not* come up with the idea of individuality—it was an inspiration. Inspiration always comes from some greater function of oneself. This does not exclude your Creator from coming up with the idea, as it were, but it includes Creator and many *other* beings. That is the nature of inspiration.

Inspiration, of course, also utilizes need. If there is a need for what you might do, you are given the inspiration not only because you could perhaps fulfill that need, but also because the need exists in the first place. Why fulfill a need that doesn't exist? Obviously there was a need for the expression of individuality. In any event, your Creator was born out of an inspiration of others.

You might ask, what others? Those others are basically the Council of Creators. The Council of Creators was very uncomfortable with the fact that something existed that had no benevolent uses – negativity. The Council of Creators was very uncomfortable having a creation that had to be put in a bottle and corked. It made them kind of wriggle about in their chairs, figuratively speaking.

So they put it out there in the world of inspiration that they would like someone to come forth and find a use for negativity. Of course, it was a *young* Creator who answered the call, because when you are youthful you feel you can do anything. That is quite a universal principle, to my understanding.

Negativity As an Outgrowth of Adventure

Negativity was something that had sort of sprouted or arisen, and it was put aside because they didn't know what to do with it? They couldn't find a constructive use for it, right? This goes back trillions of years.

That's right. Negativity was the accidental outgrowth of adventure. Adventure initially started as a joyous experience, but because it sometimes has pitfalls, negativity sprouted from it unintentionally. In the future you will have adventure with 2% negativity, you know – you don't fall down and break a leg; however, you might stub your toe, but you get over it. At the 2% level negativity will enrich adventure, and perhaps the most beneficial aspect is that it will be fulfilled. It will no longer be a prisoner living in an unfulfilled world, constantly reflecting upon itself and being, naturally, unhappy. It will be itself and be of benefit; it will be loved and appreciated because without it growth would not be possible. Thus it will feel loved and appreciated for being itself, which is a requirement for peace and stability throughout all creation.

How do people grow in these other creations?

That's the problem: *they don't!* They don't grow at all. They have reached what I call stasis, having achieved a certain level of perfection within their own culture. Although they appreciate perfection in other cultures even if it does not apply to their culture, they've achieved a level of perfection so they don't grow at all!

But that's in this *creation, isn't it?*

All other creation created by all creators anywhere anytime achieves a certain level of perfection and then *stops!* Needless to say, the discovery of a level to apply negativity that stimulates growth (you have a *very* strong growth curve here on Earth) is a revolution! It is going to change all creation elsewhere. In all creation in all universes, they won't, of course, immediately slap in 2% negativity. In some places it will be an infinitesimal (decimals out ten thousand points) bit of negativity, and in other places it will be the full 2%, but it will absolutely and totally initiate growth in some places.

Needless to say, the Council of Creators pays a significant amount of attention to this little old planet upon which there are very few individuals living, when one considers all of creation everywhere in all universes. What you are doing here is unknown anyplace else, so this is where the action is. That is why you have so many ETs here in their ships. This place is a school; you are simply living here in your school. People come from all over, some even traveling fifty years to get here just to observe your culture in action and take samples of the air of emotional extremes back to their culture to be studied (safely, of course) in isolated laboratories. Most of them don't take fifty years to get here, but a few of them do.

Are there certain places that ETs from specific areas go to on Earth, like one race might go to South America and another to Africa? How do they know where they need to go?

They are allowed to come here if you on Earth are resolving one of their problems. Even if their makeup might resemble one race or another here and they might drop by to see how people are doing who look like them, they go to the areas where their unsolvable problem is being solved by you. They are allowed to come here on that basis.

Eighteen Civilizations Experienced Negativity

How many very high civilizations on the Earth destroyed themselves in this attempt to find a use for negativity? How many overall got to the level where we are?

Oh, about eighteen civilizations got to this level, some of them significantly beyond yours technologically and a few even spiritually. For example, one civilization on Earth got well beyond where you are spiritually, but the obvious took place: They simply moved on up in dimensions and were gone! They didn't hang around to become the Explorer Race because they had the capacity to *leave* — so they did. It was at that point that Creator decided, "Wait a minute — I don't want to let this happen because it is self-defeating." That's when Creator clamped down a little in terms of applied limits.

But they weren't aware of the purpose of why they were here. They didn't say, "I don't want to . . ."

No, they weren't aware. They thought they were here to achieve spirituality at its highest level, so they moved on. Only *after* they moved on were they informed why they had been here. By then they were sufficiently expanded so that it didn't upset them. [Chuckles.] Creator and they had a good laugh together. It was like, "Oops!" — one of those oops! laughs, you know; it is kind of amusing at higher dimensions.

So it was prohibited to us, the opportunity to expand?

Certainly! You *are* expanded. You have to contract to come here. You cannot possibly expand beyond what you are in your natural state, okay? In order to come here you have to compress yourseif significantly. You are not some lowly little being – that is a joke. You are compacted, as it were; you are compressed in order to explore and apply actual principles of creation in the creation lab. Here if you create something that doesn't work, it doesn't simply become a theoretical function. If it doesn't work – such as the initially promising idea of atomic energy – you are immediately or soon after confronted with actual examples of why it doesn't work, which you then get to apply in some other way.

Am I telling you that you cannot expand and rise up off of this earthly plane? Of course I'm telling you that, because it is your nature. You are *already* expanded. You have risen up well off of this earthly plane, and you have to struggle to squeeze yourself into what you are now in order to forget who you are, so that you can create something entirely new. What's the point of rushing back to being what you *were* when you had to stand in line to qualify to come here to become *more* than you were? But I do understand your motivation to do it.

So the irony is that we're trying to remove the negativity on the planet when that's the one thing that we actually have that's of value.

It's not the one thing you have of value. You now have about 46% negativity on Earth, and you do not need more than 2%. So feel free to reduce it.

I want to get back to those eighteen civilizations. Was it the first one that expanded out of here?

The ninth civilization expanded and went on up. The eleventh one expanded and went on up, rethought it and moved elsewhere. In discovering what they had done, they decided to try out Creator's idea elsewhere. But it didn't work.

The civilization on this planet has been in a constant experimental process. Can you say that we are the ultimate result of all the learning that took place? Is it like a laboratory experiment where we keep refining and refining it – this doesn't work, that doesn't work? Are we the ultimate result, or have there been finer examples of humans before?

Oh, no, you are the ultimate result. You have shown that limits, when applied judiciously, can achieve the desired goal. The key for Creator was learning just the right amount of limits, just the right amount of inspiration and so on. You know, it's trial and error even if you are Creator. When Creator creates something that doesn't work, He can't just throw it back into the pot. She has to live with it until it goes on and takes its next step. So Creator just sort of sits by and watches it, and when it takes its next step, then Creator goes on and starts again. Creator chose to do this in a basically linear fashion so that Creator could assimilate the full personal lesson for Herself.

Of course, Creator's idea of linear might be a little different from your own. Creator can have a multiplicity of things going on at the same time and would consider that linear, whereas for you linear is one thing after another. But that's another story.

So our focusing on creating, looking at consequences, refining our lesson here is like the Creator's – slowing it down to look at it and then learning from it.

Yes. Originally, Creator was applying in the second dimension this attempt to achieve what you have achieved. Creator thought it was necessary to do it in the slowest dimension, but in time He discovered that it wasn't necessary to slow it down that much and that it was possible to achieve this in the third dimension. Creator doesn't like to slow things down any more than necessary, so after a few flubs in the second- and third-dimension applications, He eventually spat *you* out.

Will we be able to continue in the fourth and fifth dimension?

You do now. Why should you be denied it? You have to remember that time here has *appeared* to be one moment after another. But it's being curved now, which is why it's possible to do things much quicker than you used to do them, not only technologically but on the applied level. You learn and achieve things faster than you used to since time is beginning to curve, which is its true nature. If you are asking whether you are going to be able to move onward and upward, I'd say this: If you took a big, thick rubber band and held it above your head, stretching it down to your belly button, and you let go of the portion that was at your belly button, would the rubber band then spring back up above your head? That's where you are! You are naturally at a very high dimension. You have to stretch to come down to do what is necessary here, which is to do things very slowly. At a higher dimension where you naturally exist, you might conceive of something and pouf! it's there. *Here* you get to understand what happens in that moment of pouf!

Is there a value in our looking into those eighteen civilizations and learning from them?

There is not very much value, because most of them were significantly different from your own. The eleventh one (that achieved significant liftoff in the spiritual momentum) actually went to Sirius and became the negative planet that blew itself up recently. They tried to do it on their own even though Creator said, "No, no, don't do it, I've got it going here." They decided to try it without Mom and Dad – Creator.

Oh, they had to go back down to third dimension, and they're going to come up again?

Yes, they're going to come back up again. They discovered the fallacy of attempting to do something without sufficient guidance. When they went to that other planet, you see, their civilization inaugurated negativity, but it didn't have the checks that Creator would normally put on it, so it just went hog-wild.

Aren't we processing Atlantis now?

We are really processing the final days of Atlantis, yes. This whole genetics thing is very Atlantean, but this also preceded Atlantis, you know. Atlantis is blamed for doing a lot of things, but it was basically imitating other cultures. They had the capacity to fly around in ships, which you call UFOs. They picked up a lot of little things from other cultures in space, which they applied. They didn't invent genetic research; they just picked it up and misapplied it.

Did it go to past-Orion?

Yes.

Are we processing on all the different dimensions at the same time right now?

Certainly.

Do we change our focus to the fourth and then fifth even though we'll still be on the third? Is that the way it goes?

No. There is no linear progression. You don't progress from the third to the fourth, from the fourth to the fifth unless there is a reason to do so. For example, at the end of this natural cycle of yours, you might immediately spring off to the ninth dimension on the other side of the universe, because the ninth dimension is everybody's natural place. Some of you might be higher, some of you might be a teeny bit lower, but not much. It's your nature to be at the ninth dimension *at least,* and you have to stretch to get *here.*

So no, there is no numerical progression. You are not small beings who are gradually achieving expansion. No, it's not that—you are large beings who temporarily compress what you are so that you can learn and become more.

The Eighteenth and Last Civilization

Does the eighteenth civilization begin with the seeding, with Jehovah bringing the souls here?

Well, I don't like to limit Creator by referring to Creator only as Jehovah.

I meant the ship that brought the souls here—was that the beginning of the first civilization, or was that . . .

No, that was at the beginning of what you now know as *your* civilization.

Everything that has happened since that seeding, that's the eighteenth, that's us?

That's you.

It's important to understand that the Explorer Race's ultimate destiny is to express yourself as creator. You will go out and be the Explorer Race for *x* amount of time, but then you will re-form as creator in all beingness. Your universe, which Creator has stimulated or created, will contract (I'll bet you didn't know that) and you will become all that Creator has created. You will then take over the application of what is left of your universe to guide it and stimulate it; and you will guide and stimulate other universes.

What I meant when I said you will contract is that you will become more connected with all beings and all life. You will appreciate the focus of beings, and that beings now considered animals in your world are not animals at all. They are not second-class citizens or even third-class citizens, but beings who have very specific skills and abilities and could be said to simply be different from human beings. You will become very connected to all life in your universe. And you will take over for Creator, give Creator a coffee break and allow Creator to buzz on up to what is next. Your Creator really has no idea what's next, but knows that there *is* something next because Creator has absolute proof of something beyond Itself—*inspiration.* When something comes to you from, say, the Council of Creators and you are at the creator level, where did *they* get the idea?

At some point you want to go on up and see what that world's about. Your Creator wants to do that. That's why you are here—so that you can not only replace your Creator benevolently, but give your Creator the feeling of pride, because you are not only as much as She is, but more. And that, as any good parent knows, is the ultimate desire for one's child. Good night.

Visitors from Orion

The White Race and the Andromedan Linear Mind

April 9 and September 10, 1996

Tonight I will speak about a subject I will discuss for the next few weeks. To understand where you all came from, you need to understand who populated Earth in terms of what races came here, where they are from and how you can benefit from knowing this.

As a preamble, understand that those of you who have been on Earth more than once will be different races and different nationalities every time because your soul always seeks variety. The only way your soul can truly understand something is by being it. Your soul does not have the capacity to think or rationalize. Thought, regardless of how many people believe it is so, is not the means by which Creator manifests or functions.

Thought, as you now know it in the linear fashion (from point A to point B), is not even native to the Creator. As a matter of fact, it was some significant struggle to create linear thought and linear manifestation, which means to create on the basis of one step after another. The normal way of creation is to perceive a need, then whatever is needed immediately or soon comes into your life. You do not need to go out and tear down a mountain to make it. You are here to understand moment-to-moment creation, how things move along in a way in which they can be examined. Creator believed that the application of linear thought as well as linear creation would help you toward that end.

I'm going to discuss tonight a little of the history of the white race. This has a wide range because some people might have other influences, other nationalities in their family tree, no matter how lightly or firmly you shake it.

I want you to understand that people of the white race are recent arrivals on Earth. The first arrivals were from Sirius, and their skin was a little lighter than the color of milk chocolate. But recently the white race (in terms of your societies and cultures) arrived, largely from Andromeda, with some aspects or minor arrivals from the Pleiades.

You arrived here because Earth was the last place on your chart, as it were – the chart that's entitled, "How to Find Your Way Home without Getting Hurt Too Badly." So you went to all the other places first because they were more benevolent ("you" meaning those of you who are experiencing that particular race as your own in this life).

Earth was the last place on the chart because everyone knew it's a place of struggle, a place of polarities, a place of great joys but also of significant sadness – in other words, a place of growth. While we all like to grow, we don't necessarily like to do it twenty-four hours a day every day of the week, which is necessary here on Earth. Thus this was the last stop.

Gifts: Polarization, Goddess Religion, Linear Thought, Curiosity and Impatience

The white race brought with it primarily a need to discover its true heritage. It also brought full understanding of its relationship to God through the feminine. When beings of the white race arrived on Earth, they had already become polarized. They had very clear divisions as to who did what. Women were the ones who did the praying, were the medicine people, the mystical people and (should it be necessary) the priests (or priestesses, if we want to be using that word); whereas the men were (if necessary) the warriors, the gatherers, the hunters. There was a very distinct delineation that was not a great factor for other people on the Earth at that time. Thus you brought with you a prescribed-before-birth way of expressing yourself.

The religion that came with you had to do with the stars. This largely grew out of your meanderings through the stars one generation after another to try to find a place that truly felt like home, yet a place that reveals all there is to be known about who you are. As a result, the relationship to God was a functional relationship through the stars, which were considered goddesses. This is not at all unusual, since women had that role. So when the white race arrived on Earth from Andromeda and the Pleiades, God was feminine. It is only in very recent years that the idea of a masculine god has replaced the idea of a feminine god – within the last thirty thousand years. The idea of the goddess (or the feminine god) was tapering off during that time, so they coexisted for some time.

In order to fully appreciate why different races occur on Earth, you must understand that Earth is school. It is creator school, but it is also a place that is meant to be challenging, so school must necessarily have some challenges built in. Therefore one cannot have a continuum on Earth. That means you cannot have a shared, unified culture and still have a challenging place. This is why Earth has had such a colorful history, to say the least.

I'm condensing this for the purposes of our talk tonight. The white race arrived here, largely from Andromeda, bringing with it a more practical application of linear thought than any other race here had before that time, as well as significantly more polarity. This is because of the generational motion from star to star, from planet to planet, and the constant, restless seeking for who you are in relationship to home. It is interesting to note that because you come here, you naturally (it is a given) forget who you are to have the gift of ignorance in order to re-create. Because of this you have the ongoing experience of continually regenerating your cultural history.

One of the problems with linear thought is that it is not based in something permanent. Linear thought, linear communication, linear books and so on are necessarily based on assumptions, presumptions or statements that have been made by others before you. This creates a separation from the natural world and allows for the generation and continuation of mistakes. That is an important function of the school.

You might say, "Why would Creator encourage, much less allow us to make mistakes?" But of course in reality, Creator knew that since you would forget who you are (everybody forgets when they come here), one of the best ways to learn is to make a mistake that you survive and then attempt to resolve. Thus mistakes are a gift, even though they don't feel like it at the time. So within this culture of linear thought and linear creation, entire volumes (to say nothing of cultural identities) can be built on something that is either a mistake or a lie.

This is now drawing to a close. As I said in the first Explorer Race book, the Andromedan thought structure, the linear structure, has been processed by you sufficiently on Earth, and they are reclaiming it. This will allow you to reclaim your natural "thought" function, which is vertical.

When the white race arrived here from Andromeda (and the Pleiades, to a minor degree), most of the people on the Earth at that time were utilizing vertical thought, otherwise known as instinct, but more than just gut instinct. Vertical thought allows you to know what you need to know when you need to know it. And when you don't need to know it anymore, you don't. It literally comes and goes on the basis of your personal need, your culture's need or your clan or your tribe – whatever. So it's with you while you need it, but not when you don't.

Vertical thought allowed these other cultures to have a very close relationship to nature, because all of nature other than man (at least currently) functions with vertical thought, or instinct. This is how animals know how to do things, how plants know when to shed their leaves and when to bud in the spring. It is instinctual; the instinctual body is based upon a reservoir of all knowledge generated by a given species throughout all time. But you're not stuck with it all the time; you have it only when you need it.

So when you arrived here bringing linear thought and in the process speeded up the growth cycle on Earth, you not only brought with you accelerated growth (because linear thought would allow you to make more mistakes [chuckles]), but a need to know who you are.

Before the arrival of the white race, the other races on the Earth did not have that desire. By now it has permeated literature and gone into every corner of the Earth, but even today some tribal peoples who have managed to avoid most contact with modern mankind still have enough identification with the land and the energies of nature and spirit that they don't need to know who they are because they know who they are in relationship to all other life.

The Challenge of Linear Thought: To Stimulate Conflict

The big challenge associated with linear thought is that it allows you to become separated from the rest of living beings. Thus you attempt to discover who you are based only on yourself, not upon your whole world. You see, in reality you cannot discover who you are based solely upon yourself. You can discover who you are in relationship to all other things – that is the nature of Creator. Creator is who She is based upon His relationship to all things. You are children of Creator and you are this also.

Linear thought forces you to make mistakes from which you must learn through trial and error. This allows you to create things beyond what you have accumulated in knowledge in your soul's experiences. In other words, you have to work by trial and error, and in the process you might invent something new to your soul. Because only your thought stimulates that, it necessarily follows that the farther away you are from the Creator, the more linear you are.

Creator knew this was a hazard, but by the time of your arrival it was necessary to infuse into Earth a faster, stronger, more powerful, more distinct growth curve. So you might say that the arrival of the white race on Earth was designed to stir up the pot to stimulate to some extent internal conflict, even external conflict, and to bring about re-creation based upon the gradual move away from emulating Creator.

Now, the other races who were already here had already evolved into an emulation of Creator. In other words, they were primarily wrapped up in their instinctual body, their physical body, their emotional body. They were not at all involved in their mental bodies. In storytelling and in imaginations they could be involved in linear communication, but in terms of what they needed to know when they needed to know it, they were primarily instinctual or vertical.

The white race meandered through space because of its desire to leave no stone unturned to find whatever was under that stone. This has largely led to and supports the technological era in which you find yourself. Although it arrived about a million years ago and civilizations have risen and fallen since then, I need to bring this up because this is your heritage. The white race brought with it an absolute need to know —even beyond reason, beyond the face of danger, even in the face of certain death —what is beyond the next curve in the road. Even if half your people were lost in a landslide and you had to peek around, it didn't make any difference —you *had* to know. Let's just say that you brought with you an undying curiosity.

Curiosity existed before you got here, but it didn't border on self-destruction. By the time the white race arrived, Creator needed to speed things up here because the timetable required for the Explorer Race (of which everyone on Earth is a part) was being moved up. With a flick of Creator's finger you could have easily been nudged someplace else. But Creator allowed you to come here. Because of your evolutionary journey you would necessarily stir up the pot and make things grow faster. You would bring about a very fast growth curve, although a little destructively —which has happened.

I want you to take this in the brightest possible light. People of the white race or those who find themselves a white person in this life (everyone on Earth has been most of the other races in other lives) need to know that you are not the troublemakers of the universe. But because of who you are, where you've been, where you've come from, why you're here and what you're doing, you necessarily are the impatient ones. You have been created largely to be impatient.

An Experiment in the Pleiades and One in Andromeda

You came from Andromeda but also the Pleiades. Once upon a time the Pleiades had a planet that experimented with polarity, and there were a few people who were for it and a lot of people who were against it. The majority did not rule in those days in the Pleiades, and they decided that the idea was a valid one to experiment with to see how it would work. So those who were for it, who represented a minority, were encouraged to go to a distant planet within the Pleiades star system that was supportive of life and try it out to see how it worked.

Well, it didn't work too well. As a matter of fact, it was a total disaster and affected life on other planets in the Pleiades. This created the need for a major cleansing of those planets. The original parties in the Pleiades who gave permission to the small minority who wanted to do this were (how can we put it mildly?) voted out of office quickly. Those who tried the experiment were invited to leave the Pleiades and find someplace else where they could express their interest in growing at an accelerated pace. (As I've said before, polarity necessitates faster growth because you've got to think fast to get out of the spot you got yourself into.)

A similar thing happened on Andromeda. Andromeda is known as the seed of thought, thus they have all the different forms of thought there. The whole thing played itself out there in a thought process. As a result of significant analysis and interplay of various thoughts, another, much larger group was invited to leave and try different places.

The white race arrived here and gravitated initially toward colder places, because you had come for the last hundred years or so through places that were unbearably hot. The idea of going someplace cool was a very refreshing idea indeed. There have been, in the past million years or so, three or four significant civilizations that came and went. Most of the history of the white race, as you know yourself to be, resides in libraries on the Pleiades. Although there is a duplicate on Andromeda, the Andromedans are not particularly interested, even though they do look this way. The Pleiades has many people of the white race and they are very interested, so they have it readily available. They come here and check up on you from time to time, although not so much lately.

I'm presenting this brief overview of the white race because in order for you to fully grasp your purpose here and understand the experience of different nationalities and apparent racial groups on Earth, you must recognize that they have all brought to Earth some very significant function that would allow the Explorer Race to percolate. The white race largely brought a degree of *planned instability* that would stimulate the growth rate, to put it in a nutshell.

What was the journey before Andromeda and the Pleiades? Were we in polarity before?

The journey was to other planets and star systems where significant populations of people who were basically human beings of the same racial context were hanging out—and they were all over the place. That's why it took about twenty generations to get to Earth. Even with ships that could travel through light and time, it took a long time because they had to hang around in each place a while to see what it was like. And everyplace they went on this journey they picked up a little more of their history. The history is deposited on the Pleiades because they have the most long-lasting, continuous, harmonious cultures of all the cultures in your immediate region. History books (or history capsules, as they're sometimes called) tend to be on the Pleiades because the chances of the Pleiades going on as it is is high. It's a safe place to leave your history.

The Journey from Andromeda

Anyway, when you originally launched out from Andromeda, you quickly picked up the folks from the Pleiades (because you were essentially on the same mission, albeit in different ships) and you wandered around. You went to other places where there were significant populations of white people to try and find who you are and where you belong.

This experience took you very far in the opposite direction of Earth. You had to travel in time because you'd never get there otherwise, and most of the ships, with one exception, could travel in time. When a place was far away, all the people would get into the ships that traveled in time and park the ship that traveled at light speed where they could pick it up later, because light speed is nowhere near as fast as traveling in time.

So your first journey, when departing from Pleiades and Andromeda, was to a very distant star system and the planet Triana. With the most powerful telescopes you could not see it from Earth; it's not in your galaxy at all. But it has a very unusual population (it's still there, by the way). It's not polarized, but it has a theme, as many extraterrestrial planets do: the reunification of all living beings in a single moment. This means that the people spend a lot of time meditating with the intention of connecting with all living beings everywhere for at least one moment during every time period (what you would call a day, but longer than twenty-four hours).

In the process, they can in that moment (which might last, in your time, from twenty seconds to ten or twelve minutes) feel utterly and completely all of the light, love and beauty of Creator without any of the chaos. Since unification is their intention and harmony is their apparent means, they have no place in their culture for chaos.

This was largely the culture of Triana. On your journey to discover your other cousins you stopped off there for about a hundred years because it felt good, because you were approved of and appreciated and because you liked those moments of meditations. (And, I might add, that is where you picked up the pleasure of experiencing a high.)

You stayed there for roughly the end of one generation and the beginning of another. The new generation felt that not enough was happening, so, ticking your way through the list, your next stop in a distant galaxy was the planet Obay-tch-tosk-nah.

In all these planets (being benevolent places, of course) cultures were of a continuum—no polarity. You didn't have a chance to experience polarity and they didn't want you to, but they were okay with your living with them for a while.

On this planet the people were involved in something more than in the last place – the recording of the histories of all the races they could contact. This was your first contact with the idea of the Explorer Race. There you had a chance to read a little about yourselves and you had a tremendous amount of time and capacity to study and to encrypt (within the logs of your ships) data and understanding of other cultures elsewhere. This was the beginning of the idea of the library, I might add.

You stayed there about 600 years. You were quite happy there for a time, but after a while you wanted to discover things for yourself (which is your nature, being a portion of the soon-to-be-developed Explorer Race), so you left. You were at your farthest point from Earth, and you started heading back in its general direction.

Your next stop was at a planet that looks from a distance similar to the planet Mars. As you approached, it appeared to be somewhat red because of the soil. On that planet the population lived underground. You hung around there only about ten years because you liked to be in the fresh air. The population was very small; when you got there it was about 600. But the people, though they were physical, were immortals. Understand that your soul, your personality, is immortal, but you had never before encountered physical immortals. (They're still alive, I might add.) They simply choose to function in the physical the way your soul functions in the nonphysical.

This was fascinating for you, but ten years was enough. You felt that the people there had become somewhat egocentric, because they were totally involved in their own beingness. Since you had just come from a culture that had acquired tremendous bodies of history about other cultures all over the universe, the idea of bumping into an egocentric culture didn't interest you very much, and ten years was about all you could take.

You then started heading back in this direction, and fairly soon you got to the Sirius star system, where you discovered a planet that had only about three people living on it who were also physical immortals. You were ready to leave the moment you found that out, but you discovered that they were doing something that allowed them to function primarily as physical-being telephones. They could contact anyone, anywhere, anytime. You found that fascinating, and that introduced you to applied telepathy and its unlimited potential. So you hung around for about sixty years to learn that. You started coming closer to Earth then – or the place where Earth would be. (I'm just hitting the highlights.)

Jupiter and the Language of Music and Applied Magic

You spent a significant time in the planet Jupiter, which at that time had a significant civilization – about 100,000 people – of the same race. (These people are not there anymore, by the way, but they were there for quite a while. They have since moved back to Andromeda.)

When you visited these people you discovered that they had come from Andromeda a long time ago, before you left there. That fascinated you. They introduced you to music as a language, which was also something you found very fascinating – the idea of being able to communicate by sending tones and to retain knowledge through music. That introduced you to artificial intelligence and to sound structure as a means of deep communication. It also got you very involved in the structure of language and its ultimate varietal uses, causing somewhat of a fascination in creating words (which are basically sounds) and applying them in many different ways.

It was also the first time you began to learn about applied magic. Sound has always been intended to be used for benevolent magic – singing to plants and to children to encourage them to grow, and utilizing different sounds (different words) sung in different harmonies, different scales, to manifest a desired benevolent goal. That was your first real exposure to that form of magic and you've never quite let that go. It's even in your myths to some extent now – "hocus-pocus" and so on. That's part of your mythology.

Since Jupiter was reasonably close to Earth, you decided then to come on over. Those are the highlights.

What happened in the other direction? We didn't spring full-grown from Andromeda. Where was the genesis of the white race before that?

Before that, the white race had unfolded, as it were, from the expression of Creator's desire to have a race of beings who would speed things up. Creator pulled out bits of Its impatience [chuckles] and endowed Its impatience with the white race's physical expression.

Now, you tend to think of impatience as a negative trait, but it really isn't. Impatience, though it can be expressed negatively, is also the foundation for *every level of growth* there is. You might say that inventors who've come up with technological creations or even philosophers who've made leaps have done it largely because they were impatient with the technology or the philosophy or the knowledge of the time. There are many benevolent applications for impatience.

The white race was stimulated out of Creator's need to move things along a bit, so the white race is cored in impatience. So don't feel bad if you're impatient; you were not always impatient.

Were there actual lives on other planets before that?

Oh, certainly.

The Beginning on Mars

Where did they start out?

I can't give it all to you, you understand, because there is just so much, but I'll answer that one. Where you started out was someplace that will surprise you: You started out on Mars. The entire Martian culture, both on the surface (bits of which remain today) and under the surface (significant portions of which remain today), is the beginning point for the the white race. You were there about a million years. You were there, I might add, before the planet you now know as Earth was in this solar system. You started out there not only because this would be near the place where the Explorer Race would eventually be fostered and encouraged, but also because Mars has certain capacities.

Mars is a place that encourages and stimulates; it speeds you along. It is not (astrologically speaking) the place of conflict. That might be how it relates to Earth and how the planet ended, in terms of its surface culture, but it was not that when you were there.

You learned about being fragile there during your first experience of physical bodies as white human beings – any human being learns that pretty early on. You survived; you learned how to take care of yourself. You did not start out as primitive man. When you arrived you were very evolved spiritual beings, and there were evolved spiritual beings beyond you to guide you. But because of your natural restlessness toward any established application of knowledge, you have to try new things. Sometimes they're good for you, sometimes not.

For a long time that civilization thrived. Technology began to move into the fields of creation without the permission of that which was being modified. This largely came about as a result of carving stone (some of which are still recognizable today on the surface of the planet), which was done without the mountains' permission. Only then did the planet begin to exude a gas from underground, which immediately forced your civilization to move underground. (The gas, by the way, was oxygen, which was poison to you at the time. Oxygen, you understand, oxidizes, so it shortens the life of everything.) You moved underground to places that were safe for a while, but eventually the gas permeated them. You had ships, so you moved on.

The civilizations were not blown up on Mars. The reason there are just bits and pieces left is that after a million years you can't expect too much to be left standing. But you did start out on Mars.

And then went to Andromeda?

No, you bounced around the universe for a while before you went to Andromeda. What you liked about it and attracted you to it as a race of people was not only that there were a lot of white people there already (you would be accepted – that was necessary), but because the Andromedans had something that you liked very much – different types of thought. At that time the idea of having all knowledge did not appeal to you so much. You were beginning to get a glimmer of the advantages of ignorance, because when you have *all* knowledge, you do not have the joy of discovery at all, ever.

You could walk into a chamber they had there and totally forget everything you ever knew. You could then rediscover in bits and pieces, over as many years as you chose, all of the stuff that you had known and have the joy of discovery every time you rediscovered it. This was an unknown experience to you and was a powerful learning tool. That's why you hung around Andromeda as long as you did. You were there for about 1.5 million years. You were happy there for a long time, but eventually you wanted to have more physical experiences.

The thing about Andromeda and the planets upon which you resided is that the population was stabilized at a very large number. Thus there was a limited amount of what you could do physically on the planet. You couldn't go to wide open spaces because there weren't any. Mostly you could enjoy going into these booths (not too much bigger than a fairly comfortable telephone booth), sitting there comfortably, staying as long as you wanted. You could be supported with whatever you needed, discovering, forgetting and rediscovering things. This might sound on one hand exciting and on the other hand boring, but when you experienced it you could stretch out the moment of joy in that discovery so that it was like a fantastic experience not unlike a high.

If they started on Mars, then how did they get to Andromeda and be already there when they came from Mars?

Well, you wanted to know where they were before Andromeda, didn't you?

Yes.

You got there because the civilization was pretty evolved. To get from point A to point B, all you have to do is create ships that will haul your people, and they did that. Creating ships is not difficult, especially if you have a technology that utilizes the cooperation of materials. Granted, toward the end of the civilization cooperation was being overlooked, but they still had the technology to utilize the cooperation of materials. And when you have the cooperation of materials you can build the hugest, most gigantic, most complex machine that travels through space in about a week, because you don't have to hammer and nail things. You just have to conceive of them, and the materials basically jump into position. So it wasn't difficult to move your populations.

I might add that to this day, the so-called motherships can be so gigantic and go on forever because the materials are cooperating with what they are doing. If a material wants to be something, it will hold that position and be happy with it.

Maybe I'm just slow tonight, but if the white people started on Mars, then how could they have been on Andromeda when they got there? I'm losing it.

I am separating out the Explorer Race version of white people. You have to understand that on Andromeda you have white people, but they do not look like human beings. They look like humanoid beings and to some extent like human beings, but they are very tall. An eight-foot-tall Andromedan is normal; an Andromedan wouldn't know what to do with a five-foot-tall Earth person. Even today they look like a variation of a white human being. You looked enough like them to be acceptable.

In this day and age, no matter how liberal we claim to be, there's still an underlying division between white and black and yellow and brown. Has that always been there?

Arrival on Earth

No, the division of the races (good question) – the structure, the tension between the races – really didn't occur until the white race arrived on Earth. I'm ruling out struggles between planets and referring to an actual division based on race alone. It really didn't happen until the white race arrived on Earth. That was because everybody else on Earth was using their instinctual body; they were fully connected to nature (as you would call it) and were functioning largely with tribal instinct (or animal instinct). They did not experience impatience in any way. They were not fully realized beings per se, most of them (some of them were), but they were not easily herded into impatience.

When you arrived you did something that most other races did not do when they came to Earth: you literally spread out all over the place. You were initially attracted to the cool places (the extreme north, the extreme south, in terms of poles), but because of your desire beyond reason to find yourself, you were in constant search and constant contact with other races of beings. Every time you met someone you would try to understand who you were through who they were. And you would sometimes even emulate the philosophies or manners and cultures of other tribes. You'd try on the shoe to see if it fit or even portions of it fit. If some of it did fit, you'd take that into your culture and move on.

But you also inadvertently passed on impatience and the need to have a self-identity at all costs. Sometimes you would find yourself at war with other cultures; you would do something you couldn't imagine would offend them – emulate their culture – because you did not understand their religions.

You initially created a reputation for yourself as the takers. Sometimes you would assimilate things into your civilization because in needing to understand yourself, you tried to find yourself in the cultures of others. You did this from one star system to another. You were still doing it here, not realizing that your race was created to have that restless need so you would stimulate other people to want to know who *they* are, where *they* came from – all of these things. You were designed to be messianic in nature, to spread your traits around. And in order to spread them around you had to want to go all over.

I know this seems very convoluted, but Creator needed to shave about nine million years of experiential time off the evolution of Earth and the Explorer Race. In order to do that, He needed to input impatience into civilizations.

The primary thing you did was spread impatience amongst other races. True impatience is at the core of racial strife: If you don't have time to hear the other man's explanation of how he got to be the way he is, or if you don't have time to hang around with him for five years to notice, you're eventually going to build up a lot of resentment in him. That's how that happened.

Now, you might ask yourself, how, as a racial group, are people going to discover who they are? Creator does not create something without making it whole. Creator has built in somewhat of a reward for you to discover who you are. That's why Creator has caused your last place to come and look to be a place of so much racial, national and ethnic diversity. It is planned that by the year 2012 publications will appear as fictional accounts written by people of the African race. There will be one, two or three books (it's in the works now) that will help you to understand that your root race, before you were cultivated into being what you are now, is associated with Polynesia.

The Polynesian Connection

The Polynesians have long been associated with other races, but they are a unique race in their own being. They are the white race's direct predecessor. You might say that Creator scooped up a little bit of Polynesian ethnic stuff and stirred in a little impatience, and voilà! —out popped the white race.

For those who are curious and want to discover your true roots, read books associated with Polynesia, Polynesian religion and Polynesian philosophy, and you will discover those roots encoded in them. In Maori culture there is a reference to *waitana*, which has to do with a culture, an ethic, a knowledge, but it also has to do somewhat with the white race. On the surface it is Polynesian in appearance, but it is the springboard from which you came.

As I said before, at least three different novels are planned, which will be produced by people of the African race that will guide you to discover this en masse and allow you to see yourself in a better light.

I have been somewhat blunt with you tonight because I want you to understand that some of your traits that you might consider negative have a positive side as well. I want you to understand that impatience has a purpose and is not something to be rooted out at all costs. And I also want you to understand that in the long run everyone is really united with everyone else, and that surface appearance is strictly a mask you wear so your soul can understand life from a different perspective. I assure you that no soul ever chooses to be one race alone. It will constantly try different races, different types of people, extraterrestrial civilizations that don't look even remotely like human beings—everything. The soul remains restless until it manages to express itself in every way it can.

Do you mean that our souls inhabited Polynesian bodies before inhabiting these Earth bodies?

No, not your souls directly. Let's just say Creator . . .

Oh, I see, He just used . . .

He used the Polynesian ethic, the Polynesian emotional body, the Polynesian spiritual body—even the Polynesian instinctual body—to springboard the white race. This tells you, of course, that the Polynesian race goes way back, and it does.

Before this planet?

Oh, yes. The Polynesian race goes back to the beginnings of what would be identified by the casual observer as a culture on Sirius, as does the African race. Their inception as Polynesians is connected with the definition of the feminine values, virtues and cultural applications. They also perpetuated their culture through storytelling, sometimes called myth. They are the root race of other cultures as well, including many Native American groups, and their constitution (that which makes them up) allows them to apply themselves with significant variety due to their core feminine energy.

Is there something before Sirius?

Well, there's something before everything, isn't there? But I'm going to leave it at Sirius tonight because, while they existed before the galaxy of Sirius existed, they were combined with something else at that time. They became a distinct culture (a distinct ethnic group, if you would) when Sirius came to the point where it could be populated.

If they're our root race, as we get back our vertical minds will we be more like them—more intuitive, more sensual, whatever?

You will be more in tune with the natural world, because Polynesian culture and religion is very much involved with the natural world, meaning life around you—rocks, water, elements, animals and so on. It will allow you to communicate with all life, which you desire, so that other people can tell you what they think of you (usually complimentary, I might add). It will allow you to accrue knowledge. Polynesian culture is about the acquisition of knowledge and its application to support wisdom (knowledge that works) and the willingness to allow in their culture others to use knowledge that does *not* work.

As we go on into the galaxy, then, we will have the technology that the white race represents. Would you say that the masculine part will be integrated more with the Polynesian roots by then?

Yes, integrated more with the feminine, because in order to be welcomed as visitors you must be more than tolerant of other cultures.

You must be interested and supportive of their cultures' perpetuation as they are, even though you want to share your own. You will not be quite so messianic then.

Not so gross as here.

Well, you will not be knocking on their doors trying to convert them, although that is not a racial trait.

With all the evolvement we've gone through, it seems like we haven't really evolved very much. Will we ever know harmony on this planet?

Thank you. You will know it, although it will not be the benevolent, total unconditional love/harmony that one experiences beyond the veil. It will be the experience of such a small amount of discomfort that it aids you to grow but does not provide pain or punishment as it does now. At the fourth dimension Earth will experience only about 2% of negative energy, just enough to stimulate growth by being, at the very most, annoyance. By comparison that will feel like harmony. If that stands as a model of *almost* harmony, then you will have that.

Third-dimensional Earth (you are not there now, but at 3.47) will live on as a school—a school of polarity, a school of extremes—in which theories can be put to the test to see whether they actually work or whether they just look good on paper, as it were. Although third-dimensional Earth will remain a school, the people of Earth (as you now identify yourselves) will not be living there. You'll be on fourth-imensional Earth and reincarnationally elsewhere as well.

Of course, in your immortal personalities harmony is your natural way. Only when you are here at school do you get the experience of disharmony so that you can be forced to grow. I'm not trying to paint disharmony as a good thing, but to explain its purpose.

People are finding their own way now instead of being involved in large groups. Is this disharmony that seems to be scattering us right now meant to make us grow as individuals, as opposed to being part of a large group?

Yes, the level of disharmony that exists right now is largely designed to force you into each other's arms, as it were, because an organism cannot function if it has a disease so prominent that it is overwhelmed by it. Right now your societies are sufficiently diseased that you can no longer turn a blind eye—you have to do something about it. Whereas religion in the past functioned to teach a certain amount of morals and ethics, it also (in the beginning, at least) inadvertently taught separation and elitism, which have been perpetuated to encourage people's political devotion to a particular religion. But this is changing.

Mass education and mass communication allow you to see how similar you are to everybody else, and it breaks down the barriers of elitism because you are immediately informed about what's going on on the other side of the world. The reasons are not always explained, but at least you can see how similar everybody is. These kinds of stimulations invariably cause people to notice their common ground, and the moment people start to notice their common ground they are less attached to being separate or elite and more open to being a portion of the whole.

If the Earth is going to continue to function as a school, there will be waves of Explorer Races graduating. Is that the plan?

Exactly. That is the plan, because at some point in time your souls will be satisfied with what you have done as the Explorer Race and on Earth. At some point in time you will want to go on and do something else. Nevertheless, the idea of the Explorer Race is a good one, and it will continue to have new graduates. Explorer Race individuals will go out, stir up life on other planets and encourage growth as the Explorer Race. So yes, there will be further graduating classes to come.

On that note I want to say this in closing: I discussed a little bit about the white race tonight. I also want to talk about other races. Many core races have truly changed, influenced and added spice, as it were, to the Explorer Race. As I said, without the arrival of the white race, it would have taken another nine million years for the Explorer Race to evolve—but it would have evolved.

The Nordic Polynesians and the Orions

I will speak for a short time only about what has been referred to as the Nordic race. The Nordic race is not a root race.

They're part of the white race?

We've already covered the white race. But I will allude to the Nordic race because publisher is a member in good standing of the Nordic race (though she does not wear a helmet with horns). This group of individuals is largely rooted in the white race and also has strong connections to the Polynesian race. You could say, "Well, take me to Sweden and show me the Polynesian features." I would say, "To be perfectly honest, there's not much left of the Polynesian features except in the bridge of the nose. There is a slight connection there, and in the lower spine there are few nerve bundles that directly relate to the more adventurous members of the Polynesians (specifically, the Maori people associated presently with New Zealand). The Maori people absorbed the culture in New Zealand, which still exists at a higher dimension, and, being durable and determined people, were involved with some individuals who emigrated to the north country. The Orions came in later, but originally the Nordic race was primarily Polynesians who went north.

The Orion Colony and Its Adaptation

When certain elements came from Orion, the Polynesians who went north had become fairly well established in the Nordic country near the seacoast. Several Orion ships offloaded quite a few years ago, and these people had lighter-colored skin, features identified with the white race. These individuals were set to do battle with the already established Polynesians.

However, the ships were unexpectedly called back to Orion, so were able to offload sufficient equipment for the Orion beings to survive only about five Earth years. Cooler heads (if I might make a pun) amongst the Orion expeditionary team realized that if they had a shred of hope to survive beyond that point, they had better find a way to make peace with the Nordic Polynesians.

Well, they had to really stretch their philosophy to do this, because it was the Orion philosophy at that time to establish a source culture that was strongly biased toward Orion. It was a culture they believed in to the extent that nowadays would be referred to as a religion, but it didn't have the religious trappings of today—no churches, no specifically identified singular deity. It was very deep and entrenched rather than rigid. But here they were in the not-too-enviable position of knowing . . .

It was survival, or your beliefs.

That's right. Of knowing that they could live without the Nordic Polynesians for five years, but that would be it. The Nordic Polynesians were crystal clear that they could just wait out the Orions. They had figured out by that time how to get along, how to survive, how to make it work.

So there was an outreach from the Orions to the Nordic Polynesians that said, "How can we bring our cultures together so we can live in peace and—oh, by the way, how can we survive here?"

What kind of numbers are we talking about here?

Not a huge number really, about 700 or 800 . . .

Orions?

Orions.

And how many Polynesians?

Oh, by that time I think it was a balanced civilization, perhaps 3500 Nordic Polynesians. Granted, the Orions had weapons; they could have leveled the Nordic Polynesians, but that was not their intention. And the Nordic Polynesians realized that they didn't have to do battle with these "invaders" at all. They took one look at them and laughed to each other and said, "They'll never make it." So they thought they would ignore them until they went away.

But when they were approached by the Orions, they decided that they would be willing to educate these people, but there could be no barriers. There would have to be total integration or nothing. After much consultation with leaders of the Nordic Polynesians, the chief scientists of the Orions, who were the leaders, basically burned their bridges. They looked at the equipment they arrived with, including food, and said, "If we use this, we'll miss it the more when it's gone." So they burned their bridges: they destroyed their equipment. This was not done as an order; it was a vote taken among the Orions, who agreed they would do it. They knew they were taking a tremendous risk, but they said, "We're not going to be able, with what we've lived with (advanced technology), to be able to turn our back on this without missing it terribly, unless we simply do away with it."

So they did that. It was quite a brave act. And they didn't lose any people. They had to adapt to eating things they'd never thought they would eat, but they also kept enough basic necessities so that they were able to help out initially in creating shelters to support and sustain the primitive survival culture.

The Orions had one thing going for them at that time. They were unusually potent, speaking physically, meaning that the idea of birthing twins or triplets that would survive as healthy babies was common. Multiple birth was quite rare before these beings arrived. I believe they brought with them a genetic bias toward multiple birth. This is how their population grew so quickly.

The Nordic Polynesians took on these newcomers as a wise grandfather or grandmother would teach a child. In a way they became shamanic teachers to these new beings. Eventually —within a few years, actually —the Orions (cum Nordic race) began to develop a deity that grew out of the best and most nurturing qualities of the Nordic Polynesians. The deity had to do with strength, survival and, ultimately, endurance. In a very significant way it was a tribute to the wonderful qualities that the Orions admired in the Nordic Polynesians. This was the actual beginning of the conceptual practice of "looking up to Thor."

Over the years some of the Nordic Polynesians emigrated back to the South Seas, and because of the blending of the races, eventually the appearance of the Nordic races became more predominantly white as it is today. It is interesting to consider that the first Nordics, if we can extrapolate a little bit, were from Polynesia and their skin was significantly darker than . . .

The Orions were white?

Yes, *those* Orions were. There are many Orions who have darker skin, which is probably one of the reasons the Orions who arrived thought that these Nordic Polynesians looked rather like many other beings, especially those associated with the old royal classes on Orion, whose skin was darker.

So there was no god named Thor who came down in a spaceship; it's that they built the myths out of their culture and the Nordic Polynesians?

It had more to do with their admiration of the Nordic Polynesians than visitors from space, which they had from time to time. When possible, the Orion ships would come back and check up on their colonies, but by the time it made its first trip back several thousand years had gone by. Of course, nobody remembered where they had come from because the people felt by that time that they had always been there.

So when the Orion visitors from space arrived, they realized quickly that the Nordic people were shy, not hostile, but apparently had no recognition or memory. Remember that the Orion colonists basically burned their bridges and did not pass memories of home on to their younger generation because they felt they had to give the young ones the will to survive, which had to do with feeling they were in their homeland. A colonist who is not in his homeland always has that backdoor feeling that they can always go home and forget it. But if you burn the ships, the crew tends to forget about going home.

That's beautiful.

Many people on Earth do not realize what a tremendous impact the Polynesian race has had. I thought it might amuse you to realize that even that far north they had an impact. This is largely because the Polynesians are very adventurous, and at their root (specifically, the Maori) have no fear. This does not mean that individuals are not fearful of things, but that at the root of their culture fear does not prevent forward motion.

And the Vikings were totally different?

Oh, I am including the Vikings.

The Vikings are an offspring of these Polynesian Orions?

I'm including that because the Vikings *followed*. The Vikings are Nordica; they followed. Certainly they were driven by that "can do," if I might use that philosophy.

That's a wild genetic combination – Orions and Polynesians.

It's an interesting thing, you know. But even in families that have managed to keep their lineage reasonably intact, if you looked at the facial features of an average Maori person you would say, "Well, these features are not that far afield from what you might see in the Nordic peoples even today." There might be some slight difference in skin color, but the features themselves are very similar.

It's important to know how the different races contribute to the makeup of the Explorer Race – the ingredients, as it were. And also, to some extent, to remind those of you who will be reading this of traits that sound similar to yours or that feel like portions of you so that you will remember your lives as other races a little bit (maybe in dream time) and those experiences. This will further increase your experience of your mutual common ground here on Earth. Good night.

Return to Andromeda

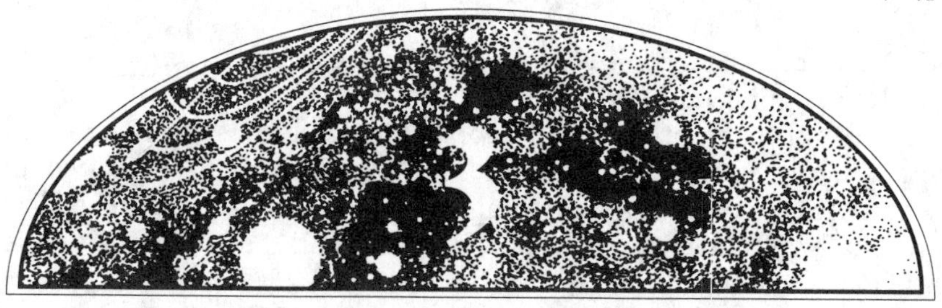

The Asian Race, the Keepers of Zeta Vertical Thought

April 16, 1996

Tonight I am going to talk about the Asians, or what Western people sometimes call Orientals. These people are very pivotal in your world right now. This is a race of people who owe their heritage, past and future, to a most significant star system that is involved with you in intimate ways. The people from this particular star system did not know until recently just how intimate those ways were. But now they know, and even day by day (or time period by time period, they might say) they are accumulating more knowledge of the intimate connections between them and you. These people are from Zeta Reticuli.

These people, by the way, are not "Grays." That term is really insidious. It is a very clever way of identifying an extraterrestrial group of citizens as a racial group on Earth. The reason I am using the term "race" is that I want you to get accustomed to the idea that race is not so much a factor of genetic heritage as an extraterrestrial point of origin.

The reason people look different here is that their point of origin has to do directly with other planets. That's why very often on other planets people tend to look the "same." They don't *really* look the same, but since you are not one of them, you do not notice the nuances of difference. Even on this planet individuals from different racial groups will sometimes say, "They all look the same; I can't tell them apart." Well, of course that's a flip remark, but in reality it doesn't take too much effort to tell them apart.

Now, the Asian people are truly prototype-oriented. They were some of the very first human beings here on Earth. The Native American people and many of the tribal people, including those in Alaska and other places, even Siberia, are all connected. Their roots go back to an extraterrestrial origin and are strongly rooted in the Asian race. (I realize the cradle of humanity has to do with Africa and the African race. We will talk about them soon, but not tonight.)

We want to talk about the Asian race because Zeta Reticuli beings have so very much to do with that race. Do you know that even now many of you on Earth (*at least half, if not two-thirds* of you now on Earth in terms of soul lineage) have had past lives or will have future lives on Zeta Reticuli? That is an unusually large percentage, so you must understand more about that race. (In the book *ETs and the Explorer Race* Joopah talks more about them.)

I want to talk to you about this because your linear mind owes its direct connection to Andromeda, although your *actual* mental body is more directly associated with what I've called the instinctual body of vertical knowledge, and vertical knowledge is the *exact form of knowledge* native to Zeta Reticuli. A Zeta Reticulan being has a very massive brain in comparison to the rest of the head; this is not so much to store knowledge (as one might store it in a library), but to access vertically all wisdom that exists.

The Zeta Mental Focus on Science

Zeta Reticulans have chosen for many years to be mentally and scientifically oriented. That is not only because they have specific and personal interests in those areas, especially science, but also because they have access to all wisdom that has ever been experienced as well as all knowledge, which is a considerably vaster field. Thus they must deal with how to balance all of that knowledge. One of the ways they have chosen to do this is by choosing a specific field—science and the knowledge associated with science—which allows them to remain in a fairly narrow field. Compared to the humanities, for example, or even languages, not that much science knowledge has been accumulated over the millennia.

Few races of people or beings have done much with science compared to the arts, humanities, languages, history and so on, which have been more interesting areas for beings. Science is a small enough field that the Zetas are not overwhelmed by the knowledge they can tap into. Their large brains are like circuit breakers designed to fend off (that is a term opposite to your own mental concepts) too much knowledge and to screen out, as a filter might, everything except that which suits their current specific interests.

That is the nature of your true mental body in a little more intimate detail than I have previously given you. Before, I have always given you the nutshell description, "you know what you need to know when you need to know it." You are mentally attuned with the Zeta Reticulans' experience because there is so much knowledge (consider the origin of the universe and beyond) that a mental being who is focused through the mind and living for and applying mental concepts to life as an individual or society would have to put forth a tremendous amount of effort to *not* be overwhelmed by this ability.

Can you imagine the incredible distraction you would have if you could sit down and have all the mental knowledge that ever was? You wouldn't have it all at once, of course, because there would be chaos, but you could have it in sequence. It would be so fascinating, so incredibly seductive in its splendor and magnificence, that you would be very hard pressed to get anything done in your life. You would spend your days listening and seeing with the mind's mental imaging a panorama of amazing thoughts and actions based on mental contrivances.

Now, this tells you that the most important function (if you are going to have a sane life in which you can accomplish something) is to filter out everything other than what you are currently dealing with. When you have this knowledge available to you, you naturally need to be specific. To some extent the soul functions as a filter in your life here on Earth. But when you have a more unlimited life at higher dimensions such as the Zeta Reticulan beings have, the soul does not perform that function much because it gets to be involved spiritually, sometimes concurrently.

You might, for instance, have a physical life (and a mental and emotional life) while simultaneously bilocating as a single individual (as the Zeta Reticulans can do) to your spiritual plane of existence. For Zetas this spiritual plane is rather like another planet where they function in their spiritual tasks, lessons, duties, applications and so on. All Zeta Reticulan beings can bilocate this way. Their ships can also bilocate (and have been seen and even photographed doing that) because they are made of the same organic material as the beings themselves.

The Asian people – those who exhibit those traits in this immediate life although they have been other kinds of beings in other lives – have a past and a future that is directly associated with Zeta Reticuli. It is only recently that the Zeta beings have become fully aware that they have a direct racial connection with Earth. They have learned fairly recently that you are their past lives, and that was quite startling to them. They didn't know until *very* recently that they are also *your* past lives and that the people to whom they can minutely relate are those of Asian racial heritage.

Now, I have stated that Native Americans are connected here (some Native Americans and native Alaskans) through certain traits. In order to truly understand what the different races are doing here and why they are here, you need to know that they are connected to different parts of your universe and that they have different jobs.

Relinquishing the Job of Holding the Vertical Mind

The Asians have had a job for many, many millennia, which is to hold as a mass group of souls the actual mental body of all human beings incarnated on Earth (which, as I've said, is correlated with the Zeta mind). That means that all men, women and children who have Asian roots or even direct traits and who are born, living on, living out their final years and everything in between, are performing a function within their lightbodies. This function has a direct connection to the Zeta mental body and to the human being's (for that matter, your soul's) actual mental function. To some extent this job or trait has so occupied the Asians that they sometimes have a need to expand their lightbodies beyond the limits imposed on human beings by Creator (to quickly learn the limits of theories when they are applied).

So this extreme challenge, which has absorbed 40% of all Asian beings' lightbodies, is now being lifted. As your linear mind passes back to the Andromedans, who are reclaiming it, you are now beginning to move into vertical consciousness. You are becoming forgetful, and while you don't necessarily remember past-oriented things, you still have your imagination for potential future-oriented things. But that will also change. You will eventually be able to project into the future things that you want to happen. You will actually be able to tap into energies in the present to *stimulate* their happening! This is your true nature.

The Asian people are being relieved of their responsibility to hold that mental body, and what once absorbed 40% of their lightbodies now absorbs slightly less than 37%. As a result, the Asian people will begin to notice over the next thirty to thirty-five years a sense of greater capacity, and everywhere they are concentrated (in Japan, China and so on) will become the commercial, philosophical and artistic centers on Earth. These people have accomplished a great deal in spite of the handicap of 40% of their lightbody having to carry your true mental bodies. As they are relieved of that responsibility they will be able to exercise their full capacities and talents. If you had to walk around with a forty-pound pack and suddenly were relieved of it, you would feel light on your feet and would be able to do many more things. It will be just like that for them; they will become very accomplished.

This does not mean they will be better than you. Their handicap is now being removed and it will be completely gone within the next nine years. There will be a transitional stage as your true mental body eases back into all of you. During the next nine years the Asians, long operating at a handicap, will become lighter on their feet, have more applied acumen and be able to do more while seemingly not working as hard. So it is a good time to be an Asian [laughs]. Because of this extra work they have taken on, they as a people will be able to communicate much better with each other. For a long time there have been tensions between Japan and China and between China and Thailand and so on. A lot of these tensions will melt away. Because of the increased availability of the lightbody to accommodate and assimilate so much more, the people will see themselves very clearly as one people. They will form philosophical, political and (perhaps more to the point) economic alliances that will literally allow them to become the new world in many ways.

Ironically, those who have emigrated will want to go home; even the children of those immigrants will want to visit and take up residence there. It might be hard to imagine that some of these places will change like that, but they really will. Most of you will live to see quite bit of this, and some of you in this room and in the readership will live to see all of it.

I wanted to tell you this because Creator rewards those who do Creator's labors. Understanding the connection to Zeta Reticuli must tell you something else if you are Asian: that the Zeta Reticulan beings have in the past made some spectacular errors, not unlike human beings. They had a time when they experimented with emotions, and because of their tendency to put their total focus on one thing, they became too emotional. After this they put their attention on the mental body, where it has been for some time. They are, in many senses, too mental now.

I have to tell those of you who are Asian beings that because of your direct connection to the Zeta Reticulum planets, these traits are somewhat genetically present for you, although as a mass of beings you do not all have these traits. So guard against becoming obsessive. I don't think it's something to worry about. I think that one can easily become obsessive no matter what racial group you are. But as you become more influential as a world power, please try to guard against the desire to have everyone emulate your better traits. Let them be themselves, and that will enough.

I am talking a little about the different races because the more you understand the roots of the Earth physical bodies, the human-being forms (you are much *more* in your spirit bodies), you will understand that certain traits are literally programmed into you genetically. Zeta beings have traits programmed into them, so if you are born as a Zeta Reticulan being (as many of you have been in the past or will be in the future), you will have certain traits besides being a humanoid: You will be brilliant beyond present human conception. You will have a tremendous opportunity to function in a community that accomplishes great things by working together, although you might not have as much opportunity to be an individual. Although there will be things that you do as an individual, you will achieve a much greater sense of purpose and happiness functioning in groups.

For you to fully experience yourselves, to truly feel the nature of who you are, you must quite well understand where you're from. That is most important. Although you have different lives on Earth (one time you will be an Asian, another time an African, another time an Englishman and so on), all of these experiences are, racially speaking, preparing you to understand that the actual shape and presence of a person *do* affect to some minor degree and occasionally to a greater degree (comparing a Zeta being to an Earth person) your actual approach to the world. Genetics does make a difference, and appearance and function also make a difference.

I am not nominating Asian people to become the brains of your world, though with application some of them have managed to become just that. Your labors and services have been duly noted, and you will receive your rewards in this world. You will certainly enjoy your talks with your teachers in other worlds like everyone else does, but you will receive your rewards *here*. I want you to think about that so nobody passes over before their time. It is a good time to be an Asian, so don't kiss it off just yet.

More About the Zetas

In order to fully understand how you are connected to Zeta beings, one must truly understand the Creator's plan for you. In many senses you are on loan here from the Zeta Reticulan people. You know, they've always been connected here on Earth. That's why they have been coming here for so long, why they have such respect and regard for human beings. That's why they send expeditions here, although they don't come here as much as they used to. They have accumulated a great deal of knowledge about human beings even though they do not always understand what you do and how you behave. They are fascinated because they see something in you that they as a society have been unable to accomplish. They once experimented with emotions, and that experiment was a failure (in their own eyes, at least). Looking back at it from my perspective, I can see where they might have considered it a failure. And yet you *are* their past lives (those of them who travel through time to these times) and their future lives (those who travel through time from their past to your present).

They have the utmost regard for you. They are trying to understand where they went wrong and what genetic traits were added to you that allowed you to succeed where they failed so utterly. They have not fully understood that by choosing to focus so completely on one thing, that focus functioned as a filter to exclude many other things. You have been able to accomplish some levels of balance with your emotional body here because your spiritual, physical and mental bodies have a great advantage that Zetas do not need and have not needed for a long time – the instinctual-body function that serves survival. This function heightens the senses and sharpens you to inspiration. Simply put, it allows you to change course in midstream and do something more constructive as an individual or society.

Sometimes societies make instinctual changes, but sometimes it is individuals or small groups. Because Zeta society is so stable and has been all along (except for their experiment with emotions, which was a very short time), they do not need a function in their instinctual bodies to survive. Focusing on emotions is largely what caused them to fail when feelings were their primary trait. They had never been at risk before, and focusing completely on emotions exposed them to self-destructive behavior, which eventually ended that experiment.

They did not have the capacity for survivability within their instinctual body, nor did they develop it after that experiment, since the experiment was regarded as a failure and the lessons from it were integrated only as far as to say, "Emotions aren't safe, so let's keep them restricted to calm and peace and we'll be all right."

Because that was the only thing they got out of the experiment, they didn't learn how to gradually assimilate the value of survivability as an important factor in the instinctual body, where the emotional body communicates with the physical body. They have not acquired that to this day or even into the future, including all of the futures of Zeta beings who come up to that time.

Humans can be marching along in the woods and suddenly hear a noise. Without thought or analysis you will immediately take an action that protects you. That protection usually proves unnecessary, but occasionally it is very necessary and helps you and others in your family, clan, tribe or society to survive.

The Zeta beings have been testing you for years to try to find that genetic trait, but because it is actually an instinctual trait, they didn't find it. Thus their obsession with mental knowledge has excluded the possibility of discovering by scientific means why you have been successful with emotions and integrating the emotional body here.

What Asians Can Do to Help Zetas Achieve Creator Status

Because the Asian people are so integrated with the Zeta beings, I am going to give you Asians a special task. I would like you (those of you who would care to do this) in your meditations, in your thoughts, in your imaginations, to help your Zeta brothers and sisters. Imagine them being able to react for the sake of their survival and the survival of their people. Picture them walking in the woods when a sudden storm breaks out with thunder and lightning, and see them recognizing that some threat is present and immediately seeking safe shelter and protection. They are so closely connected with you that you as individuals (to say nothing of meditation groups) can be heard by them on the mental, the spiritual and even on the feeling level. They can hear you, and you can change their lives and help them to change their society.

It has always been the Creator's intent that the Zeta beings become able to demonstrate a full range of emotions and instincts as well as utilize their magnificent brains. When they have this one lesson – which Asians can help them get by reaching out to them – they will then be able to fulfill their purpose for being, which is to function as a living, breathing, feeling, spiritual computer.

Right now they are so close to becoming their own creator that they need only a little help. Should you give them that help, they will be able to reach their ultimate destiny in the next ten years, transforming their culture into a combined spiritual culture such that their separated, bilocated spiritual life will take on a form of its own, become its own creator, achieve creator status and begin creating its own universe! And you Asian-culture people can help them do this *in ten years* instead of the three million it would otherwise take!

I offer this to you as a worthy thing to do and bid you good fortune with it.

Does this mean that the Zetas will transit from the fourth to the fifth dimension?

I don't think the Zetas have a culture right now that is functioning at anything less than the fifth. As a matter of fact, I think they are at about five and one-half. They haven't been in the fourth dimension for a very long time.

Will the Zetas be making a particular effort to connect with Asians because of this possibility of becoming a creator within the next ten years?

Thank you. That's the question I have been waiting for. Yes. They will be hanging around areas that have a significant racial connection to the Asian culture. You will see their ships; you might even see them in your dreams. They will be around. They know there is something you can do for them. *Now they know exactly what,* since we're revealing it to both of you at the same time. They will not be annoyingly present, tugging on your skirts, as it were, but they know that enough of you will do this so that they can make this quantum leap.

This does not mean that their physical civilizations will cease to exist. It means that their separated spiritual life that has been bilocating, creating a separate planet for itself, will have a birth of its own and combine itself. Even though their physical life will go on, their bilocated self will literally become a creator and create its entire universe. To be a Zeta person when that's happening would be an extraordinary event, because then your spiritual life, instead of being a singular effect or even a cultural effect on another planet of a higher dimension, will literally become a creator. It is a wonderful time to be a Zeta being!

Because of a one-mind race, will that mean a part of them is our Creator?

Well no, not that creator. There is more than one creator. They are not *really* one mind; they simply have the capacity to be unified and work in unified ways toward a single objective or from a single objective applied to different things.

They have achieved complete and total mental unity, but they do not have to be this all the time. I don't like to say they are one mind. It makes them sound like automatons, which they are not; they do have an individual life. But because they are so *good* at union, they do it.

Does that mental unity help them in this extraordinary experience that's coming up for them when their spiritual self becomes a creator?

It will help them. Those Zetas who do not readily understand the conversion of their separated spiritual life to creator status will be able to assimilate it more easily because of their ability to have total union mentally. Individuals who might not otherwise understand what's going on will be able to move beyond the feeling of apparent chaos and randomness in their spiritual level and understand through a mental model of what is actually happening.

Next time we will continue talking about core races of people on Earth and I will discuss at length those aspects that affect you, your planet, your people and your purpose now as well as in the future. Good night.

Water From the Heart of Earth

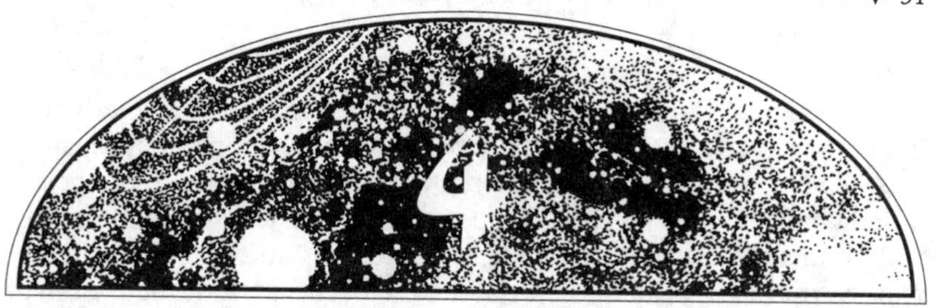

The African Race and Its Sirius/Orion Heritage

April 23, 1996

Tonight I'm going to talk about the African peoples—if there weren't African peoples on this planet, *you wouldn't be here!*

This planet Earth, as I discussed in *The Explorer Race*, has its origin in the Sirius star system, where there is a significant group of individuals who have a milk-chocolate-colored skin and variations thereof. Not all of them look like human beings; as a matter of fact, most of them don't. But they are humanoids, at least the ones I am talking about. Sirius is the origin of the darker-skinned peoples, and this planet Earth, in its third-, fourth- and fifth-dimensional apparitions, is also from Sirius. (When you get to the sixth, seventh and eighth dimensions and beyond, the planet that used to be in this location but was destroyed is still here in those higher dimensions. Thus Earth does not need to be here at those dimensions.)

In *The Explorer Race* [chapter 11] I told about the origins of the male body and the female body and so on. I like to give you little bits and pieces [chuckles], so I've got another bit for you tonight. I didn't tell you earlier that the prototype male body that was intended to be the prototype male human body here was also created on Sirius. It's not the body you have here now, but its origin was in Sirius. That body is an important factor in your history.

The Sirian Male

The original human-being male body did not look like what you've got here now. There was very little body hair except on the head, and it was dispersed similar to the hair on a woman's body. There were other differences: For instance, the musculature was less significant. The male human body can become heavily muscled and rounded with age and overindulgence. This could not have occurred with the prototype Sirian human-being body because that body had a lot of functions that were different. One of the most significant was that the reproductive system was entirely internal, emerging only when it was actually necessary to reproduce the species. Thus sex for enjoyment did not exist the way it does now. It existed largely ceremonially and without the drive you now have. There was only one testicle, which was fairly small, and semen production was about one-tenth what you have now. As a result, an individual could reproduce only once or twice a year tops.

That was the body that was intended to be the male physical body on Earth. The feminine body that exists on Earth right now is the same body that was intended to be here. [See in the appendix "The Body of Man" and "The Body of Woman."] This history is very important; I'm mentioning this in passing, but later you'll see why.

Going back in time to about a million years ago plus about 10,000 years before that, from time to time there was a lot of tension between the Orion constellation and Sirius, sometimes erupting into actual hostilities, although not that much. In the early days of human population here the actual body was that star body from Sirius – that's how people looked. The male was largely *protected* by the female! [Chuckles.] Yeah, a little different. And even to this day some of that tendency exists. The genetics are still sufficiently similar that this protective attitude, even fierceness, is present in the woman to protect the man even today. Many of you have seen this.

That is because the man lived largely in a position of a ceremonial entity. He was not as physically grounded as the female of the species. He primarily provided a very powerful spiritual connection to Mother Sirius and also acted as a link to all other stars, times, dimensions. In a word, the male of the species was, in personality, dreamy. I want you to understand where certain of your traits come from. Are they an accident? No, there are no accidents here; nothing is by accident.

That circumstance prevailed for some time, where that version of the male existed alongside your now version of the female of the species while life on this planet was relatively benign. (We're talking about life at higher dimensions, where the prototype versions of the Explorer Race were being demonstrated.) This was happening *before* the planet that was in this position was destroyed. What does that suggest? It suggests that the original prototype human population that was grown and encouraged and nurtured to become the Explorer Race *did not start out in this galaxy* [solar system], *but in Sirius*. After all, if you're going to create prototype bodies of human beings and an entire civilization, you're going create them in one place. You're not going to kick them across the other side of the universe and have them exist there if all the support systems are in *your* area. Thus the Explorer Race started out in Sirius and you were there for some time, where there were relatively benign circumstances.

At the time the masculine prototype changed there was a significant conflict going on between Sirius and Orion. The Warrior League from Orion was both fascinated and revulsed by the Sirians' male human prototype. These feelings come into play later; that idea of simultaneous fascination and revulsion is a theme even today amongst many men. (I'm mentioning these things on purpose.)

That fascination/revulsion had to do with the fact that the prototype Explorer Race male was dreamy and was not involved even spiritually in repelling an attack. Thus this planet where the Explorer Race was being nurtured on Sirius was fairly vulnerable to invasion. The Sirians were fairly strapped as to what they could do because they knew that you were supposed to establish your own culture, evolve in a pattern different from their civilization, have the opportunity to experience polarity so that you would grow, and then create your own outcome to events.

If it had been one of their ordinary planets, they would have come to the rescue and beaten back the Orion ships very quickly, because they had a weapon (really a defensive instrument) that takes the energy being thrown at it aggressively and, not unlike judo, uses it to repel the attack. If, for instance, a weapon beamed a hundred-million-gauss energy at someone, the energy to repel it would return by that same amount and possibly slightly more. This kind of weapon functions defensively and tends to return to the aggressor exactly what the aggressor sends. It is very effective, because the enemy is faced with their own weaponry. It's a pretty advanced defensive system. But they could not use it on behalf of the prototype Explorer Race because they knew you were supposed to evolve on your own.

These prototype beings looked human. The females looked like the women on your planet today except that they had darker skin. Several males were captured by the Orions because the Sirians couldn't exert their normal means of defense; they weren't supposed to interfere with you too much. The geneticists on Orion did not harm the people who were captured, but they analyzed the means, methods and function of their genetic and physiological systems. They decided to add a little something to the stew because these members of the Warrior League were personally offended by the dreamy quality of the male. So they added some of their own genetic stuff, and in the process of adding testosterone, aggression entered the physiology of what you now know as a human being.

If one of the members of the Warrior League from Orion were to pop in to model for you, you would see someone who was quite tall, unusually muscular, and who would, in certain versions of itself, have a significant amount of body hair. Members of the Warrior League at this time functioned largely out of their will. They were converted later by emissaries from the Creator in the guise of a feminine warrior [see *The Explorer Race*, chapter 17], but at this time that had not yet occurred. So they did what has happened many times in the history and evolution of various species in the universe – basically, they interfered. You could say, "Did they really interfere, or did they just add a little flavoring to the soup?" You could look at it that way and say, "Well, they were a portion of the universe. They came, they saw and they changed things." Certainly one might say (at least with the male human being) that this describes in a nutshell technological man – he came, he saw, and he changed what he saw, thereby changing himself, too.

The Orion/Sirian Male Version One

This brings us to the next evolutionary version of the present masculine being that you know as human-being male. It didn't take the Orions long to create their version, because their understanding of genetics was pretty advanced. They knew that in order to make their hybrid version of the prototype male they would have to use the foundation of what was there. It took them about twenty-five years to do it.

They used samples retrieved from those they captured, which was not painful; they took only little bits and then put them back. (They were sufficiently scientific to know they weren't going to interfere with things too much.) Retreating, they created their own prototype, which was made up of the Sirian prototype but dominated by their own genetics. They decided that they would need to do something with it but knew they couldn't just take this version of the Orion-Sirian male back to the Sirian planet. For one thing, they wouldn't be allowed in.

What should we do? they thought. Well, they knew what the plan was. They had an idea about what the Explorer Race was going to do, because this was a universal creation you all held – it wasn't secret information. They asked themselves, "How can we influence this experiment and create a male that is something to look up to and that can not only survive but help others to survive – have strength, courage and bonding?" (This later became a Roman ideal, but started out as strength, courage and the bonds between men.)

The Orions wondered how they could create that, influence the experiment for good and cause it to continue in some benevolent way. They had to think about that for a while. The heat between Sirius and

Orion had died down at that point, so they thought, "Well, we could ask Sirius if we could continue the experiment by seeding a planet in some other galaxy and try to do, with *our* prototype male human being and volunteers of female human beings, what they are trying to do in Sirius."

Normally one might expect this to be abruptly refused, remembering that Sirius and Orion had a long history of struggle. But cooler heads on Sirius thought, "Well, where's the harm? It will bring us together and create a connection between us. Maybe this will be all right; it will create a foundation for peace." It actually did do that, I might add.

So the Sirians said, "Okay, but it can't be in our galaxy, and we'd just as soon it wouldn't be in your galaxy because we'd like to see how they evolve on their own." The Orion scientists agreed, because they understood that the purpose of the experiment was to create something new (although they wanted to include *their* genetic structure within it). The Sirians figured, "Well, we've got our own experiment going on in Sirius here, so let them do what they want as long as the beings they're working with are willing."

They got a few volunteers of the women from Sirius and buzzed on down to Earth with a few of their male prototypes. (The Earth they buzzed to, however, was the Earth that used to be here.) The people then evolved in a kind a truculent fashion on the third planet from the sun in your solar system. They evolved in a not-very-comfortable way because they had been on an evolutionary cycle that had been receiving a great deal of nurturing from Sirius. But here all of the genetic structure of the feminine and about 48.737% of the masculine structure from Sirius was no longer getting that radiated energy from Mother Sirius—it was too far away. And the Orions backed off, not wanting to mess around here too much and not having the capacity then to provide nurturing. So you were kind of stumbling around in this early version of yourselves.

The Orions saw that your civilization wasn't going to make it, so they asked for help. They said, "We don't want to interfere too much, but if someone wants to come in and give us some ideas, let's put our heads together to see what we can do." A call went out and various individuals from different civilizations came down. For starters the Andromedans said, "Let's give them our mind, our linear mind. It'll help us out and it'll help them to work in sequence" (which you hadn't been doing).

So they gave you their linear mind—that helped a lot. Then a few other beings from other places came and added their stuff. The main thing is, they added too much, because that race of beings on the third planet from the Sun started reproducing the history of other beings. (This is what tends to happen when you mess with an experiment too much; you overinfluence it and it tends to reproduce previous experiments, as any scientists knows.)

Well, that's what happened: You accelerated, got very technological and blew that planet up! Although the Sirians felt badly about the people, the planet and everything being killed, sitting back from a distance their tendency was to say, tauntingly, "We told you so. Well, we've got our prototype going *here*."

Right around that time a couple of synchronistic things were happening: The Warrior League on Orion was being visited by emissaries from the Creator in the form of God's daughter and the beings she brought with her, and Sirius was experiencing what amounted to an age of enlightenment. So the Sirians were able to get beyond the taunts and rise above their attitude about the Orions and say, "This is a noble experiment, even allowing for our former feelings about the Orions meddling with it." They thought it was something worth doing, so they changed their position.

This was happening over a few years, during which time the Warrior League was being enlightened, so both the Sirians and the Orions were on the fast track toward further enlightenment. Because of this and the foundation that was building between them (which is now a solid connection – they really get along well), their approach became: "What can we do together to reproduce this version of the Explorer Race experiment?"

The Sirians said, "Well, we've got this planet over here where we've got our version, and we like the way it's going. However, we must admit," said the Sirians in all candor, "that the civilization has not evolved at all. Although the women are protective, nothing is happening. The civilization has reached a status quo, very much like civilizations in the rest of our system and all over the universe where they've created beautiful, loving civilizations and cultures that are so good they don't want to change." They realized that they had inadvertently re-created themselves, but that the addition of the genetics from another culture – one that might be overmotivated to do things – might be added to their prototype to create the motivation for growth.

This is why a planet from Sirius came to your system, which is close enough to Orion and Sirius that their ships could travel here to check on you, and also sufficiently close to Zeta Reticuli so that those individuals could be emissaries and have direct contact with you.

At this point the beings from Sirius realized, "One of the reasons the civilization we tried to create on Sirius to be the Explorer Race and prototype human beings didn't work is that we were contacting these people all the time. They saw how we were and emulated us. What can we do about that?"

Therefore later on they would come and visit but wouldn't have direct contact with you. The Orions did that, too. They asked for volunteers: "Who will come and do what needs to be done for these human beings?" The Zetas, because they were so interested in genetic experiments that it was almost a religious tenet in their philosophy, said, "We'll do it; we'll stay with it. We'll enjoy it because of who we are and what we like to do." The Orions and the Sirians said, "Great!"

This explains why (from their point of view, anyway) the Zetas have been with you in this experiment for a long time. It also explains why the foundational elements of the African peoples exhibit a greater lung capacity to a minor degree. On Sirius the atmospheres and the size of the beings there required a significant amount of breath to incorporate the gases that would support life. The appearance of the core African individual and other core features are all directly related to the tropical planets on Sirius.

The Orion/Sirian Male Version Two

These initial beings that were seeded by both Sirius and Orion became a little different in the second prototype. The first Sirian-Orion prototype had about 48.7% Sirian genetics and the rest Orion, but the second prototype had about 50% of each. They said, "Let's try it that way and see how it works. That way it gives a chance for the Orion traits to be highly influential but not dominant."

Well, civilizations rose and fell, and they tried variations. The female body remained the same – 100% Sirian – but the final outcome of the male body (the one you've had for the past million years or so) has about .0022% more Orion genetics than Sirian, just enough to create a significant influence but not so much as to dominate. That's what you're functioning with. Down through the years, going back a ways, there have been cultures where women have been the warriors, but since the creation of that .0022% greater influence of the Orion energy, all subsequent wars that have taken place have been stimulated, carried out and largely fought by men.

I give you that history so that you would understand the whys and wherefores – why the female body is 100% Sirian and the male body isn't (which is important) and why the true differences between the male and the female lie not only in their extraterrestrial point of origin but in the fact that the male body is more of an experiment than the female body. That's why the male body has a wider range of potentially self-destructive or self-constructive energies.

So the origin of the male is traceable to both Sirius and Orion. The root race is *all* Sirian, so the original prototype beings here on Earth had the darker skin – everybody basically looked African. This went on for some time. Then you had a change in policy directed toward Earth, which amounted to speeding up the experiment. You had proven sufficiently durable to accept more external influences to increase your Explorer Race growth curve.

The Arrival of Other Races

After that, different races were allowed to people the Earth. (I'm talking about the surface; underneath the surface is a wholly different program. When we focus underneath the surface we'll be able to talk about other beings.) About a million years ago white people as we now have arrived on the surface. We also had some short-lived species that were here for a while, variations of human beings who came from other planets, saying, "We'll have a colony here and see how it goes. The colonists who come here will live and die here, leaving their genetic substance. This will add to the stew."

You had several races that didn't make it, and one of them was a race from a galaxy *very* far away that you don't have name for, but who basically refer to themselves as the Siu-ka-tanz-da. These people are highly involved with a curiosity that goes beyond reason. One might say that that's why they didn't survive here, because they just *had to know.* (As you all know, there's a point at which prudence provides a significantly greater contribution to survival than curiosity.) Although they didn't survive, they did make a big impact, bigger than one might expect for that number of people. They started out with a colony of about a thousand people. Because they were so curious, they traveled all over the Earth, exploring all the peoples and animals of the Earth. Thus they touched the lives of everyone while they were here.

They lasted for 10,000 years, not very long in the evolution of a culture. They never had more than 1500 people, but they really made a big difference while they were here because their level of curiosity rubbed off on your culture. As curious a people as you were before they arrived, you were about twice as curious after they no longer existed here. The Creator let it happen because the Creator realized you needed to be more curious. If you're more curious, you are likely to accelerate your growth curve, which you did. So He allowed that; yes, She did.

The origin, then, of the darker-skinned peoples is so rooted in this planet that they truly represent the energy (in walking and talking form) of the connection with Sirius and Orion. Over the years the Earth humans would rely on the visitors to do things. Sometimes, for example, there might be an area that was becoming a desert. When no people were around the extraterrestrials would land, interact with underground spring water, bring it to the surface and create a lake, stopping the desertism. The animals would come, the trees would grow, the cycle of nature would begin, people would be nurtured and so on. They would do things like that.

So over the years there's been a lot of energy going back and forth, but it couldn't have happened if we hadn't had the darker-skinned peoples. They are the prototype human beings (now, I don't want that to go to your heads, dark-skinned people!).

I'm presenting the origin of darker-skinned peoples on Earth because they are *so unique* to the founding of your culture here. They are really functioning as a living, walking, talking connection between this planet and Sirius and to a lesser degree to Orion. If anything ever happened and they decided to return to Sirius and Orion (and had the capacity to do so), over a period of about 8000 years (10,000 maximum, but probably 7500 to 8000 years) the energy coming from Sirius and Orion would gradually halt. Because it takes a while for the energy to travel here, it would take about 7750 to 8000 years for it to cease after they left. That energy is what allows you to function, to be alive here on a planet borrowed from Sirius (which is why it looks so different from the rest of the planets in this solar system) —in short, to exist not only in this solar system but in this galaxy.

To put not too fine a point on it, the dark-skinned African peoples are what's keeping you alive here. Their origin is so rooted in extraterrestrial cultures that the ability for civilization to exist at all on the surface is directly related to them. The other races who started out their cultures on the surface such as the Pleiadians and the Polynesians (the Polynesians are the root race of the white race) would not be able to maintain the planet's energy and keep you here intact.

How much of the Orion genetics were put into the original prototype?

Well, they started out with a little bit of this and little bit of that and had different versions. Originally they wanted to put in a dominant force, but they put in more than 51%. Then they saw that when a civilization got to a certain point, the overaggressive nature of the being would be too individualistic, having a little too much Orion energy and genetics, and couldn't make the leap from the individual to the group.

The 100% Sirian feminine being could easily exist within a group for the group's sake. Females could expand the group, be involved in extensions of the group and more easily adapt to networks. But the masculine being was too focused in individuality, and while the Orions wanted that, they could see that that focus was too fierce, too truculent. It would not, could not change, so they experimented with different variations. That is why the version of the masculine being you have here now is only .0022% more Orion than Sirian. They discovered that this version of the body can make the move, albeit with some difficulty, from the individual to the collective and then to groups, networks, tribes, world families and so on. That is why the specific experimentation went along those lines.

That was after the planet was moved to the new location?

Once the planet was moved out of Sirius, it was no longer getting that direct Sirian energy, so the experiment went on full-blown status. So yes, most of that happened after the move.

The female body is still 100% Sirian; does that mean that the small amount of

testosterone that is in the female body is also Sirian?

Yes. The version of testosterone that the average woman has is also native to the Sirian culture. It allows growth, change; it allows the fierceness that can come up naturally within a woman. It allows various traits of protection and so on to take place. That is native to Sirius. There is enough similarity in the genetics and the hormone system that Sirius and Orion could together create a combined being.

You said that we are dependent on the Sirians sending us energy right now?

Yes, the *planet* is dependent—the surface culture.

Could you expand on that a little bit? I'm having a hard time understanding it.

The Sirian Heart Energy and the Female Africans

Well, I will expand on it within terms that make complete sense. If you take a baby away from its mother and father and family situation and put it into a circumstance where it is not getting love (yet still getting the food it needs to survive, the vitamins, the minerals, the diaper changed and all of that), the survival rate for that child is not much better than 43% tops. That is because the energy of love, nurturing, friendliness, happiness, the whole business within the family, is what keeps most babies alive: simply put, a familial sense, even if the love is not particularly there—brothers, sisters, some context of culture for the child to gravitate to. But if you take that child out of its family and put it into a laboratory, the death rate climbs right up.

I'm using that analogy because you can understand it very well. That energy is what keeps most beings alive and, by nurturing you and allowing you to thrive, *want* to be here.

That still doesn't answer the question: Is it the genetics in the black race that is drawing that love? How would you describe it?

It is not the genetics per se. Between lives people are spirit—etheric bodies and so on—and any person of the African core race now has been many other things. But during the time when a soul participates in the African core race here on this planet in all of its variations, the genetics activates a level of heart energy that is not naturally dominant (meaning it is present in everyone but not a *dominant* force in all human beings). It will activate that heart energy of African core-race persons and they will have a greater sense of expression. When a soul is a variation of an African being, there is naturally a more pronounced emotional body, a more pronounced love center. As a result, they have about 1% greater connection to the spiritual realm; in other words, it creates a slightly higher vibration that connects the individual strongly with Sirius (and to a degree with Orion). For a man, the connection would be roughly 50% (fifty-fifty Sirius/Orion), but as a woman the connection is 100%.

It is not an accident that your reproductive cycle births more females:

It was actually intended that women would comprise about 73% of Earth's population. If cultures did not interfere so much through various rites at birth, you would have 73% (plus a few decimal points) females here and as a result, the energy from Sirius would be more powerful, allowing the planet to be more balanced physically. But because your population is closer to fifty-fifty masculine/feminine, you're not getting enough energy from Sirius (and to some degree from Orion) that supports water resources, love, spirit, emotions, feelings and so on. You're getting just enough to squeak by, but that's about it. Ideally you need to have 73% women.

There has been a great deal of mixture of the black races with other races in our country and elsewhere. Does that help?

Does that have an influence? You know, I want to support your line of thought here, but I can't. It doesn't work that way. You need to have *core* root-race people—African people. Now, there is a minor benefit when core root-race African people marry or mingle with other races. Let's say that the core root-race African being has a *ten* connection. When there is offspring from marriage between races, that offspring will have anywhere from a *one* (available to all other peoples) to a *three* connection. It's a good question.

Obviously, if you're going to improve your quality of life here—[chuckles] I love irony—you need to nurture core root-race African peoples, especially women, because the more root-race African peoples you have, the more energy potential you'll have to receive that nurturing energy that supports, sustains and allows this planet to manifest certain elements within herself that you as technological people have pumped out of her.

Female Energy and Material Manifestation

Say, for instance, that you had what you once had—about 73% connection to Sirius (right now it's about 50%). Mother Earth could reproduce to a much greater degree the oil that you have pumped out. (She uses it, by the way; that is part of her body. It relates to your lymphatic system. Those of you who understand medicine know what that means.) If you had a 73% connection to Sirius, Earth could reproduce her oil, her minerals—gold, silver, copper and so on—at the same rate you take them out. She would be totally in balance, and all the beneficial things she does for you would be very present. Even individuals would be able to manifest what they need much more quickly, and technology would be influenced much more benevolently.

Let's say you want to build a house. You would be able to walk up to a patch of ground and invite the water to come to the surface rather than drill a well. Most likely women would do this. Standing, they would say, "This is the spot where water could come up." (It's a form of dowsing.) At least three women would basically triangulate that spot and touch the Earth. They would go into receptive energy and focus on drinking water that comes out of the ground. They could pull that water up to the surface and it would take them only about five minutes, because that water would *want* to come up and touch these beings.

You wouldn't have to drill a well because the water would pool at the surface. You could use it for whatever purposes you wished. But you would have to have those three people (or whoever helped bring up that water) come and touch that water every day or the well would dry up. That water would come to those original people because of love, and it would need to touch that love at least once a day.

There are a lot of tribal peoples who know this, but that information doesn't get out to the westernized world because it does not have its basis in the mind. I tell you this because if you understand that you stand to *gain* from the inclusion of these feminine ceremonies (and to a large extent these African feminine ceremonies), then maybe you'll do it, at least eventually.

Are they doing things like that on some planets on Sirius?

If it's needed. Although on one hand their cultures are stuck, on the other hand they're focused in a totally loving, balanced place, so they don't need to do that anymore. Initially they did do it, but now they have all the water they need. Their population is balanced, fixed to a certain point, not because someone says, "This is how you are going to do it," but because everybody knows and feels how many people can live on a given planet with that planet's full consent. Everybody there can feel it, because their planet is an intimate portion of themselves. Just as you feel it when someone tickles the bottom of your foot with a feather, they can tell how the planet feels because they're directly connected to it.

That's what is intended for you here. If all of you were directly connected to the planet, you would know what she felt at all times. Some of you can now, but if there were 73% Sirius energy here, you wouldn't even have to focus on manifesting balance and comfort and nurturance for Mother Earth – it would just happen.

It comes to this: Sometimes some of you wish, "If we were all just alike, we'd get along fine; and if those people weren't here we'd have a lot more room." (Lebensraum, maybe? Those of you who are historians know what I am talking about.) You'd have more room, but if "those people" weren't here, *you* wouldn't be here, either!

There are countries where they literally can't afford to raise their children.

Well, they think that. But the more girl children there are, the more Sirius energy you get, and that nurturing energy encourages plants to spring up. Instead of an apple tree dying in the front yard no matter what you do to save it, that apple tree will *want* to be here – it will thrive, dropping apples on the ground, and those apple seeds will eventually turn into trees. Before you know it you've got an orchard without even working, simply because you've got enough Sirian energy here.

The female body, no matter what race she is, brings in Sirian energies. It's true that the core root-race African woman brings in more Sirian energy, but every woman brings in a significant amount of energy from Sirius. No, you can't afford not to have women. The population is suffering because they don't have enough women! We need to get this information out there because people are shooting themselves in the foot. They think they're doing something to help themselves, and in the process they're doing something to hurt themselves and others.

Things don't happen here by accident. Creator didn't set it up to have more women born than men because Creator is capricious. Creator does things for reasons, and just because you cannot figure it out mentally does not mean that there are none. And let's not say, "Oh, that's the mystery of God." You're past that point now, kids; you need to be responsible for your own creations. You're being tested by Creator. Creator whispers, "Can they accept what I've provided for them? Can they accept the fact that women are born more often on purpose?" To the extent that you do not accept that, you reject Creator's wisdom.

Reducing and Now Increasing the Heart Energy

In the past you talked about the heart and how we could have at the very most 10% of the heart energy that was originally put aside for us. Many people are now trying to expand the nurturing energy and the heart energy, so how much do we have right now?

Let's say that every human being on the face of the Earth was core root-race African female—no men. For starters you'd have 100% of your heart energy right there. Even if the men could accommodate 100% heart energy as in that original version, what inevitably happens is that the culture would reach a certain level of happiness—of love, cooperation, nurturing, joy to be alive together wherever you are—and stop growing.

The reason Creator allowed the Orions to step in and say, "Let's throw in some of *our* energy" is that Creator realized it would stir the pot. Of course, with the hybrid Orion/Sirian male you weren't able to have 100% of your heart energy, but your growth curve was accelerated, along with the risk of extreme polarity. But since Creator trusted you to find a good purpose for negativity (discomfort), Creator figured, "The chances are that I cannot trust a civilization totally rooted in heart energy to find a good place for negativity and a means for negativity to express itself and be loved in at least some small degree for it" (which was originally intended to be 2% and later will be).

Creator said, "If we had a civilization 100% focused in heart energy, it would not be able to tolerate, accept or experience negative energy or discomfort of any kind because its natural tendency would be to evolve toward complete balance and love. I've got to make them more durable and stir up the pot sufficiently with the Orion Warrior League energy so that they will be motivated. Yes, they'll make mistakes and fall short of their objectives, but they will discover the value of negativity in stimulating the growth cycle. Ideally we want it at 2% so they don't have any more than annoyance, because that will stimulate growth; we don't want it at 49%." But you're evolving toward that wisdom.

How much of the heart energy do we have at the moment?

Right now you've got about 13%.

Okay, so more is being released; we're progressing.

You wanted a statistical answer, but I like to give histories. Numbers do not necessarily mean anything unless we understand their meaning, you know.

Let's look at the history, a little bit of the male and the female. At one time they were in different galaxies, at least on different planets, and they were deadly enemies. Can you put this in the context of how we got from there to here?

Emotions of Discomfort and the Female Race

Well, I wouldn't say deadly enemies; I'd rather say that they were not compatible. At least, the feminine being was not compatible with a being that could be self-destructive. The core 100% feminine being (basically an instrument of love and creation) cannot abide an instrument of destruction in any way, shape or form. This is why most civilizations (if not all) on other planets that have achieved loving, beautiful cultures cannot abide discomfort of any kind. (It actually makes them sick.) The feminine being could not be around any kind of being who could be self-destructive, even if that self-destructiveness was for a higher purpose—for instance, laying down your life for a friend: Here comes a speeding truck; your friend doesn't see it coming. You grab the friend and pull the friend out of the way, but in the process you fall in front of the truck and get killed. You saved your friend's life, but you were destroyed.

Now, that's bravery—self-sacrifice—and admirable, yet the 100% prototype feminine being (referring to the Earth being you know the woman to be today) could not accept that, could not be around it.

The refinement of understanding here is about the ability to tolerate anything outside a very narrow range of expressed feelings ("narrow range" means that the primary feeling is love and all that relates only to love; even sadness or loss would not be comfortable). The 100% feminine being couldn't live in a civilization where ants crawl around on the surface because someone would invariably step on one, and if one ant were stepped on and suffered (they do suffer, you know—just because you don't feel their pain doesn't mean they don't feel it), the females would feel it, being totally connected to all living things through the love connection, and would themselves suffer.

So they couldn't live on a planet that had much variety of life like that; they couldn't even live in a circumstance wherein life survives off other life. They couldn't eat an apple because that would be destroying something to sustain themselves. That's why it was necessary to mingle a little Warrior League energy from Orion. That increased their durability and also the range of emotions. In the long run (more revelations here), *the actual Explorer Race is women!* The Explorer Race is not men; men are here to help the women. Men are here to help you to expand by needing you, by being a full 360° of emotions – love and joy, yes, but also anger, rage and action.

What do those emotions do in a woman? She can't live on Sirius in the benevolent, beautiful circumstances of nurturing life and love. No, she is forced to react. She's got to bring up something within herself that not only allows her to live or even just survive in those circumstances but to *thrive* in those circumstances. *She must be here!* If we don't have at least 40% women on this planet, this planet will die. All surface life as you know it in terms of human beings would be gone.

So this drives the woman – she knows she's *got* to survive. A woman drowning at sea will hold her baby above water, but a woman out in the desert would survive just to keep her children alive. A woman's will to survive is stronger than a man's. The man is here largely to serve the root-race Explorer Race person, and that at its very core is female African root race (at the very least, female).

Someday you will begin fulfilling your function of sending out astronauts to other planets (eventually, not right away). In the beginning your one-world government will do this. (The one-world government is coming, by the way, so don't fight it too much; at first it might be a little uncomfortable, but eventually you are going to like it.) Initially, the cosmonauts will be men and maybe a woman (perhaps a government will deign to send one woman along). The government will get more enlightened and become more nurturing and supportive for all people – which means that it will not tolerate hunger or misery or suffering. As your planetary Earth personality emerges, you will not be an American or Chilean or Argentinian or Egyptian, but an Earthian, and it will be more feminine. (As you know, the feminine being can feel that energy; she's got a strong emotional body.)

As that happens and as you become more sensitive, you are not going to be able to say, "Oh well, there will always be hungry people." That's a mental concept, not a heart concept. If there were always hungry people, then you would always be suffering whether you had enough to eat or not, because you would *feel acutely* the suffering of others.

The Feminizing of Earth

As the planet feminizes (which it's doing now) and that one-world government becomes more feminine (you're going to love it), cosmonauts will gradually have more women and fewer men. Eventually they will all be women.

I might add that Pleiadians, who have a very strong feminine society, have mostly female cosmonauts. This is because women can, want, will and *must* bring the new version of the feminine being out into the stars to initiate people. So although the core root-race feminine person is rooted in the feminine energy, as a result of her Earth experience of being tested by the masculine being—*really* tested, some more than others—they've had to change and adapt their emotional body. The female's emotional body cannot express only love; it must react to anger, hate and so on. It thus discovers bravery, courage and to some extent self-sacrifice for the preservation of loved ones.

It is *that* version of the human being that can be most easily welcomed by extraplanetary culture (ETs), because ETs in general will react better to the feminine being than to the masculine being, and they communicate more easily with the average feminine being. (I'm not talking about people who are spiritually evolved.)

So the true Explorer Race is women! That's right. I want men to understand that you are not here as some extraneous variation of being here to train women—you're here to *challenge* them, and what a wonderful job you're doing! This does not give you license, however, to be obnoxious. It is a refining process, you know.

At some point in the far-flung future, this version of Earth (or maybe a version around dimension five) will be peopled by about 93% females, and the original Sirian prototype male will comprise the rest of the population. By that time the Orion energy will no longer be required to test, refine and hone the expansion of the expression of feelings and (to a minor degree) thoughts in the feminine being.

At that time the dreamer will return—the ceremonial masculine being. But that's quite a ways off in your future. None of you in your current bodies are going to live to see that.

We've talked about a lot of things tonight. I haven't talked in any great physiological detail about the origin of the African peoples, but I've talked at length about who they are to *you*, what they mean to you and how they affect a significant portion of your population (women and also men). I needed to give you this information now because you're going to have the opportunity in coming days to be even more responsible for your creations. You will need to understand yourselves better, and that is the whole purpose of this book: to help you to understand yourselves, your point of origin and your motivations. Good night.

From Creator, through Fairy, To Child of Earth . . .

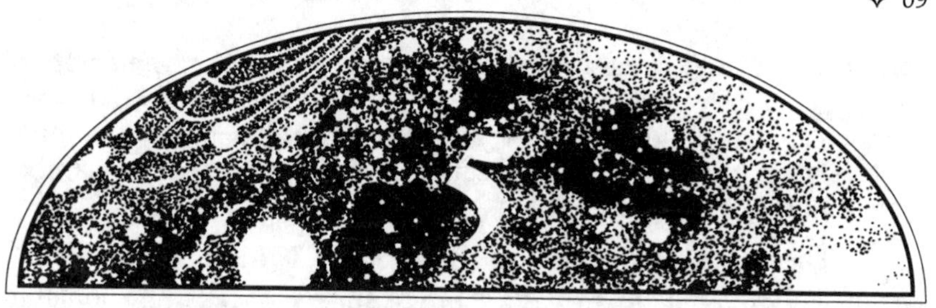

The Fairy Race and the Native Peoples of the North

April 30, 1996

I want to talk about Native Americans tonight. But there's another group of beings I want to talk about first.

Underneath the Earth there are many groups of lightbeings, many higher civilizations at different dimensions. But amongst you now today, although not as numerous, is a group of beings that relates well to those lightbeings who are underground and who have a connection to you. I'm going to include them among the races because they have something in common with the Native Americans: although they are aware that they come from the stars, they're related to a culture that has done a great deal to train many native peoples for their stay here on Earth and for their experience in life.

The first beings I want to talk about are those in the world of the *fairies.* These beings can be seen physically or remain unseen. When they are unseen it is because they are in a different dimension. They can move between dimensions and travel through portals into higher-dimensional Earth under the surface as well as many dimensions of the planet that was once here. They can travel to all planets, all galaxies, all dimensions – everywhere. They are the divine representation of love in a physical form and are readily available on the planet for all to experience. You might not be aware of the experience, but it is available.

These beings are truly the link between the angelics and the Creator Itself. The angelics cannot be on the Earth at all times walking amongst you, as it were; they come and go, they are nearby, but not here amongst you. The fairies (an apt name for them) can be amongst you. If your energies are uncomfortable, they go to a higher dimension or simply move – when too much industrialization comes to places where many of them live, they move on.

The Work of Fairies and the Work of Nature Spirits

Once upon a time there was a more pristine surface population and Mother Earth was more like she once was (about 4000 years ago she was in pretty good shape). At that time there was a population of fairies here exceeding 50,000, more than were necessary for plant and animal species and so on. (Although there were more plant and animal species then, you don't need one fairy for each.) They were here primarily to act as a kind of a connection, a means of passing the energy of the Creator right into physical Earth. They would pass it into the plants and animals and occasionally into children, if the children were still innocent and not jaded by worldly ways. They did this with the intention of helping you to gradually resolve your past lessons.

But when the civilization began to blossom and become basically overpopulated, as you are now, there was less room for fairies on the surface. So they retreated to higher dimensions of this planet inside the Earth. Many are the times they interacted with underground civilizations to prolong the education and the pleasure of knowledge about the surface for those who were being trained there.

Because of their interdimensional quality I can't say that fairies came from a certain place and they're going to another place – they're divine beings. With that thought in mind, there is only so much I can say about them. I'm mentioning them because there is another race with which they closely interact, what I call the *nature spirits,* who dwell all over the universe. These beings exist primarily to teach receptive, sacred-oriented individuals. Here we're bridging into certain Native American populations.

North-Country Native Tribes and Nature Spirit Teachers

I have alerted you before to the idea that some Native American populations are rooted in the Polynesian race, but I want to discuss others more by region, those located in the northern United States – Montana, North and South Dakota, Minnesota, occasionally Wisconsin, basically people of the plains – as well as people connected to them, such as some peoples in Siberia, Iceland and Alaska.

These native people arrived on this planet from many different points. Some are from Sirius, such as the Algonquins; although they are an eastern tribe, I mention them because their philosophy has permeated the highest levels of your United States government and even other countries. And there are the Lakota (I won't say where they're from).

They did not come to the surface but to the Inner Earth or the inner planes first, where they studied for a long time with the nature spirits. The fairies come in at this point because they are corded directly to the Creator, and by just being in their presence, you feel the divinity. Fairies would often be present when the nature spirits were teaching these recently arrived ETs (soon-to-be-native peoples). It was not so much sacred-education basics, because they already had that, but about what it would be like to be a physical person and how they needed to behave so that Earth would produce food and water in the right places to nourish and shelter them and so on.

Primarily the native peoples were taught the abilities of manifestation within their sacred rites. Some of these peoples have been here for millions of years, coming and going through different civilizations and occasionally retreating underground when the surface civilizations got too warlike. So they've had lots of training.

The main training for a nomadic tribe, for example, was in manifestation abilities wherein they went from place to place and blessed the land. (Today nomadic tribes move about primarily because food and resources are used up in a given area or cold weather drives them to a warmer place.) The original purpose for nomadics (the Bedouins are included here because they're also from Sirius) was to bless the land with these sacred rites. They were trained for their individual areas in how to go from place to place and *be human* on the land.

You might ask, "Well, isn't that natural?" But you have to understand that these people were primarily the prototype surface human beings that set the standard for the needs of the human being. There were people who came and went and were looked after by different extraterrestrial civilizations, but the nomadics came from extraterrestrial sources who were not really going to look after them. They would have to do it on their own, so their job was to cover as much territory as they could in the beginning, basically touching the Earth as they walked (barefoot) and being as human as they could.

On the face of it it sounds like a given. But remember that these people would have recently arrived from extraterrestrial points, and in order to be human, which was new to them, they had to practice, to actually explore humanity. They had to discover the range of human feelings, emotions, thoughts, capabilities, inspirations and so on.

When they felt like they were getting too meditative or dreamy, they would stop and do something physical. They were encouraged in their sacred rites to dance, because dance is not only a means to celebrate humanity and spirit and form—and Creator, of course—but it would inevitably involve them in physicality, in visions and in using the divine mental through the imagery system you call imagination. They also had to be emotional, feeling and instinctual; they had to be fully human because they would become very involved in feelings and action.

So sacred dances were initially encouraged: "Go out and be human, because the humans who follow will have needs very similar to your own physical needs. They will need to have food grow, and Mother Earth is perfectly willing to manifest what you need. Some of this will be provided by other civilizations, but some will not. Sometimes you'll go to a place where there won't be any water, and you'll have to dance for it, making rain, making a lake. If it rains long enough, there'll be enough water—if not for you, then for somebody who comes after you." They took their spirituality and their training in sacred rites very seriously. For those who were nomadic (such as the Bedouins, some Plains peoples and Siberians) this was very important.

Thus we had a circumstance where people actually had a specific spiritual duty to do less for themselves than for others. That began the idea of applied shamanism on this planet, where a person would act out a sacred rite for the good of self, family, clan or tribe, but also for the good of all beings, possibly even for animals or plants. They were trained by their spiritual teachers, and they simply went out and did it. Many generations came and went doing that; and even today many tribes perform certain sacred rites for the good of all beings. (This is a very important thing to keep in mind.)

Understand that native peoples are literally on a mission; most people with sacred training know this. Some of the tribes have been decimated due to wars, invasion and so on, and they are no longer here to offer what they have in the past. But some tribes have returned from whence they came. This was always promised. The people who are now established as native peoples came for a while. They were assigned to do something, but it was never on a permanent basis. They weren't required (nor encouraged, for that matter) to integrate, to blend in with the rest of the people and become part of a homogeneous community. They were urged to remain amongst their own kind, keeping up their civilization and sacred practices. So that the surrounding civilization would not accept them offhandedly, they were made to look somewhat different, and their sacred practices also made them stand out.

During the past 900 years some tribes have been gradually disappearing. We know from conflicts and histories, as vague as they are in written form in your time, that some tribes disappeared because they were wiped out by some plague or were invaded by somebody and so on. But that has not been the case with all people. For example, some Pueblo peoples went north and joined other tribes (this is well-established in archaeology and various histories and tribal legacies), but certain Pueblo peoples simply vanished.

Cliff Houses for a Future New Race

You might ask: "Where did they go?" Well, some peoples (not just Pueblo peoples) were called upon to perform a duty and leave relics (ruins and so on) to establish a civilizational pattern needed more in the future than by themselves. A certain tribe of Pueblo peoples established many of these high cliff houses. Ofttimes you couldn't see them from below, but they've been discovered by archaeologists and explorers, and some have been preserved. These people were establishing a basic form of architectural archaeology.

You could say it was for protection, because it could go unnoticed for years; certainly it was very safe. It was also rather inconvenient. If your house was 200 feet above the source of water and you had to climb down and bring up water a bucket at a time (though they didn't have buckets), it would take quite a while and you had to be pretty careful bringing it up. If there were two or three hundred people living there, it would take a lot of climbing and coming and going. And with all that coming and going, the secrecy was lost, because anyone could see all that activity. So that theory has a slight hole in it.

There were other people who managed to catch rainwater, but it doesn't rain conveniently, especially in the desert where it comes when it wishes, usually not when you want it. No, they would have to go someplace to get water.

These high houses were established to suggest something to people of the twenty-first century in the third dimension. Briefly, I will simply say that at the time of this dictation you are now at 3.47 traveling to the fourth dimension (4.0). At some point people will begin a civilization at 3.0 dimension; it will not be human beings as you understand yourself to be. It will be humanoids, but not you, because you are on your way to the fourth dimension while you are alive on this planet. [For further information about this race, see *The Explorer Race*, chapter 22.]

The civilization that's going to come in the 3.0 dimension will inherit a lot of the predictions that have been made about Armageddon and disasters and so on. In place largely within 3.0-dimensional Earth – the soil, the rocks, the air, the energy – is the idea of a great flood a-coming. The flood has been here, and there is scientific evidence to show it, to say nothing of stories in various civilizations. That story has been planted strongly in the third dimension.

It does not actually affect you here now where you are in 3.47. But in your myths and stories and religions (and scientific facts, for that matter) you are perpetuating the idea of great floods coming – "California's going to fall into the sea" and so on. You are at 3.47 marching off to the fourth dimension while you're all still alive, and here old Zoosh is saying that the "I Am America" map, however valid and relevant in its issuance to those who have received it, is primarily meant for 3.0

dimension. And it's happening at 3.0 dimension as we speak.

The water's rising for them [the negative Sirians], although it appeared that it would happen here. You've got about a 98% chance that it's not going to happen to your civilization. You were supposed to put that idea in the ground so that those who arrived in 3.0 dimension would, shortly after they got there (and they're getting there regularly now), have an anticipation building to an anxiety that there might be a flood coming. They are not going to be highly technological at first They don't know much about living in 3.0 dimension, although they have a few teachers, so they need to have significant input on where to put their houses. New arrivals on a planet don't know much about its geography or weather, but they will have to put up some kind of shelter. They might not know about flood plains or have a great deal of geological understanding, to say nothing of visionary spiritual insight.

So here we have cliff dwellings. These people who are arriving at 3.0 have not had much experience in extreme weather conditions, but they are experienced in needing to be secure, in fearing things, in being afraid of invasion. [See chapter 22 in *The Explorer Race*.] Thus the idea of having a cliff house and being unseen to their enemies will be highly appealing to them.

In the past we had a civilization built high on top of a mountain, as if to say, "Look at this; you could live up here. If there's a flood you'll be safe. Build a boat. If you have to walk out your door and put the boat in the water, that's fine. If not, you can climb up and down and be safe. Bring up plenty of food; your enemies will never get to you." That's basically the logic when you look at a cliff house.

The Pueblo peoples of the past have created an architectural archaeology that is meant for the future in a denser dimension. When they built it, they were in 3.0 dimension, but they were building it not so much for their own convenience but for convenience of those who would follow and who they would never meet. In other words, they were performing a sacred rite for someone else, in typical native fashion. In the native culture and teachings, they were taught, and it was even put into their genetic structure, that to perpetuate their own lifestyle, to be fed by the Creator, to be helped and assisted, it would be necessary for them to do things for people they would never meet.

It is very important to keep in mind this theme that runs through all aspects of humanity: cooperation with those you have never met, doing things for people you will never meet. You'll be there for a while, then somebody will come along and use what you built. In a more modern version, you put a quarter in the expired meter of a parked car (a random act of kindness, it is said), doing something for someone who will never know who did it.

The Disappearance of the Pueblo People and Preparations for the Departure of Others

Where did these Pueblo people go? Ten percent went to other tribes, but most of them went back to their home planets. Native peoples all over the world have been expecting to do that for a long time. You could say that when they die off, their souls go back there. But I'm not talking about souls; I'm talking about *physically* going back. Archaeologists have been puzzling about sites that look as if the people had suddenly just left: "What happened to these people? Where did they go?" Some of these people did go back to their planets.

Some of these off-planet people are back here now to prepare the way for the return home of other native peoples. If you're going to be taught how to return to your own planet physically, you also need to be taught how to let go of your Earth duties and embrace your other planetary duties.

A lot of people who returned to their home planets several hundred years ago (basically the Pueblo people) are back here now in spirit form speaking, whispering to shamans of many tribes and sometimes also leaving messages (sometimes cryptic, sometimes obvious) and signs that it will soon be time to go home. A few tribal people have even spoken about this publicly, but the average non-native person doesn't know that the native peoples are getting ready, feeling that something is going to happen.

A lot of the rest of you also have this feeling of anticipation. People are going to come from the stars—yes, you've been expecting that—but these original peoples are also going to go home. They'll go to different points, some to different times. Some will return to the time on their home planet when the original people came several million years ago so that, as the heirs of the people who left, they can carry on life there. They will be prepared for that.

For example, some of these Pueblo people's spirits are preparing a tribe (which will go unnamed) in the Pacific Northwest as well as some people on the Plains for the time in the next generation, maybe the generation after that, when a vehicle will come (a light vehicle that cannot be interfered with) and take them to different points in the galaxy and perhaps to different times. This particular tribe in the Pacific Northwest (and possibly the tribe in the Plains) will be returning to Sirius and their native planet. It's a beautiful place, not dissimilar to here—many mountains and mesas, but very green, with lots of trees, some of which look like bushes and weeds you have here, but as big as full-sized trees.

As the people are being prepared to go there, they are given new stories to tell the next generation coming up. Sometimes the stories are coming to the children. To any of these children reading this: If you have new stories, write them down or tell them to your friends if you feel they are of value and connect you with the brothers and sisters from the sky.

The people who have prepared Earth for the Explorer Race will gradually begin to leave. First it will be the native peoples, who have been there for a long time. Eventually you, the Explorer Race, will go out. In time everybody from fourth- and fifth-dimensional Earth will gradually head back home so that the population of the Earth will then be mostly people who are core-connected to Sirius. By the time everybody goes back you'll be roughly around 4.5 to 5.0 dimension.

Why won't the Lakota tell us where they're from?

It's not that *they* won't tell you where they're from; *I* won't tell you where they're from.

They don't want you to tell us?

That's right, because they want to preserve their stories, their wisdom, their knowledge. It's not someplace scary; it's just private. I'm being vague not because I can't tell you but because I want to honor their special knowledge. I tell you where other people have gone because for them it's all right. They're gone; they don't care.

The Mayans and the Mimbres

What about the Mayans? You didn't mention them, but they were only here for a while.

Those who survived basically migrated. Some went under the Earth and stayed there a long time. Later they came up farther north and integrated into what is now northern Mexico, becoming the basis for one of the current root races there. Some of the others migrated north along the Sierra mountain chain into Canada.

How about the Mimbres?

These beings are still amongst you. Few are known by their original name and most of them also went underground. However, the original people, those who held the original tablets of knowledge, went to other planets. However, they did not go to their home planets right away; they journeyed to tell other planetary cultures what they could expect. Most extraterrestrials have no idea what day-to-day living on Earth is like, what it's like to breathe as much as you breathe here. Some of them live in atmospheres, for example, where one or two or three breaths an hour is enough to sustain life, so to them you're breathing like crazy here.

They had to tell about these things. When you're talking to a race of beings that is related to you but of a different type, you talk slowly, you tell stories, you talk about who you are and they talk about who they are —you know how it works. Eventually you talk about what life on Earth is like. The reason why many extraterrestrial races who come to Earth and visit have so much respect for people here (even though they might abhor some conditions) is because they have heard about you from these people who've gone all over the universe (and are still doing it, I might add) talking about life here and its challenges.

Adapting Native American Ceremonies

Is it inappropriate for non-Native Americans to use the ceremonies and the sacred teachings to bless the Earth?

It is inappropriate if you're using the ceremony actually practiced by the native people because you do not have the training or the discipline or the genes (and I'm not talking about Levis). While it might feel good, I would recommend that you change the ceremony to adapt it to something more appropriate. You can easily adapt ceremonies to something else—and don't be shy. Bring it into your own culture; that is better. Doing their ceremonies as they do them doesn't work.

What do you mean, it doesn't work?

It doesn't do what it does when *they* do it. I understand the desire to perpetuate this wisdom and to serve in this way. But it's very easy to adapt—to be stimulated by, build on or do something similar. I cannot support doing the exact ceremony and expecting the same result, no.

Is it the genes that are so important? I understood that most of the white people here have been Native Americans, and some present Native Americans were the ones who earlier killed Indians?

The Significance of Genes, Inspiration, Training, etc.

Yes, but you have to understand that the physical body relates to the auric body in terms of its actual physical representation. So yes, you could very easily have been Native American at some time. And you could have been Bedouin, for all you know. Your genes *do* matter, because the way you appear has to do with what you are *now*. And I'm not just being polite here or statesmanlike—*it makes a difference*. That doesn't mean that people who are doing medicine-wheel ceremonies should stop. What I am saying is this: When a spiritually trained, prepared person (Native American, for example) does the ceremony, even if they are alone, the energies of the full unifying force of the lightbody, the spirit, the inspiration and the physical body are at a level (on a scale of one to ten) of *ten*. If somebody else does it who was not born to it and it's not in their genetic structure or their lightbodies, even if they do it exactly right, it's about a *five*.

You might say, "If enough people did this, it would have a cumulative good effect," and that would be true. I would rather see more native peoples doing this, but it really doesn't need to be done so much anymore, because the evolution of consciousness is moving a little past that point.

Who is actually a Native American—five-sixteenths bloodline, seven-sixteenths? 84%? 22%? Where do you draw the line? Is it because they're raised in that energy or have

the genetic heritage or what?

If you want to draw a line genetically, I'll say half.

So if you have 50% native blood . . .

Being born to it makes a difference. For example, say you're a native person and your family was doing a particular ceremony for generations, and for some reason someone died (you were away or something) before you got the training. Let's say you returned, but you can't do that ceremony because it wasn't passed on to you completely. You don't know what to do. Then some other ceremony is brought to you from some other family where someone is dying and the grandfather teaches it to you. You're doing the ceremony you weren't raised to do, but you've been trained how to do it completely as a full native person. On a scale of one to ten, the good you can do for all beings is a *two*.

Does that mean don't do it? No, but being born to it, training for it, and expecting it and being involved and immersed in it does make a difference.

So it's the experience, not necessarily the genetic line?

The genetic line helps.

On a soul level many of us have done that many times.

Yes, but it's what you're doing *now*. This is the *physical* world, and physical-world things are different. We're talking about the core native peoples and what you do just by *being*. This is important to remember, because when you're an extraterrestrial sailing around (a Pleiadian, Andromedan or whatever), your relationship to your people is fairly fixed. You are born to a culture; that's who and what you are. Let's say the Pleiadians land on some remote planet in Andromeda. They put on full-body paper bags and walk out, waiting for the Andromedans to say, "Hi, who are you?" But the Andromedans know that Pleiadians are underneath those paper bags *because they exude that energy*.

To some extent the physical plane trains you to understand that what you are born to and what you become, you tend to radiate. For example, when you draw the same experiences to yourself over and over, it's because your energy body is exuding that, you're broadcasting a signal. So you learn more about the *broadcast signal of being*. When you go to other planets you will understand better that adaptation here is very much an illusion. On Earth you adapt, you learn, you become the best you can at something, and yet you remain what you are and tend to transmit that signal. (If it weren't for that, obviously psychoanalysts would go broke.) The main thing is, you broadcast a signal to attract what you need to experience and learn.

You learn more about that here because on other planets you take it for granted. A lot of what you learn here is designed to be applied in other places. You're learning something here about creationism, because the signal you broadcast will communicate, draw to you or even project certain things so that when you go someplace you're going to affect that place without saying a word.

The Broadcast Signal of Who You Are

Let's say an astronaut from your space program goes to a distant planet and lands. He doesn't say a word or do anything but just walks out and stands there. Even if he's wearing a full uniform, radiation shield and so on, he is broadcasting who and what he is, and the people there are going to be affected by it.

Are they going to broadcast that they are Terran, Earthian, African root race or Native American root race? Is this culture more diversified? Is it like "a Pleiadian is a Pleiadian"?

Yes, he's going to broadcast *all* of the things; he is going to broadcast the masculine energy if he's a male, the Earth energy and everything he's ever done, thought or felt. He's even going to broadcast things that people said to him—even if he didn't agree with them, even if he is outraged by those things, even if he has totally, completely forgotten them. He's going to broadcast *all* that stuff!

Let's say he stands there for five minutes. People gather around and look at him: "Who's that?" (Or in many cases, "What's that?") He stands there for five minutes, gets back in the ship, flies back to Earth and that's that. Well, that five minutes is going to affect those people. Those who are particularly receptive (and most extraterrestrial civilizations are *highly* receptive) will begin acting out and saying some of that stuff; all of that astronaut's projected energy is going to affect them. That's partly why extraterrestrial civilizations are very careful when they bring somebody from another planet.

I'll give you an example of this. Let's say a surfer is picked up in Malibu, California. His main concerns are about that next big wave and whether the weather is good enough to surf (not that that's all surfers think about, but I'm trying to simplify here). He's taken to a desert planet for two weeks (two weeks of extraterrestrial time, meaning about twenty minutes to an hour of Earth time). The people there cannot even relate to the idea of surf, much less vast quantities of water, because they are not a water-based civilization; they have liquid crystals surrounded by a flexible shell. (There *are* beings like that.)

So our surfer gets on the ship. The first thing they do is talk to him, tell him what's it like and so on. They ask, "Would you like to tell us what your life is like?" The surfer says, "Oh, yes," and begins to rattle on about himself. They say, "No, no, no," and lay him down on a table for five minutes. They attach a few instruments to him and basically (like you do with a computer) "download" him. They don't take out everything in his head and leave him blank; they just duplicate all the information he has. By doing that they know all the information in his body genetically, experientially, emotionally—everything. They can identify the patterns that might harm their civilizations. They leave the other patterns there because the rest is interesting stuff—stories he can tell. They see patterns basically in colors, some of which need to be altered.

Above the table there is a panel with different colors, and they basically deprogram the surfer from experiencing that aspect of his personality whose colors do not fit in their civilization. They cancel out that programming in his personality so that he won't exude it even on a lightbody level. Of course, that alters the surfer's personality, so that while he's there he's very benevolent and will not exude any energy that would influence the civilization other than to educate them about what he's been up to, what the surf's like in Malibu, and everything else he knows and likes and thinks about.

When they take him back, do they put those patterns back?

Of course. That's why the same ship takes him back. They lay him back on the table and put his patterns back in again. By putting in those exact same patterns, the experience on that planet is omitted. From their point of view it helps their security, because the surfer remembers little about the experience, which basically becomes subconscious. (The information is not actually taken away; he just doesn't remember it.)

It's back there in a file that says "read only"?

Yeah, it's back there in the sleeping files.

You have just explained in a way that everyone's going to understand why we were pushed up dimensionally to continue to the fourth dimension so that we could embody and exude those experiences. It was a full and beautiful explanation.*

Creator School: The Importance of Physical Life

Thank you very much. I needed to explain this so you can understand some of the fantastic innuendos of physical life here—what you've already learned and things you can learn here that you can't learn anyplace else.

The big thing is to understand why we have to stay in the body to get to the fourth dimension.

You have your memory and your thoughts, yet there is so much more going on. As psychology says, you have a subconscious, and it postulates that you have an unconscious (which of course is true). Thus you

* Humanity was saved from extinction by disease, plague and Earth cleansing in September 1995 by being pushed from 3.37 to 3.48 dimension because the rest of the universe needs the wisdom that humanity will gain by living in the body during the transition between the third and fourth dimensions. See the November 1995 issue of the *Sedona Journal* and *Shining the Light III*.

have all of this information inside you, and a lot of is processed beyond the level of the brain's function (meaning beyond thought). You're constantly processing, learning, understanding, balancing and weighing physically and emotionally all that you're experiencing so that when your soul passes out of the body it does not have to remember what you learned consciously or unconsciously. Your soul, your lightbody, has *enough compatible physical matter* so it can take with it the lessons you learn biologically, microbiologically and, for that matter, subatomically without even thinking about it.

This is obviously a requirement for creator school, because to be a creator you need to know *beyond* what you know. And in order to know beyond what you know, you have to fully access the memory structures of the physiological self. For example; someone seventy-five years old might gain 35% of his knowledge through the mind for the sake of his lightbody and soul to perpetuate the knowledge of the third-dimensional creationism school; adding the subconscious mind to the conscious mind will bring it to 37% of that person's knowledge. After that, add the emotional body, the spiritual body itself and the physical body, the rest of it.

Here is something that will shock you: Fully 40% of that remaining number (63%) is learned physiologically, microbiologically and so on! Your body is processing stuff like crazy all the time —you're breathing in pollens and matter all the time from different things, and each cell in your body passes on its knowledge to the cells that follow.

And eventually to the soul?

Yes, it eventually permeates the soul of the lightbody. The knowledge gained from all this physiological stuff that's going on does not just die with the body —what's the point of that? *That* is the basis, the foundation, of your understanding of physical creationism. And once you're beyond the veil of ignorance, you can tap that knowledge so that you can re-create reality here!

About exuding: Isn't that why when you come through Robert you exude through him in a way that's beyond Robert?

Yes, and for that matter anyone who interacts with a spirit —even of a mountain or plant or tree —will exude a certain amount of the energy of that being. And that being (the mountain, the tree, whatever) can exude more or different energy through *you* as a creation-school student than it could exude by itself. The combination of the two of you creates something "more better."

There's a difference between being in the aura of a spiritual teacher and hearing that being talk on a tape. I feel pictures coming from you, like I'm there and living it, which is different from reading it or hearing it on tape.

That's true. The printed word might not be quite as "exalted" as my actual presence, but it wouldn't be too bad.

It's as if you lift us up into a space where we're sort of there with you, like any spiritual teacher does who comes through a physical being with a lot of their energy. We're getting it on other levels besides the words.

Yes; as you say, the biology of your bodies is getting something out of this interaction with energies that you are not getting consciously, nor will you *ever,* because it's intended that that function be performed by your physiological self. You don't have to be concerned with it. It's really highly technical.

Physical Body-Lightbody Connection

For example, a physiological unit of your genetic structure will broadcast signals to appropriate mates. This appropriate mate might be someone you're going to learn something with. Maybe, to your dismay, it will be something you've already learned, but perhaps it'll be something you haven't learned that might be beneficial. In any event, this all goes on physiologically and through your lightbody. This happens long before you even consider that you might *want* an appropriate mate, or even gotten over your experience with the last one.

The main thing is that your physical body is working on these plans physiologically, even genetically, long before your conscious mind has to grapple with it. When your conscious mind begins to deal with the idea of bringing in the perfect mate, because it's a vagary within your culture (how to do it, where to find the person, do I really have to go to a bar, etc.), your mind is at sea how to do this. Your physiological self performs its function with your lightbody involved (your soul body, if there's a programmed lesson for you).

Once your mind becomes aware of needing a mate in your life, most likely your physical self has been working on it—broadcasting various signals back and forth to your lightbody and the general population—for at least two years. I'm trying to use a practical example because sometimes you'll hear about a relationship breaking up, and the next thing you know, a week later one of them is in the arms of somebody new, "the greatest person I ever met!" Well, this didn't suddenly just happen. Your conscious mind did not send out a telegram telling everybody on the unconscious level that you are now ready for your life mate. It had been going on for some time.

When any spiritual teacher comes through a human, then all of that is obvious to the teacher from our aura, our broadcasting, everything. You see. hear, feel, know all of this stuff even before you talk to the soul, right?

Yes, when I do that.

We're just like one great big open book.

You have to remember that I'm the end-time historian. I can be anywhere energy is, but very often I'm in the future so I can examine the past. Your own scientists examine something from the past. They might not have all of the data, but if it was something that happened within the last week, they'll have a lot of ways to examine or sample it.

By the time we have this conversation I've already had a couple of million of years to digest it. Sometimes I might not say something that would follow what you ask, but when that happens usually there are other things at play – maybe you need to look someplace else, and soon.

All right, I'll let you get back on the topic.

Well, we've gone so far afield from the topic, I don't know that there's any point in getting back on it now.

I'll simply say this: Native peoples, be aware that many of you are being prepared to return from whence you came, but it won't be a discomforting place. If your originals came from a place or time of discomfort, you will not be expected to bear that burden. You will be taken to a time and a place upon your planets where you can bask in the comfort of a job well done, if not by you, then by your ancestors.

You will be, even now as I speak, having whisperings and many dreams of people speaking to you, sometimes showing you pictures of places where you might be going. If you're seeing pictures of something unpleasant, don't worry about that; that is something else. You will all be going somewhere pleasant just like all souls do, but you will have the special honor of going to a place in time that your ancestors came from.

You might stay there for a short time or a long time, but you will be able to see and smell and taste what they saw and smelled and tasted before they came here. Many of the old stories and symbols will become very clear then and make sense. You'll remember the old songs and what they mean. You'll stand around and sing them with all your ancestors, and you'll hear them singing, too. Good times are coming. Good night.

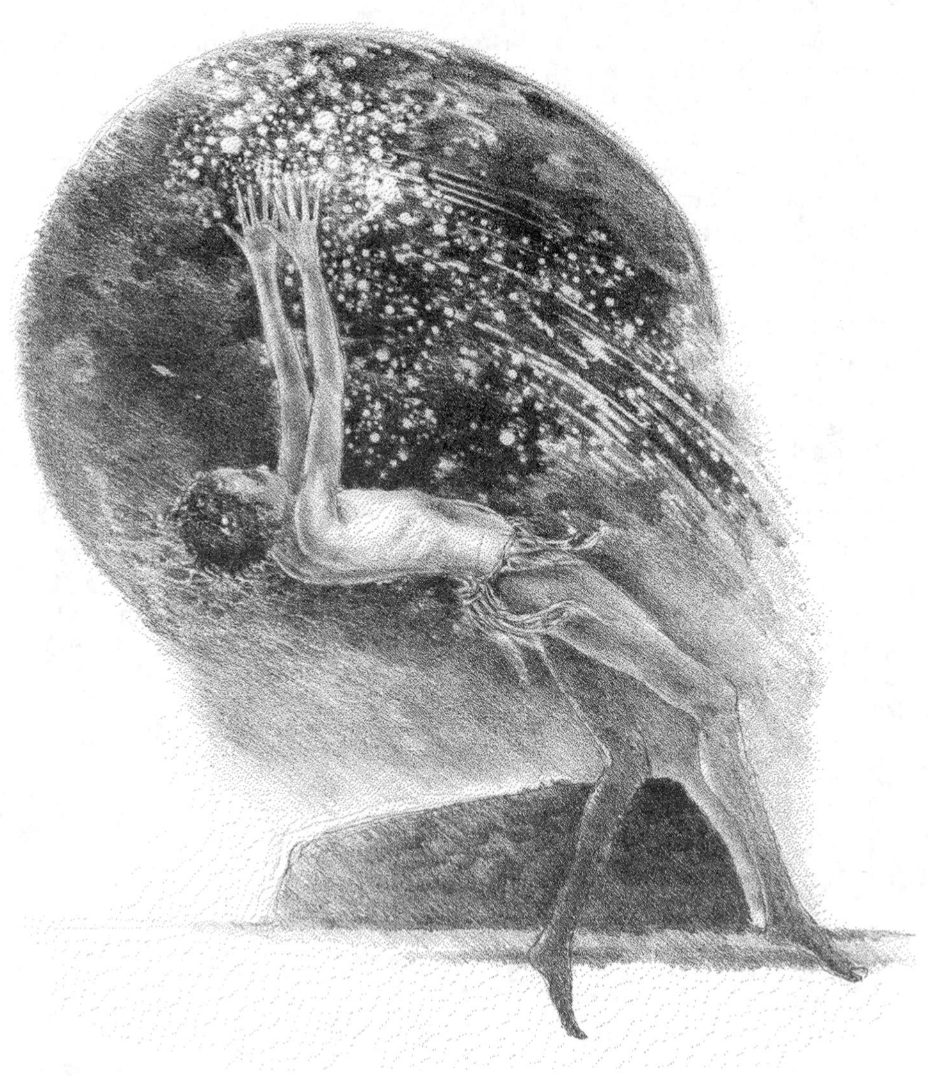

Seeders of Myth Over All of the Earth

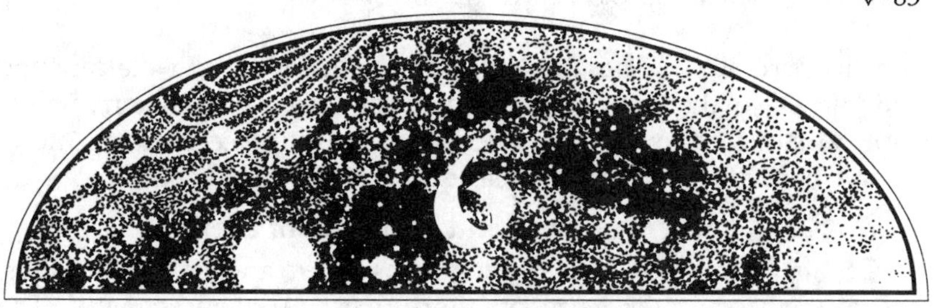

The Australian Aborigines, Advisors of the Sirius System

May 7, 1996

No discussion of the races would be complete without some mention of the aborigines of Australia. Now, the native peoples of Australia are profoundly mystical. Unlike many native peoples who have been trained extensively under the Earth before peopling the surface of the planet, these beings came here with the planet. They had a civilization underneath the surface well in advance of any other underground civilization, and though their numbers be few on the surface in Australia, their deeds be many.

Well back in time when this planet was in the galaxy [system] Sirius, these beings occupied a position of some trust and responsibility, and their ancestors still occupy this position in that system. They are the advisors (long ago some might have said the soothsayers), the guiding lights of the entire Sirius galaxy. Because their job has always been to oversee the development and maturity of the species human being, they must have representatives here on Earth although their core colony remains in Sirius. That is also why, having a vision of the future, they began and maintained an underground colony when Earth was still a planet in Sirius. That colony was begun about 45 million years ago.

These beings are working directly with Creator's representatives, the same goddess beings I've referred to before [see *The Explorer Race*, chapter 17, and "The Body of Man" in the appendix] who came to visit and initiate the Warrior League on Orion – specifically, God's daughter and her associates. These beings are very involved with the function and purpose of the human being, which is to discover and benevolently apply worthy ideas by using limits as a challenge. Of course, one knows what is benevolent by how long-lasting it is in its many guises.

Seeders of Myth and Touchers of Earth

Certain things have been developed as a core application on Sirius, one of them being the legendary story that you know as myth. The function of myth is not only to perpetuate wisdom worth remembering, but also to create the future along guidelines. True mythical stories do exactly this. This was perhaps one of their finest contributions. One ofttimes does not understand the truly legendary proportions of stories. You yourself know that to live without stories would be difficult indeed. Your most beloved books are basically stories, and even day-to-day tellings of "this happened and that happened" (to you or others) is very storylike in quality.

This was the primary contribution of what you know as the aboriginal people. It was their job to perpetuate the energies of legendary myth, seeding it over the surface of the Earth. They had to supersede Earth's arrival in this galaxy by living within the planet and radiating outward the golden rays of legendary myth, and the moment the planet arrived in its present orbit they had to go to the surface and spread out, touching the surface with their bodies.

As many of you know, material mastery on the physical plane is accomplished physically. It cannot be done in any way mentally, because the mental body cannot produce anything tangible. This is not a putdown of the mental body; its job is to acquire, to identify and to classify, but in its own right it has no capacity whatsoever to create, certainly not anything lasting. It does take direction very well, and in that light it is a good student.

So these first aboriginal people emerged from the place now known as Ayers Rock, and they walked (if necessary, swam) all over *every single point* on Earth. Now, you might say, "But Zoosh, how does a physical person swim the ocean?" These original beings did just that! If you consult your Old Testament, you will note that people lived for hundreds of years. This is not only legendary myth; it is factual. These beings who emerged from Ayers Rock, as it is now called, had life spans in those days of 1500 to 2500 years.

They also had abilities that are not now allowed you. They could not literally fly, but they had a capacity for buoyance: They could divest themselves of 10% of their natural body weight. Gravity still affected them, but there was less strain, less friction. I might add that this is how people could live so long in the third dimension – by functioning with less gravity.

Now, even though in those days there were a little fewer than 250,000 aboriginal people, it took them only about 5000 years to touch every part of Earth. Again you might say, "But Zoosh, how did they touch the spots in the deepest oceans? How could that be done?" It was done because they had special powers. If you are an aboriginal person today, you will hear stories, sometimes from the grandfathers and grandmothers, of ancient times when people could breathe underwater and fly. This is where these stories come from.

The people could fly in the sense that they could jump *very* far. For example, an aboriginal person of that time might be able to stand or take a running jump from the end of, say, an Arizona mesa, fly fully sixty feet in the air and land easily on the other side of a canyon with perhaps twenty feet to spare. Any one of them could do this; they did not have to train for it.

They could also breathe under water. They did not use gills as one might expect, like fish that process the oxygen molecules in water. In those days atmospheric oxygen was present in a far greater concentration than today and had much more life force, so if a person simply took a deep breath, he wouldn't have to breathe again for an hour or an hour and a half tops. But that wasn't even necessary, because the aboriginal people could draw oxygen from the undersea plants and even stone.

I do not intend to reveal the mystical secrets that exist even today amongst some of the initiated of these aboriginal people. But I will say that these are things they could do, and because their bodies were not fundamentally different from yours, it is something you can do also, given a high state of spiritual integrity.

Creating the Future and Recreating the Past

So these people walked, swam, climbed, jumped all over the surface of the Earth, and with their physical bodies interacting with Earth, they planted the functional aspects of legendary myth so that all those beings who came after them—including other native peoples who had developed foundational stories underground before they came to the surface—would be able to use these stories to tell about what had happened and also what *would* happen. The function of legendary myth is not so much, then, to remind the child who might hear this about who he is and where he has come from, but to literally *create the future.*

This is why your world is in such trauma right now. Some of the myths that you are creating for yourself are frightening—a casual observation of television and the movies will tell you this, although there are also some programs and movies that are filled with the stuff of legendary myth (compliments to *The Lion King,* for example, which stands out in recent time).

As you know, many wretched excesses have taken place in the form of discomfort and violence. That is why it is important to understand that materiality is created through the interaction of the physical body with the source that one is living upon (the Earth and to some degree the Sun, Moon and stars); I'm not mentioning Creator because that is a given, but certainly that as well.

These aboriginal people literally dropped the seeds from their bodies onto all portions of the surface of the Earth for you who came afterward. They understood, and were trained to understand, that one of the functions of the human being is to create the future and the present, but the most important function around the end of times (third-dimensionally speaking) would be to *re-create the past.*

It is for you to take the lesson of these aboriginal people and experience and fully understand through permission that within your physical bodies you can transform the Earth not only by what you think, but by what you feel while you're touching the Earth.

An Exercise to Re-create a Benign Past, Present and Future

So here is homework. I would like you readers to go out upon the land, ideally barefoot (during winter it's not necessary, but go barefoot when you can). Before you step foot upon the land I want you to understand what your job is. Are you going to re-create the present, the past or the future? Don't do them all; pick just one. You've got to crawl before you can walk.

Re-create a more benign present, past or future. You have the power within you *right now* to alter the worst suffering that is going on now in other places on the Earth – even in the stars! Suffering in the stars takes a different form. It is mostly frustration with what you would call stuckness ("this far and no farther"). On Earth it is more dramatic – pain, suffering and so on.

I want you to place yourself in a full-bodied emotional state of *benign calm*. (All you meditators out there, this is the goal of the meditation.) Those of you who have an ability to interact well and directly with your emotions, put yourself in a state of happiness, bliss, joy – any emotion you look forward to – and use thoughts or images necessary to maintain that feeling and no others. If you can maintain the feeling *without* thoughts, all the better, because it's better for the mind to rest during this exercise.

For one to five minutes (or longer if you have the ability) focus your energy in walking, touching, sitting on the Earth with your physical body only. Ideally, walk barefoot so you are contacting the Earth. If you wish to touch it with your hands, all right, but not with a clothed part of your body, because that greatly reduces the effect. While you are touching the Earth, maintain that emotion (even of safety or security) and project it toward the past, the present or the future. This is vital homework, and I want you to do it as soon as you can after you read this.

This exercise basically involves coming into the feeling of safety (or love or any emotion of benevolence) in some relaxed place where you can do it without distraction, then moving your body around while maintaining that benevolent feeling. Then continue the feeling as you gradually open your eyes. You might have to do this in steps over a few weeks; that's all right. (Better to do it right the first time than in a state of confusion.) Learn to get up and walk around and touch things while you maintain the feeling, then practice bringing on the feeling and going out and interacting with your physical surroundings.

Relieving the Aboriginals of Their Job of Responsible Creatorship

You will be seeding these benevolent feelings for the past, the present and the future—wherever you choose to aim it. But it cannot work unless you are in *physical contact* with the Earth! It can also improve the conditions of your life (take note, especially those of you who are stuck in buildings). When you go out and touch the Earth, aim these feelings toward the past, the present or the future. When you do that, you will be employing true aboriginal abilities and you will also be taking one of those first important steps toward responsible creativity. Creativity does not simply mean the creation of something worthwhile; perhaps I could call it *responsible creatorship.*

Understand that the aboriginal peoples are still here today because they are waiting for others to perform the same duty they performed when they came here. Their job is to stay with Earth until the other peoples of Earth begin to take over this duty. Only then will they be allowed to return home. Some of them will return inside the planet through Ayers Rock. (They cannot be stopped, so those of you who might want to stop them, don't even try.) Others will simply access portals available to them, return to their native point of existence in Sirius and continue their valuable work.

Those readers who are writers, engaged especially in fictional works, please begin to integrate more and more lessons, myths and stories of value. Fear breeds on itself, and the more you create fearful stories, the more you breed cynicism. So do not overlook this creatorship. Accept the consequences of what you are doing and (perhaps more important) the mantle of responsibility. I do not expect the publishing industry, Hollywood and so on to change overnight, but I do encourage you to begin. You have within your hands now the ability to change your world, and there's not a moment to spare!

Those who have lives that you feel are not influencing others—you're not a writer, a famous actor or author—you still have the power to effect change. Do the exercise, if you would.

Those of you who don't have feet or can't use them or who are in wheelchairs, roll outside, touch the ground with your hands or any part of your body that is not clothed. If for some reason you cannot do even this (and there are some who cannot), then generate those emotions of value so others can use them to amplify their own when they do the work. Let *everybody* participate!

Are there any questions about the aboriginal peoples?

What are one or two of the most meaningful legends they planted?

No, no. I'm glad you brought this point up, because it gives me a chance to clarify. They did not plant any legends. They planted the *material* that would support and sustain legends. They planted *the energy of legendary myth,* but they did not plant any legends. Remember that many of the native peoples who came here and originally peopled the Earth on the surface were trained underground. The legends of their people were ofttimes adapted to Earth conditions, but sometimes those legends did not need to be adapted. The teachers (who were underground) just listened and said, "Fine. No need to change that for the Earth. Go up and perpetuate your stories." No, the aboriginal people did not put stories in the ground. They put into the physical Earth the energy to support legendary myth, which can affect and transform all life. Thank you for bringing that up, because I want that to be clear.

Their Current Work: Holding Energies of Continental Shift

The aboriginal peoples have much of the skills that have been written about (basically, shamanic abilities, which many cultures also have); we take that as a given. You might ask, "What are they doing here? Are they simply waiting for other peoples to perform this activity so they can go home?" Even now they are doing something very important. They're helping to keep Earth from toppling over, in a sense.

You might ask, "Topple over onto what? Earth is not standing on anything. Earth is a free-floating spherical object that maintains its integrity by motion." This is true, yet one must understand that the magnetic energies of Earth are somewhat out of balance. Many of you know that in the past the Earth had other axes, but there are new ones to come.

One must understand the function of poles. The continents look like they do today because the polar axis that goes through the Earth balances the magnetics. The poles and the axis connecting them are primarily places of radiated influence, not unlike the function of the soul within the physical self, wherein the physical self tends to sprout around the energy of the soul (lightbody, if you like). The Earth's axis performs much the same function, and continents drift in relation to its placement through the Earth.

The next North Pole will move (it's beginning now) to north by northeast Siberia (you can figure where the South Pole would be). It will take place over a thousand years or so, giving plenty of warning to the people who live there. That area of Siberia will gradually break up, and land will be created at the new South Pole. This isn't going to happen overnight, so don't worry about it. Over a period of 10,000 to 250,000 years a significant amount of continental shift will take place. The Australian aborigines are holding this energy; that's why I'm talking about it, although it might seem to be off the subject.

This continental shift will take certain forms. What you now know as Europe will move very close to Africa, and the seas will become much smaller. There will be more lakes and rivers in what you now know as Africa. Australia will move north by about four hundred nautical miles, but not before it's joined by New Zealand, making New Zealand look rather like a comma hanging underneath it. South America will bend somewhat and move up at an angle, gathering Cuba on its way and easing toward Florida, where it will make contact and fully enclose the Gulf—in short, northeastern Brazil will collect Cuba on its way to contact Florida.

The Hawaiian Islands will by that time be growing like crazy, becoming about the size of Victoria Island off the coast of Canada, but not because of volcanoes. These and many other changes, including the connection of Greenland to Iceland, will happen as a result of the reorientation of the poles. These things will happen more easily and benevolently to life because of the aboriginal people's ability to support and sustain energies that affect all peoples. In time they will return to their point of origin, or at least their point of recent residence (point of origin is perhaps someplace else), and they will continue their work monitoring, supporting and otherwise nurturing the species human being.

I speak to the aboriginal people now: The work that you need to do for the world at large does not need to be anything other than mystical. Maintain your lives as best you can. Try to entertain yourselves more with your own stories and less with the stories of modern mankind, because these modern stories distract you from your primary function. Feel free to make up new stories, be they creation myth or pure entertainment that supports and sustains all life. Your work does not go unnoticed, and you will be rewarded accordingly—never forget that.

Could you comment on the book about the aborigines—**A Mutant Message from Down Under** ***by Marlo Morgan?***

I will fully support her story. I must, however, say a few things about it. First, it was actually intended to be published in Europe and the U.S. and so on, but not in Australia. Yet these things happen. Second, the incidents of the story did not happen with aboriginal people who are normally surface people. I must protect the people, but I will say this: Some of the people she interacted with were surface people and some were from an underground city that can still be accessed by aboriginal mystical people. Those of you who read this book, disregard the controversy. Any further question about it?

I heard a talk by the author in which she said that one of the two remaining sacred sites was dynamited by miners, with the consent of the government, and the place she had visited was the only one left. Is that pretty much correct?

Yes.

Hopefully that will remain secret so that it will be preserved?

Certain things are preserved regardless, but that's another reason to change the past.

Are the aboriginals also changing the past, just like we are supposed to learn to do?

No, no, they've done their job—at least the ancestors have.

It's up to us?

It's up to you, yes. Any other questions?

How can we change the past?

More About the Exercise

In the moment (and this is very important) that you go walking about barefoot, don't be thinking about how it can be done. If you do, you'll immediately put doubt into what you're doing. Whenever you walk barefoot, you are seeding all your feelings into the Earth. *Your feelings are literally seeds*, and the more that people feel those feelings, the more likely those seeds will sprout.

Let's say you take a walk down a nice, smooth trail and it's a nice day. You take your shoes off and walk barefoot for a while, having a good time looking around enjoying the plants and animals. You sit down on a rock and contemplate your life. I don't have to tell you that you will eventually come up with something unpleasant. You think about it and wonder what you can do about it. Maybe you come to a solution; maybe you just let it go and continue your walk. All the feelings you had about whatever was bothering you remain there as seeds.

The rock is a natural place to sit because there there is a good view from there, so the next person coming along sits on the rock, feeling good, feeling happy. But soon he has a sad feeling, as if from nowhere. Then somebody else comes down the trail feeling depressed. She sits on the rock. Those seeds of discomfort will sprout within that person, and her depression might become worse as long as she focuses on it. Equally, all the people who have sat there (to say nothing of the energy of the rock itself) and have put good feelings into that rock because they felt good—those seeds could also sprout in that person.

Here we have a situation where people are amplified by what has preceded them. This is not only a link to the past (having to do with precedence) but also with your capacity to change what you want to change when you can (free will).

Let's say that depressed person gets worse and worse, but it's getting toward sunset. She stops feeling miserable. She looks up and says, "Whoa! There's a sunset. Fabulous!" She stops feeling miserable. "I've got to get going before it gets dark, but I've got to sit here and watch that sunset as long as I can; it's just beautiful!" While she's enjoying the sunset, all the feelings of beauty and good energy from the people who sat on that rock are now sprouting inside her.

Feelings As Fertilizer and Creation Tool

You have to remember that plants and animals sprout from seeds (or eggs). So do people. You all have that in common. And fertilizer has to do with *feelings.* When a child is born and is loved, nurtured, cared for, kissed, caressed—basically brought along comfortably—the chances are it will have those abilities to pass on. If a child is born and people are not happy to see it and don't treat it right, then it will have *those* things to pass on. It might not continue it, because human beings have the capacity to change, but it *might* do it, because that's seeded within it.

You are fully involved now in material creation, and material is created through feelings. Molecules come together because of the energy of love. Molecules come apart because love energy is not dominant enough. This is the foundation of the aging, or (put scientifically) the oxidizing process. You create and re-create through feelings interpreted through form—that's the material level. So if you want to re-create the past and don't know how to do it because you have avoided thinking about everything that makes you feel bad, just walk out on the land feeling these benevolent feelings and say "the past." Then continue that feeling, because to maintain the feeling or continue to re-create it, seed it and radiate it, you can't get too much into thought. Just say "the past," or "the present" or "the future" out loud. Say it just once, that's enough. Good question.

So it's mainly having the intention to re-create the past or the present or the future while we generate this good feeling and make contact with the Earth, not getting specific about anything? Is that right?

Not quite, because that description is too complex. You don't have to have the intention, because intention involves the mental body and sufficient repetition to maintain discipline (as is done through Eastern meditation techniques, for example). No, that becomes too much. All you need to do is maintain an emotional feeling within you, period. Then when you walk out on the land while you're maintaining that feeling, you say "the past" or "the present" or "the future"—whatever you want. But you don't need to have "the intention." I'm not being picayunish with you; I just need to make it clear so that we do not sign people up for meditational practices, because these practices have to do with consciousness. What I am talking about is *body* consciousness – what you *feel,* what happens *physically.*

Let's make it simpler. If we say it's your intention and we leave that in print, do you know how people are going to receive that? They're going to say, "It's what I *think.*" But it is *not* what you think. That's a good question to clarify the point.

I have a question about the aboriginal people. Are we talking only about the few who are left in Australia? Are they on any other continent called by other names?

No, we're talking only about the people in Australia. We're not even talking about those who have emigrated to other continents. Once they emigrate, they're no longer involved in that commitment. It happens only if they're in Australia.

Is it because of a connection to the land?

It's a soul commitment, as one might say. One is not *always* an aboriginal person, but when you come in as an aboriginal person you take that on. The soul might have other commitments, but that is the primary, overriding commitment. Often people who live in Australia wonder about the aboriginal people: "How come they're not doing this? How come they're not doing that?" It's very simple. Because of the soul commitment, they cannot get overinvolved in the world of material man.

I've heard that employers there deplore what they consider the irresponsible behavior of an aborigine It's said that when the aboriginal person starts a walkabout, they just don't show up for work. That's because of that overriding commitment they feel, right?

In that circumstance that would be true, or maybe they have to go to some place for a dance or a prayer. Prayer in the western world is words; prayer in the sacred world is motion, otherwise known as application. Prayer is not what you say; it is what you *do.* That is why higher forms of prayer involve motion, in which you are physically doing things. As any religious theoretician knows, prayer is primarily intention, but if one does not *act* on the intention, it is almost meaningless in its effect. So yes, they might have to go somewhere and pray – otherwise known as perform their intention.

Are they totally conscious on the physical level of their function, or is it done more on a soul level?

The Soul Commitment

Well, the average aboriginal person who has somewhat integrated into society in Australia might not be that conscious of it. However, since there has been significant effort by the aboriginal peoples in Australia to maintain their culture, most of them are aware of it. Just because they don't walk around and say, "I've got to go to church" does not mean they are not performing their soul function.

So they're all trained from birth in these rituals and motions, whatever?

They are exposed to the stories when they are young; they are, as native peoples say, born to it. They come in with that soul commitment, are exposed to the stories (guided) and are reminded of what their ancestors did. Then they are reminded of what is good for them to continue doing. They don't have to walk all over the face of the Earth, but should they so do, some good could come of it. This is a primary function of Australia.

However, if you take an aboriginal person from Australia who is performing these functions, transport him to, say, Italy or France—no matter where he goes (especially if he's barefoot)—he's still going to perform that function, although to a significantly lesser degree. If he's wearing shoes it won't make much difference. But if he's walking around barefoot even for a stroll in the park, wiggling his toes in the grass, that function is operating at 30%.

I don't want you deifying these people. At the same time I don't want the people to feel enslaved to their purpose. Their purpose is sacred. It's intended to enhance life, not to make life less pleasurable.

When we talked about the Native Americans, you said they had to be at least 50% genetically pure. Does the soul purpose supersede the genetics, or do they both have to be there?

I will give you the formula: In order for this to apply to Australian aboriginal peoples, one must be at least 20% (that's how powerful it is) Australian aboriginal.

And with the soul purpose?

Of course, the soul commitment. Without the soul commitment it doesn't mean anything.

They don't get born as an aboriginal without the soul commitment?

That's right. It still affects people who are, say, 10% Australian aboriginal, but to a significantly lesser degree. If you're at least 20% Australian aboriginal, all these things are working for you. You could theoretically be initiated into Australian aboriginal society and not be that race, yet have some minor but worthy contribution toward this work. But since you would not have a soul commitment, it would not be as effective. If your soul suddenly decided to make that commitment and you dropped everything, moved to Australia and lived with the aboriginal people (should they accept you), you might be able to develop as much as a 20% effect of being born to it.

This tribe said they were not having babies because they planned to leave the planet. They didn't plan to be here that long?

Now, we're not talking about the average person on the street here. That was just those particular beings. They appeared to be physical to the author because they could acquire physicality for a time. They have the capacity, however, to be interdimensional. You have to remember that this experience was largely to influence the author so she would influence the world with some of this wisdom. I'm not saying the experience didn't happen—it did, absolutely! And she didn't dream it; she was out there in the bush, as they say.

Is there a large contingent of these beings still under the ground, and do they come up frequently?

I don't want to say too much about that. I will say they come up frequently. There is not as large a contingent as there once was. Even as we speak they are continuing evacuations of people from underground civilizations; however, there are still some of them there.

When they all leave, will that have an effect on the planet?

It won't have an immediate effect because the Australian aboriginal peoples will still be there on the surface. Should all the Australian aboriginal peoples leave, you'd darn well have to be involved on a day-to-day basis in responsible, constructive creativity or creatorship and nurturing stories and legendary myths of value, not of cynicism. If *you're* not doing this, that's too bad, but I think they will stay there until you are ready. When I say "you," of course I mean the general public.

You might say it's hard to believe that people will begin to do that. But then again, you have to remember that the pendulum swings both ways. You might be at that arc of the pendulum where there is cynicism and violence, but the pendulum does swing the other way—and it will. So don't give up hope; after all, you largely invented, nurtured, created and sponsored the emotion of hope here on Earth as the Explorer Race. You'd better not give it up, since you pretty well brought it into existence. Anything else?

Let's step from the aborigines to the actual planet. Did the planet arrive here forty-five million years ago?

No, no. The aboriginal peoples began living underground when the planet was in Sirius forty-five million years ago.

So when did they come here?

We covered this in *The Explorer Race* material. They came here a while ago, but a lot of the carbon-dating systems used here on Earth do not make sense to some of the scientists who use them. Understand that the carbon-dating principle, while accurate, does not allow for dimensional shifts—movements in space and time. It's been here for a while; I'll just leave it at that.

The Creation of the Underground Tunnels

Were all the tunnels that are honeycombed under the planet already there before they came?

That's a good question. Some of the more beautiful and more spiritual ones were there. How do you identify a beautiful tunnel? Well, the walls are very smooth, yet you can see that they have not been carved by lasers. If you look closely, you will see that the stones appear intact. That's because the tunnels were *sung* into shape, and the rock moved and changed its form into a beautiful oval. The tunnels are somewhat flat at the top and bottom, so let's call the form a rounded rectangle. The material has a look not unlike what one sees at various archeological sites around the Earth where ancient stone-fitting techniques were used. The stones there were also sung into place.

Originally they didn't dig anything; at least ten people, all women, went into a cave and sang. They would start out with a melody, then notice certain tones. They'd go up and down the tonal scale. If they couldn't make a tone, they would imagine that tone and exhale so that the physicality would be put into that tone. They would have the vision of a tunnel – it was not a mental intention; they would *see* it. In short, they used their divine imagination and they sang.

They would start out in a circle, then the circle would open in the direction of where the tunnel would form. Standing in a C-shape, they would begin walking forward. That is how these tunnels were formed.

Tunnels that have been drilled will not last because the stone was not asked for permission, but brushed aside as if it were nothing. But tunnels that have been sung through this form of applied prayer, which is physical, will last indefinitely. It is love that attracts and holds things together, and this kind of singing is a form of sacred love whereby the stones simply move apart.

You could say, "But Zoosh, how is that possible? You have a dense layer of stone." That is because what is dense on one level is liquid on another. The stone is *alive*, and it liquefies itself in its interdimensional capacity, moving and re-forming into a shape in which it finds pleasure. The actual shape of the individual stones or boulders and the space they re-form around is something in which they find comfort, such as this particular tunnel shape. The flat spots are almost (not totally) flat, so it looks like a rectangle where someone has imposed an oval, curving the sides, and at the top and bottom look somewhat flat. I can draw a picture for you. Here you're looking into it, standing in front of it.

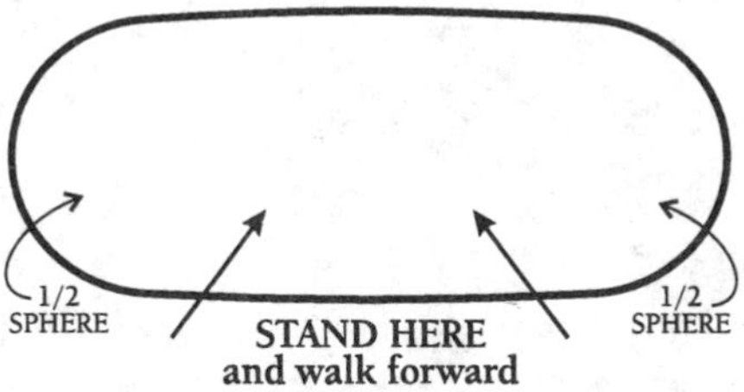

And you just walk through it?

That's right.

I have a book that has a photo of a footprint of a finely sewn leather shoe. The carbon date was over a million years ago. Is that wrong dating, or was that done when the planet was in Sirius?

Origin of Dinosaurs

Some of it might have happened in transit; some of it might have happened in Sirius. I said before that the dinosaurs are basically from Sirius – the great big ones with the long tails and the long necks that eat leaves. (Flesh-eating dinosaurs are an aberration unassociated with Sirius.) The dating systems are accurate insofar as they can date things, but they do not take dimensions into account. A machine does not easily do this unless it has been programmed by a person, and since you don't have that program available, then people will just have to add their inspired beliefs to these carbon-dating systems.

A number of years ago I remember seeing some pictures on TV taken in Australia. They said it was so hot there that the crystal miners had to live underground. It showed some of their living quarters. I couldn't help but wonder if any of them took over some of the tunnels or underground spaces that had been used by the aboriginal people, or did they dig their own?

Their mines didn't go deep enough, so they didn't happen to come upon one of those tunnels. But there have been other circumstances in which people were digging for this and that and found the tunnels. This is why the insiders in your governments know about these tunnels. That was a good question.

What is the future of the aborigines after they complete this service? What will their life be like after they leave this planet?

They will return to a point of creation, initially within the Sirius galaxy. There they will meet with God's daughter and her companions. They will pass back and forth feelings and inspirations of the future of species human being. After a time they will return to be with Creator, then go up with Creator to the next stopping point in Her voyage after you have replaced Him.

What about God's daughter? Will she stay?

She'll go. We hope that other God's daughters will be around then.

To close, I will simply say this: We understand now that the aboriginal people are very important for what they have done.

Invitation from the Seven Sisters

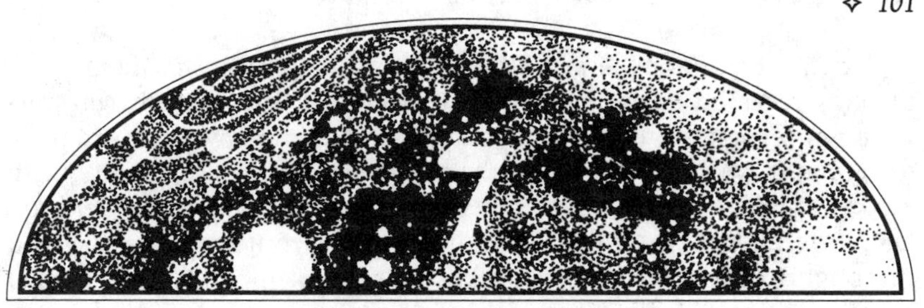

The Return of the Lost Tribe of Israel

July 30, 1996

The pace toward your future is rapidly increasing. You have been for some time now consumed with your personal experiences, and rightly so, because they were designed to encourage your growth, to test you at times and to bring forth potentials that even you do not know you have. Yet now we find that as you move speedily down your path toward becoming fully realized, you will soon experience a debt of gratitude to your past. You might find it hard to believe that you would be very grateful for your total past, some of you especially, because of the struggles you have gone through, the unexpected moments of travail and even the wonderful feelings of victory after some particularly difficult problem was solved.

As you begin to experience your future now as it unfolds, in order to fully believe in your own abilities you will need to have the courage of your convictions as well as a certain rhythm or harmony in your felt reality. For many years society has abandoned its true feelings and pursued the myth, the dream; and when you abandon your feelings to pursue the myth or the dream, you are invariably going to run smack into walls, one after another. Your feelings are designed to guide you, to show you, to command you from time to time. Now you are getting back on track with your feelings, and this is good. I want you to understand that you have help in place for this.

Once upon a time there was a group of human beings made up of all the core races on this planet, even many beings who in their own time were not recognized but whom you see all the time now. For example, the Mexican people are basically the native peoples of that country, with a Spanish influence (ruling out certain French influences and others you take for granted). I use the Mexican people – people who are well and thoroughly integrated into the public – as an example because almost everybody has a Mexican friend these days.

Several thousand years ago, before the Mexican people were known, if you saw someone who was of that nationality you might have cocked your head to one side and said, "Ah, I recognize some of your features, but not others. Where is your land? Who are your people?"

I mention this because in that time gone by a tribe, I will call it, of individuals wandered off from a larger group and went exploring, wandering into northern Africa. Some of their people settled in central Africa and some, deciding that their philosophy didn't fit, wandered farther south toward the tip of Africa. Over about a generation and a half they walked back north until they were near what is now known as the Giza pyramid. By near I mean about fifty miles away, close enough that their most spiritual seers could observe the blue light that radiated in those days off the tip toward the heavens and was directed that particular night toward the Pleiades.

About 5000 years ago this particular tribe of people had evolved into about 650 hardy souls of all different appearances. Some were unknown in those days and some were of what you now call the white race. In the nicest possible way, it was a motley group. So here they were, an Earth family before their time. Their main focus was to try to understand the benefits of variety (as was obvious by their appearances) and to integrate the best of all possible worlds they might come to. That's why they wandered about nomadically for a time. Because they started out in the Middle East, their natural direction to wander was southward into Africa and then back north again.

The night I'm describing was warm and pleasant, the kind of night when you could sleep under the stars and be very comfortable. The seer observed the blue light and noticed that it pointed toward the Seven Sisters. He felt kind of tingly and said to the tribal council, who were nearby, "This is something important, and it has to do with us." So the council and about half of the people formed a semicircle and stared at the blue light going up into the sky. (The seer probably realized that it was coming off the pyramid, but might not have mentioned that.)

They watched the light for a time. Pretty soon a ship appeared, hovering in the light. (Because they were fifty miles away, it had to be a pretty big ship.) They could see the blue and that the ship itself was radiating blue. It hovered in the light for a time, then moved slowly toward them, emanating this blue light that gave them all a very wonderful, tingly feeling. As it got closer, the seer and some members of the tribal council could hear words forming in their heads (in their own language, of course), "Would you like to come with us now? Have you gathered enough information? Would you like to come home?"

The people, especially the council, said, "Come and tell us who you are. Speak with us." With no radar to worry about in those days, the ship landed. A couple of people got out, looking not dissimilar to certain members in the tribe, but dressed a little different. They wore certain items that appeared to be jewelry, fancy-looking compared to that of the tribal members, who looked nomadic and, aside from their different facial appearances, not unlike the nomadic tribes of the Middle East today. These individuals from the Pleiades said, "We recognize many of you by your soul patterns. We have been following your soul patterns for years, and we want you to know that there is an important job for you. But we'd like you to come home with us so we can train you for it and extend your lifetimes so you can serve in this capacity.

A Trip to the Pleiades and a Long Dream

The Pleiadians proceeded to create something not unlike a holographic show, although the colors were more vivid (the three-dimensional effect was not so profound as to dazzle the people too much). They basically showed the progressing future without being overdramatic. They showed them these times you are now living in, where they had the pleasant surprise of seeing many of their own people (for example, the Mexican people and others that exist today), and the people in the tribe who resembled them said, "Oh, these are our brothers and sisters! Can we go see them?" There was a lot of excitement. The tribe at that time had gathered all they could, and since they were running low on supplies anyway and didn't exactly fancy going back to being colonial subjects, they went aboard the ship and off they went to the Pleiades.

They've been in the Pleiades for all this time. You might say, "Now, wait a minute, Zoosh, that would make them immortal. It would also mean that they would have outlived the Pleiadian citizens who picked them up—and the Pleiadians do not have that type of technology," says you, who are very well informed. And I would say, "You're right." However, the technology the Pleiadians had even then was the ability to place people in stasis (a form of suspended animation) whereby you could sleep for thousands of years without experiencing any negative side effects. It would be a benevolent sleep rather than a stupor (as in a drug-induced coma). It would be something wherein you would have full capacity to interact with all of your spiritual teachers, as you do on the most pleasant dream levels. (This was considered a necessity.) So they were experiencing physical life on the Pleiades for only about twenty years, long enough to have their bodies completely built up, becoming well-nourished and as healthy as they could be – transformed somewhat in their physical appearance but basically looking rather the same. Then this apparently lost tribe on the Pleiades was put into a "rockabye baby" consciousness.

Now, about two years ago in your time these people were awakened by those who had been instructed that the time was soon coming when they would be sent in vehicles, some made of light. They would be in lightbodies; some would be more physical because they would be transported to Earth.

To summarize, this former lost tribe of Israel, as it is called, went to the Pleiades, got educated spiritually by Pleiadian teachers (and some Andromedan teachers), learning everything they could about you (who were living in the future, from their perspective) so they could come to be amongst you to help you understand your own abilities and remind you of who you are and help you wake up.

The Return from the Pleiades

This lost tribe is returning to you now, infiltrating in such a way that they would be absolutely undetectable by any citizen born and raised on Earth. (In point of fact, they *have* been born and raised on Earth, only long ago.) This will last seventy years, but the bulk of your waking up takes place over the next twenty-five to thirty-five years. As I speak, three members are amongst you. Since there is zero possibility of their being detected, I will tell you that two are in the Northern Hemisphere and one is in the Southern Hemisphere. Over the next eight to ten years they will all have arrived – all 650 of them, including some who are quite youthful (in their twenties) and some in their sixties and seventies.

These former Earth people who are returning to you are among many who will come to help guide you to remember who you are. These individuals are uniquely qualified. They've been trained to speak any language they might need, including obscure and lost languages. (For example, they can speak and understand clearly Sumerian, not a language used in daily conversation these days.)

These "lost" ones will soon be amongst you in greater numbers. You will not recognize them as great spiritual leaders, however. Most of them will choose fairly innocuous backgrounds, although there will be a few who have to infiltrate into the higher echelons of power simply because sometimes everybody needs to have guidance. At least half of them will be just regular folk; another fourth will appear as down-and-outers, homeless people – drifters, as you call them. Another significant portion will simply blend in all over the world – maybe your neighbor down the street, maybe a friend you just met.

They will not have antennas or anything obvious, but in moments when your consciousness and hearts are open and fertile to suggestion and guidance, they will say just the right thing or radiate just the right feeling to steer you toward rediscovering your natural abilities. It is in your nature to be true magicians. What might be considered benevolent magic today? A benevolently warm rain with falling flower petals would certainly be magic, and the smell of roses on the air a pleasant experience indeed. These individuals would be able to help you remember doing this.

Integrating Benevolent Magic

Your problems are not unsolvable. Granted, some of them are not solvable through technological or even intellectual means, but there is no problem I have ever seen any people come up against that cannot be solved by benevolent magic. Understand that benevolent magicians is *who you are.* You discarded that ability when you came here so that you could learn consequences and take actions and be spontaneous and even impetuous. But you have learned that now. You have assembled enough of that information. Now it is time for you to be benevolently magical.

As you now choose that path, I want you to begin to integrate some of this benevolent magic, to actually acquire the vision to know what is benevolent (for it is not always obvious) and to know that *the way you feel* when you are asking for something has everything to do with what will happen. Progenitor and I and perhaps even a few others will talk much more about benevolent magic, but I'm mentioning this now in this context because I want you to know that there is not one individual amongst you living now or who will be born in the future on Earth who does not have the capacity to fully express all levels of benevolent magic. While your world remains polarized, this magic could of course be used in nefarious ways, so choose the path of benevolence, as I know you will, and you can make your own lives and the lives of all others around you only happier.

Healing: Attention on the Comfort

I was wondering about healing. How will we experience self-healing, and what techniques will we discover?

It is interesting that you mention healing, because it is based upon the polarized notion that when one is injured, suffering, sick or even emotionally confused or twisted, the best approach is to work with the symptomology—heal the symptoms and bring back balance and the natural beauty of the soul, as expressed through the physical self. That is the long, hard way, as many of your physicians have discovered, regardless of the letters after their names. Rather, true healing does not focus any attention whatsoever on the discomfort but rather total attention on *the comfort to be attained.* You *invite* that comfort.

If you can remember how it felt to be comfortable—whole, as you might say—that is available. If you cannot remember (if you have perhaps been born with the condition or have had it for so long you simply cannot remember), then you focus on the warmth, the loving presence within and around and about yourself. You use the divine part of your mind (the imagination) to picture how it might be to be benevolently healthy, with so much strength that you could literally run up and jog down the trail to the top of a mountain with plenty of energy to spare, and not only picture it but *feel* yourself doing it.

The total focus is on the acquisition through the universal cosmic energy (also known as God, which includes everyone and everything) *to bring about the desired condition within you* rather than to work on the symptomology itself. I'm not saying that providing aid and comfort for symptoms (certainly for drastic symptoms) is not worthwhile; certainly it is. Do whatever you need to do. But for a true cure, you must invite the state of being of *total benevolent balance.*

For example, a person might be in the throes of some serious disease, going to a physician for a diagnosis and treatment and perhaps to a herbologist or even a chiropractor or acupuncturist for relief of symptoms. The person could still, in relaxed moments when those symptoms have been relieved, imagine and focus total attention on how it would be to feel wonderful and well, imagining a physical activity that might not have been possible for some time, or even something more uncharacteristic like jogging on the beach, swimming forty laps and feeling wonderful and invigorated. (The actor must pretend so well that he literally *becomes* that.) And all the while ask that it be so. Just that.

Invite this wonderful state of being to you. And when you are done inviting it, let it go and forget about it as best you can. The way to have prayers answered is to speak them with every fiber of your being, to feel them, to include your passion in them, to bring the loving warmth of the loving being to you—and then let it go. Once you have informed fully the All That Is (as you call it), that is sufficient. Usually once is enough. If you want to say it again, then rephrase it in some benevolent way. But don't say it the same way again. They got the message. You'll know because you'll feel something good; maybe you'll feel warmth in your chest or body. Maybe you'll have a sense of benevolence. Maybe you'll relax and be able to have a good nap for a change.

The main thing is, don't give up the therapies that are assisting you, but include this if you choose. True healing, then, does not involve itself in the discomfort, but rather in the *comfort* that is desired. Anything else?

What was the point of this lost tribe of Israel? I heard your story and that they're here to help us. Was that the point?

That's the basic point. Who could help you more than people who have actually lived compatibly and comfortably as human beings in a modern society (racially speaking) years before their time, regardless of their varied appearances? What better teachers could you have than people who actually know and understand Earth, *plus* know and understand life beyond Earth? They would have broader horizons, yet have total compassion and understanding for those who might not have that ability or vision.

If we have ETs come down and talk to you, you're going to be dazzled by them. And they will not have the patience (that's right, there are ETs who don't have patience, either) for people's desire to hang onto the known past, regardless of its pain. You might be able to come to some accommodation of the discomforts you put up with in life and say, "I've grown used to it. I can take it. At least I know what's coming; there are no surprises." But you see, *you can't do that anymore!*

Feeling and Claiming a Benevolent Future

Now you have to focus on the actuality of your becoming benevolent magicians again, which is your nature. In order to do that you must practice benevolence, and in this case benevolence begins at home. You must learn to be benevolent with yourself. You must learn how to tack on. When you catch yourself saying, "Oh, not that again," tack on, "Well, I'm sure things will change for the better." But don't just say it; if you just say it, nothing will happen—zero. *Feel it!*

I'm asking you to take a leap of faith here and extend your rainbow bridge, because right now in order to truly integrate the future that is totally benevolent, you must begin to *claim* it—not just want it, hope for it or wish for it. You must claim it as your own. If it is your own, you will feel absolutely and totally justified in possessing it. I don't have to tell you that as children you all had things taken away from you, and you were totally, adamantly, justifiably demanding in having them returned.

I'm not saying that you should be demanding; I am telling you to understand that it is your nature to be benevolently magical. *Expect it to return in your lifetime—and claim it as your own.* These returning individuals will help you toward that end, but you can begin now. I'm not saying that you shouldn't be grounded and practical. I'm saying that if you catch yourself thinking something that is negatively charged, transform it now. I know you have done all this.

"Oh, that's new-age, airy-fairy, Zoosh." But no, this is true; the large mass of the public now, as you are reading this book, is beginning to wake up, but they have not yet awakened. They are stirring, as one stirs before waking, rolling this way and squirming about because they are being told by their teachers that they must choose. Most of you sensitives have had this opportunity before. But now the mass of the public must choose, and there are not enough of you sensitive enlightened ones, as you might strive to be now, to go around. That's right. Even all of you—hundreds of thousands, millions—is not enough, and it is necessary to have help. But understand that the help will come only slightly from those who have been elsewhere, such as this lost tribe, and their main focus will be to train the people here to fully know their potentials.

As you read this I want you to understand completely that all that training, all that wishing and hoping, all that ability you have garnered for many years and have wondered if you'd ever have the chance to use, as you read this you will have the chance to use it tomorrow, if not today.

Can you say more about the lost tribe? How will we know them? What do they look like? How many of them are there?

You won't know them. They will be completely under cover the entire time they are here. They will never reveal who they are.

Do they know, though?

Oh, they know, absolutely. They will not reveal who they are, because if they did even the most enlightened person would immediately elevate them. No, they have to say what they say and allow you to use your own perceptiveness to receive it as something valuable. If you do not receive it as valuable to you in that moment, perhaps the moment is not right for you.

These people are designed to initiate you, and they cannot initiate an unprepared person into the most insightful wisdoms and practices. So no, you will not know them and I will not give you any means to do so. I will just say that sometimes they will say little things, and you will suddenly have a feeling that it has been an obvious truth all along and you had just forgotten it somehow in the maze of thought patterns. That thought will go in and you will feel good. The person might have been just passing; perhaps it will be a customer, someone on the bus, someone you met on the airline, and you will think about it and maybe say, "What a wonderful thought!" And you will have great thoughts and imaginations branch off from this. Then maybe you'll turn around and they won't be there—or perhaps they will. Their job is entirely under cover.

How many of them are there?

Roughly 650, give or take one or two.

And they'll look just like us?

They'll look just like you.

But they'll have all of these benevolent, magical powers and they'll come in and go out . . .

You mean, will they come in like *pouf!*

Yes.

They might come in initially as in *pouf!* – that is easy and also entirely undetectable by existing technology or even potential technology in the fairly long-range future. But they will not do much *pouf!* around you. If they did, they would immediately separate themselves. However, they would encourage *you* to experience *pouf!*

The End of Disease

Will we be in a position in ten to twenty-five years to totally heal ourselves so that no one will need healing? What is the midpoint?

As usual you want to have an act one, act two, act three? For many of you it will be sooner rather than later. I should think in thirty-five years at most, disease (certainly catastrophic disease) will be significantly less known. Beginning in some circles and within twenty years the recognition will catch on quite well that death is not the end of anything, but is simply the means to exist in a different form of life where you take your personality right along with you. It is not to be feared at all. It is just like "on to the next thing. Bye."

There is no reason that you cannot, within some of your lifetimes here, develop death into something like on the Pleiades. There the body does not particularly wear out because they've learned how to interact in a sacred way with their planet, and they treat their body in a sacred way, as if it were a part of the planet's sacred self, which it is. There is no reason at all that you have to get sick when Mother Earth needs this body back. In many places in the Pleiades death is a celebrated occasion; there is no pain or discomfort; one simply steps out. There is usually a celebration, and the physical form, should it be left behind (it is not always left in all places), is placed near the village and allowed to rejoin the planet. (No fear of animals coming along – the Pleiades do not have them.)

The body naturally decomposes, and since they don't have polarity, this happens in about three days of Earth experiential time. Three is a very important number, by the way: three cycles – past, present, future. I've run into that before, but three is very well integrated in other places.

Healing Yourself

Is that all you want to say about healing techniques or ways that we'll be doing it? Will we use tones, colors, stones?

These are all transitional methods. You will use all of these and more. But eventually it will be entirely from within each individual. As you begin to feel, you will be tested, as you already are here all of the time; naturally that will continue. You are really testing yourselves. And when you feel some discomfort you will transform it with benevolent magic through the use of feeling your physical self. You will *not* simply accustom yourself to the burden. You will feel any and all discomforts, and you will ask (for example) that these energies that cause you discomfort be moved out of your body, but it will be more than a prayer. You won't ask somebody to do it for you; you'll do it yourself. And while you are doing it, you will ask that it happen. In that way you cocreate with Creator. Yes, Creator helps you and yes, you help yourself.

You cannot cocreate with Creator unless you do what Creator does as Creator is doing it. Cocreation does not take place by imitation but by simultaneous action. And as you are feeling the loving energy of Creator, you then ask that this discomfort be sent out of you to a place where it can be resolved most benevolently and imagine yourself feeling wonderful.

Begin now, as you read this book. Techniques will come to you; you will be taught, you will be shown. I will say more about it at some point, but I will say now that much of this will be have to be demonstrated in person. At some point in time Robert and maybe others will begin workshops to train a few of you so you that you can train others and send it around. Why keep it a secret?

Discipline and Focus: Homework

To become benevolent magicians, are we going to have to be a little bit more disciplined? I was just reading a book from 1935 about the yogis and the people in India who practiced mind control or yoga. It needs a lot of discipline and practice. But in 1996 we're not very disciplined.

Oh, I don't know about this. Let's consider it. I know this will sound funny, but how many people have sat on their couches staring at the television set?

All of us.

Becoming totally consumed. Are you not totally and completely focused on that? That's what it takes—total and complete focusing on one thing, to which you then add other things. Here is homework: One might take one's arm or even one's hand and gently open and close it. Of course, it gets sore after a time, so one might consider putting your hands lightly together and moving them back and forth. Do this lightly, not vigorously, focusing your total attention on that point of contact between your hands, noticing first how one hand feels, then the other.

These exercises are designed to give you the ability to completely focus on one thing, but it has to be physical. Even though you might think, from an Eastern meditational viewpoint, that people do not have the capacity to focus on one thing, I'm saying that they *do*. In the West people become seduced by whatever is on TV or the movies, *but they focus on it entirely.* Ask anybody who's ever played a game of pinball or, even better, a video game, where one must be absolutely and totally focused on what you are doing or you get shot out of the sky. You cannot let your attention waver for even a second. People who are good at this know it. It's actually pretty good training for focusing, although the cheaper way is to put your hands together and move them lightly back and forth across each other; that doesn't even cost a quarter!

We would want to at least be careful about what we focus on.

Exactly. But you want to *begin.* The foundation of any ability to acquire a condition —whether it be good health, whether it be rain in a drought, whether it be sun in downpour—is going to require the absolute ability to focus entirely on one thing and then add other things.

So you focus your attention and then add emotion to it, or . . .

You *start* with emotion—it's not mental. You start with the *feeling.*

Oh, you focus the feeling?

Yes. You begin with your feeling because feeling has the capacity to move things. Mind has an extremely limited capacity to create physical motion. It does have some ability, but compared to the feeling self, the physiological self, the emotional self, the instinctual self, the spiritual self, the mind has only one-quarter of one percent of the capacity of the rest of your being to create physical motion —changes —outside or even within yourself. Why bother with one-quarter of one percent when you could have 99¾%, and in the process also remember who you are?

Just because I want something right at this moment, I don't necessarily manifest it. Of course, things are getting a lot faster . . .

In the future that could change by tacking on that benevolent consideration. You actually commented on it yourself: "Things are speeding up, so it could change." Thank you for saying that; things *can* change. You might want something now and later it might not be so important. If necessary, you can ask later that it be changed, or if it's an object, pass it on. In this western world many of you will want something material —and that's all right. That's the world you've chosen to be born into. Perhaps in other societies people will want a change in the way they feel or perceive, a change in their senses, a change in the way the neighborhood perceives —or even broader possibilities.

Asking, Not Demanding or Commanding

Although you can ask for others, you cannot, even with benevolent magic, demand or command. You can only say, "I will ask" or "I am asking," because that is the reality. Even Creator does that. Creator does not say, regardless of various religious interpretations: "Now this will happen!" You see, Creator is aware of the fact that It might not be totally aware of all of Its complete being. Creator is now aware that more of Itself is someplace else, so Creator calls on Itself (Himself, Herself) to bring something about that Creator feels is benevolent, and with Its significant ability to see, is reasonably assured is benevolent. It might be something little, maybe something profound. But Creator asks that things be changed. Creator does not use English, but if It did, Creator might say something like this: "I will now ask that all beings be benefited by more personal joy that supports them and inflicts on no one else." Creator might say something like that.

Zoosh, he wouldn't say it; he'd feel it, wouldn't he?

Well, he wouldn't say it in English, obviously, or even Swahili, but Creator would feel it. If I could interpret what Creator is feeling, radiating, it comes out in the form of a request, not a demand or even an instruction. Creator is not even upper-level management. Creator is a portion of All That Exists.

Regardless of how others feel toward Creator, Creator does not deify Itself. It is not omnipotent. Creator's attitude about Itself is not the least bit omnipotent. Omnipresent, yes, but that is to be expected in a being who has reasonable consciousness, which Creator does. And by "reasonable consciousness" I mean It has the capacity to be aware of most of Itself and has the faith to know that all of Itself might not be present. In this way Creator includes the unknown as something beyond that which is conceptual. Good questions.

It seems to me that the answer to everything is that by the time we get up to 3.70 or 3.75, we're going to have abilities we don't have now, and everything's going to change.

Well, the most important thing to understand is that you're going to have abilities that you *do* have now. [Chuckles] It's essential to understand that. If you don't, you will disempower yourself. You have these abilities now, but because of the way you've been raised, most people on this Earth do not realize that. However, there are a few individuals out there who have been raised in reasonably sacred ways and who know they have these abilities. Because they have never had them taken away by innocuous remarks such as "that's just your imagination, dear," they have no knowledge they *cannot* do these things.

Children can be the best benevolent magicians in your society because it's difficult to go through life without running into some form of disbelief. You might find in a particular ashram or temple a child who is able to do many magical things, and no one ever tells that child (or even *feels*) that it is limited. And this child will demonstrate to those people by what it does what *they* can also do. I mention this because there are a few kids like that out there now.

So a major theme in this book is: Everything that you will be in 4.0 is awakening in you now. Honor it, start using it.

Yes, start using it as if you had it now. You can now tap into your future self (although not as powerfully as you might in 4.0) and, like learning how to ride a bicycle, when you practice and practice, after a while you get better. It just might be that by the time we're at 3.49 you'll already have it. You might not be creating planets, but on the other hand you might very well do something like the following. As you're traveling to work you might drive over the top of the hill, then suddenly glimpse a traffic jam ahead. When your view is obscured by another hill and because you know how to apply benevolent magic, you say to yourself, "I will easily be able to pass through this and get to work on time." Because you like your work, you are motivated. As you come up over the crest of the hill to where the traffic jam was, you are able to get through easily.

Now, I'm not saying that will work all the time every time, but as you get more attuned to bringing in energies of benevolence and projecting a clear space in front of you, cocreating with the Creator and using exactly the same methods of creation that Creator uses, there's a pretty good chance it will happen frequently. Eventually it will happen most of the time, and not too long after that, *all* the time.

So begin now, but pay attention while you're driving.

So we've exhausted the subject of healing.

We haven't exhausted it. We could talk for a long time on various healing methods that are coming in that might be used, but I'd rather put the focus of the gentle reader on this one point, because it's something you can do at virtually no expense, something you can begin *now*. Even if you are living under the most difficult circumstances, it might be possible during certain moments to begin.

Prayer

Remember that all prayers or requests must be benevolent. It's not going to happen if you say, "Oh, I wish she was out of my life" or "I wish that she would change." What might work in that circumstance is, "I would like or request that the circumstance with another person be lightened into something benevolent from which I will learn benevolently something new I can use to help myself and perhaps others, and that this argument will turn into a set of insights that I will then be able to integrate into my life. Thank you."

Always end with "thank you." Creator does; why can't you, eh? Creator thanks Itself, but also *all beings*, including that which Creator might not be aware of. I'm emphasizing that because it's very important to understand that Creator is perfectly comfortable with the idea that there is something going on somewhere that It is not aware of. Creator would not consider that ignorance, simply that there is no need to know.

So by A.D. 2050 are we off the planet and out into the stars with the aid of space travel?

I'd say that by 2050 you will not be to the stars yet. Certainly the more advanced spiritual people amongst you will be doing that without ships, but if you're asking whether you'll be gallivanting about in the universe freely—no, I don't think so. Because the universe, including comets and other planets, just people near you, are going to want a certain elapsed time. They might be perfectly willing for you to have a vehicle that will take you from here to Pluto or Saturn in a couple of hours; they'll be less willing for you to get from here to the Pleiades in fifty years. They're going to want to wait and see.

"They're all right now, but they've had other times when they were all right," say these ETs. "Let's give them time and see if it's actually going to last." I'd say that that bit of understandable balkiness on their part will probably be resolved by 2075, and by that time they will accept you in their midst, though in small numbers at first. That's understandable, is it not?

An Exercise in Benevolent Magic

Zoosh, I have a question about benevolent magic. You can't go around benevolently wishing for good for other people, can you? Can't this interfere with them? Should I wish or ask or put my attention to making my sister happy? That's somebody else's life. I can only do it for myself, right?

I want you to *begin* with yourself. Let's not say what you can and can't do at this stage. Let's say that I would prefer that you begin with yourself. If you're having an interaction with your sister, for example, and it's difficult to talk to her, then you might ask that that be resolved in some way that works for you benevolently. But can you simply ask that your sister's life be cleared up and that everything be hunky-dory for her? You can ask, but not in terms of benevolent magic, because *benevolent magic cannot be controlling.*

Thank you for bringing up that point. You can offer a blessing for your sister even at a distance, saying, "I will ask that benevolent energies be with my sister so that she might have the ability to see and feel the value of more loving, insightful ways of being." That's more like a prayer —you say it once and that's it. You let it go and forget about it the best you can. You can do that; it's like a blessing.

You cannot say, "My sister is stubbornly making the same mistakes over and over again, and I ask that she be enlightened and not continue this pattern." In the first place you have to ask for what you *want;* you can't ask for what you don't want. But I'd like you to begin with yourselves, because even if you do this regularly, allowing for all of the mistraining you've had in your lives, it will take many years to integrate this so well that you don't get caught up in old patterns that cause you to self-destruct (as you used to say, "run old tapes").

It will take years to get it going, so I'd rather see you get it going for yourself first. By the time you get around to asking for others, there's no harm in asking for something for the world. Let's say that you get an uncomfortable feeling in the process of asking. Then you would stop, because that uncomfortable feeling tells you that it is not appropriate to ask at that time or else that you need to rephrase it. If you feel rephrasing is needed, then do that. If you keep getting that uncomfortable feeling as you're rephrasing it, then let it go. Don't finish it; simply say thank you and forget about it. Your feeling will feed back to you whether it's an appropriate request. If you feel good throughout this request, then you can be reasonably certain it has been heard and you can forget about it. That doesn't mean that you don't take action toward improving your own physical condition, whatever it may be, but you say it and let it go.

Try to ask for benevolent, loving energies to come into you before and during your saying of this so that you have the opportunity to make it something other than simply a statement of intellectual desire. You want your feelings there. You want as much love there as you can summon, and sometimes in order to bring up that love and comfort, it might require imagining the condition that you wish to take place and bringing up those feelings. Then when those feelings are present, focus on maintaining them and say, "With this wonderful feeling of how I will feel when I have this condition, I ask that this be brought about for me in my lifetime now." You add "now" because that allows it to begin now. It might not be brought about tomorrow or the next day, but it will begin. I know I'm not telling you anything new as much as that this is now the time to do it. As it begins to work for you, share it with others.

Do you use love or desire?

Love.

What's the difference?

You understand intellectually that desire might stimulate you to ask. But the energy you use to the best of your ability is *love*. If you cannot identify it within you, then *peace* and *allowance,* otherwise known as the experience or the feeling of letting go. As I say, letting go is easy to practice.

It is?

Very easy. Yes, letting go has another capacity. Be in the feeling of *giving up* — that is exactly the same feeling as letting go.

Letting go and allowing are really hard concepts.

Self-Distrust vs. Permission to Be Yourself

Yes, it can be a hard concept when you've been taught that you cannot trust that part of yourself, but in point of fact, that teaching [self-distrust], while often given by the most benign and benevolent and loving beings (your mother or father or grandfather or grandmother, uncle or aunt) is actually the beginning of disempowerment.

I'm not saying that you should raise children without some sense of self-discipline. Certainly that is of value. I'm not saying that you should allow them to be wild and run around doing whatever they want. But I am saying that if they say something that sounds imaginative, like "I talked to the fairies today," just be interested and say (if you can be interested in that moment), "Oh, that's wonderful, dear. What did you say to them? What did they say to you? Can you tell me, or is it a secret?"

Do you understand? Say it with the feeling of true interest. If you cannot do that in that moment, just say, "That's wonderful" and *feel* that it is wonderful. It doesn't mean that it's wonderful for you. You don't have to talk to the fairies if you don't want to, but if your child has just spoken to the fairies and says so, feel wonderful for that child, because that is a wonderful thing for that person.

It's all in the feelings. Thoughts allow you to get there. Obviously, I put out a lot of my so-called thoughts on pieces of paper, with little scribbles that you can read. But basically all I am doing here and all I have ever done is give you permission to be yourselves. And if I have to do that intellectually, I'll do it.

In coming weeks I and Progenitor will talk about the future and how it will be. Tonight we did a little bit of that. We might actually talk about healing practices and what form they will take in the near future compared to the long range. I was talking about the long range tonight that you can integrate now, but I might also talk about the the practices that will begin, if only to encourage and nurture those who are beginning them. And we will talk about the future of other things, if you wish. But you must ask, and then I will know that you want to know. For now I'll simply say good night.

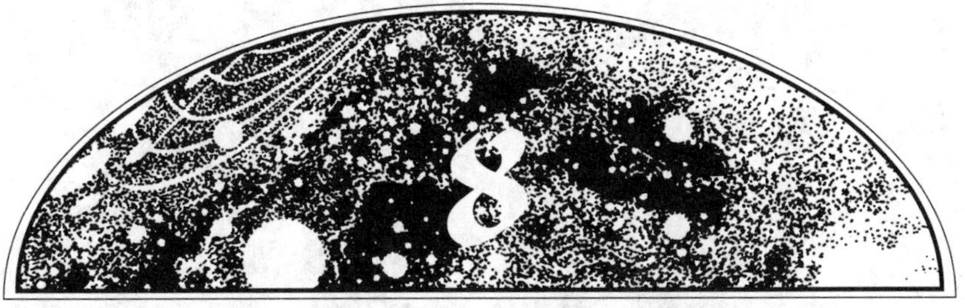

The Body of the Child, a Pleiadian Heritage*

May 14, 1996

I need to bring up the subject of the body of the child, its origins and its purposes because sometimes overlooking the obvious is easy in your complicated world.

When you start out life physically, you start out as a small child. You start out needing support (you're not exactly helpless, but pretty near) and you have an entirely different perspective. For one thing, as a baby you are still in touch with spiritual teachers, past and future lives, all forms of spirit that are nurturing, and often ETs as well (friendly ETS whom you know personally). For most children this will leave after the age of three. By that time you are beginning to learn the rudimentary language of your people and it has fully dawned on you that telepathy on the emotional level works only marginally, and on the communicative, word-based level not at all. You realize then that you are going to have to do more than make little noises to be understood. That's's when you let go of your natural form of communication.

Babies are not born stupid, but they have to narrow their perceptiveness to ease into the current level of society, thereby slowing down.

---

* In the appendix see "The Body of Man" and "The Body of Woman," previously appearing in the December 1993 and January 1994 issues, respectively, of the *Sedona Journal.*

Soul Guidance from the Crystals' Songs

When a soul is born in the body of a baby, its energy vibration is quite high, but having to ease into the mainstream of your civilization, it must densify a bit. It also needs to adjust to the world's perspective, experiencing its world from the eyes and perspective of someone small. Everything is very large to a small baby. You are born small for obvious physical reasons, then grow larger, although there are societies elsewhere in which a person is born at the size they remain. But here it is intended that you have the full range of perspectives – that of the child, the adolescent, the adult and the older person.

To adequately and reasonably see where the body of the child originated, you might assume you would look toward Sirius or Orion, since women come from here, men from there. But the whole perspective of the child – the spiritual range, the focus, the flexibility – does not come from either Sirius or Orion; here we have to factor in the Pleiades. If you put Earth people and many Pleiadian people side by side, you would not really be able to tell them apart because Pleiadians look just like you, at least when they are fully dressed.

The concept and experience of childhood and the baby as a spiritual being is directly connected to the Pleiades. When you start out life you are required to learn how to play, which is something that has been perfected as a benevolent applied science on the Pleiades. On Earth one must also move from certain means of spirituality to certain other means. But on the Pleiades children are raised very differently; it is well understood that there are two times in the life of human beings when they are most spiritual, not counting individual moments. Human beings (including Pleiadians) are at their most spiritual in their very early years and their very late years – assuming they experience what you call senility, the bridge between worlds that allows movement from the linear physical world into the world of spirit.

Spiritual Education for Pleiadian Babies and Adolescents

On the Pleiades when a child is born they don't wait very long to begin its spiritual exercises and education. When the child is less than a year old, maybe at about eight months, the mother, often with the father, will bring it to a very large room that is softly padded so that the child can roll around or crawl without damage. The child is then encouraged to communicate in its language. Even on the Pleiades a child is not yet verbally communicating; however, it is understood that when the soul comes in it has a significant memory of its most recent past life or perhaps several recent past lives, some of which might be profoundly spiritual. So the Pleiadians, for their own benefit and for the later benefit of the child, glean as much information as they can while it is still a baby.

In the room there are several very receptive people, or empaths (here

When a soul is born in the body of a baby, its energy vibration is quite high, but having to ease into the mainstream of your civilization, it must densify a bit. It also needs to adjust to the world's perspective, experiencing its world from the eyes and perspective of someone small. Everything is very large to a small baby. You are born small for obvious physical reasons, then grow larger, although there are societies elsewhere in which a person is born at the size they remain. But here it is intended that you have the full range of perspectives – that of the child, the adolescent, the adult and the older person.

To adequately and reasonably see where the body of the child originated, you might assume you would look toward Sirius or Orion, since women come from here, men from there. But the whole perspective of the child – the spiritual range, the focus, the flexibility – does not come from either Sirius or Orion; here we have to factor in the Pleiades. If you put Earth people and many Pleiadian people side by side, you would not really be able to tell them apart because Pleiadians look just like you, at least when they are fully dressed.

The concept and experience of childhood and the baby as a spiritual being is directly connected to the Pleiades. When you start out life you are required to learn how to play, which is something that has been perfected as a benevolent applied science on the Pleiades. On Earth one must also move from certain means of spirituality to certain other means. But on the Pleiades children are raised very differently; it is well understood that there are two times in the life of human beings when they are most spiritual, not counting individual moments. Human beings (including Pleiadians) are at their most spiritual in their very early years and their very late years – assuming they experience what you call senility, the bridge between worlds that allows movement from the linear physical world into the world of spirit.

Spiritual Education for Pleiadian Babies and Adolescents

On the Pleiades when a child is born they don't wait very long to begin its spiritual exercises and education. When the child is less than a year old, maybe at about eight months, the mother, often with the father, will bring it to a very large room that is softly padded so that the child can roll around or crawl without damage. The child is then encouraged to communicate in its language. Even on the Pleiades a child is not yet verbally communicating; however, it is understood that when the soul comes in it has a significant memory of its most recent past life or perhaps several recent past lives, some of which might be profoundly spiritual. So the Pleiadians, for their own benefit and for the later benefit of the child, glean as much information as they can while it is still a baby.

In the room there are several very receptive people, or empaths (here you might call them psychics). Their main duty in that situation is to receive the maximum amount of visual and tactile information by directly connecting with the baby and (if we can use the computer term) downloading the knowledge it came in with, using special colored crystals that are laid out in front of them, not unlike a large computer keyboard. As they get different levels of information they move their left hand (occasionally their right) over the crystals, acting as a conduit to download the baby's information, knowledge, wisdom, visuals and even senses of touch and motion into the crystals for storage. These are library crystals that are toned to certain colors and sounds classified by the Pleiadians to represent certain behaviors, wisdoms or levels of knowledge.

This might take anywhere up to forty-five minutes while the child is crawling around the room and playing with various toys and whatnot. In addition to benefiting the civilization, this does a lot for the baby because it's spared the worry that it will forget who it is, where it came from, who its friends are and so on, which the Earth baby forgets. ("Forgets" means on Earth that this knowledge and wisdom pass into the unconscious, or in rare cases the subconscious, to be tapped at a later time in life.)

Then the empaths relax. The crystals are coded holographically: a light ray indexes exactly which crystals that baby's information was loaded into, how much of the capacity of the crystal was used and the location in that crystal of the holographic information (very important). The average baby might use portions of, say, 68 crystals, because different information would go into different crystals. The child goes on with its play. It might, for instance, utilize something in the room to learn how to move objects without touching them (*telekinesis*, a fancy word for moving objects without touching them). This is done in a friendly and benevolent way (described at length in the book *Allies*, with illustrations).

One of the things that's missing in your civilization is a means for the baby to store its information to call on later as an adolescent. People live longer on the Pleiades, so the years would differ, but the cycles are the same. On Earth adolescence usually begins when a child becomes a teenager. That is the beginning of the cycle where chemical changes take place and adolescents seek their identity and personality apart from those they were raised with (parents, grandparents, uncles, aunts, brothers, sisters).

On the Pleiades this is when young people would go to the crystal storage room and the crystals would be laid out in front of them. They would move their left hand over these crystals (the crystals being programmed by a technician) and draw out various information. They might move a finger to touch a crystal and see what knowledge or information is in it that they brought in at birth. This helps young people to establish their identity (their soul lineage, if you would) and also connect with spirit guides, extraterrestrials beyond the Pleiades, other-dimensional beings and so on.

This is missing on Earth, but it is coming. The reason I am going into this in some detail is because you are very close to being able to do this now. To young parents or those thinking about having children soon, here is something you can do with your baby anytime from three to eight months after birth.

An Exercise for Earth Babies and Parents

Pick out crystals – say, an amethyst or two, a couple of optical-grade quartz crystals (either uncut small ones or big ones, but they must be clean; you'll know a crystal is clean when you hold it in front of your heart and you feel fine). Put them in the baby's room (someplace nice) near the baby but not close enough to reach, and leave them for two days. If you know a very clear empath, ask one to be with baby. Place the crystals near the empath's left hand and ask the person to do what is done in the Pleiades – four to five minutes is enough. (A clear empath on Earth is equal to one in the Pleiades.) Then remove the crystals and put them in a wooden box. Ideally the wooden box is compartmentalized and lined with silk, each crystal in its silk bag. If you have no box, simply put them in silk bags in a drawer, then just wait.

When your child is easing into adolescence, from eleven to thirteen years old, still young enough to appreciate their beauty, give the crystals to her. See what happens. She just might be able to pull something of her true spiritual identity from those crystals into her life.

Is it too late after they're eight months old?

The best time is between three months (the first three months is spent bonding) and eight months, when some might be trying to use words. Thought will be interfering by that time. One exception is children who have Down's syndrome. With those children this can be done up to the age of five. Most children with Down's syndrome are strongly surrounded by spiritual beings until they are at least seven, many of them during their whole lives, especially when supported and nurtured in spiritual ways. These are profoundly spiritual children.

I'm bringing up the subject of the body of the child tonight because I want you to understand that this book is primarily about origins and purposes. We've talked about origins in the original Explorer Race book, which many of you are reading now, but in this book we go a little deeper into it. In succeeding books we're not going to do much of that

What percentage of civilizations out there reproduce sexually?

Which percentage reproduce by having children, as you understand children to be, and use the same means of reproduction you have here?

Yes.

In terms of your universe, about 13%.

Are the others cloned or created, manifested or . . .

Yes, all of the above and more. The birth cycle as you know it here has everything to do with creation and lessons in creation. It also has to do with responsibility toward the feminine, because creation is largely a feminine process. That is why the creation cycle is basically a feminine activity.

It seems odd to me that tonight's topic is "The Body of a Child." It looks more like it's about the nurturing of a child or getting information and then giving it back later, rather than about the physical body.

Well, no children are going to read this book—I'm taking that into account. Men and women will read this book, so I'm talking to the target audience. I'm aiming what I'm saying here largely to those who live with children rather than to children themselves, as I did with the races. And I'm not really identifying a child as a race of beings; I'm just filling in the gaps.

The Explorer Race

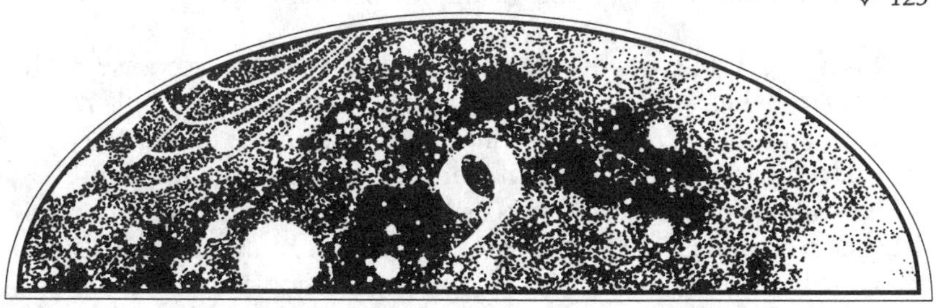

Creating Sexual Balance for Growth

September 10, 1996

All right. The races that have been mentioned represent the root races, and I don't want to go into anything beyond the root races.

Your Predominantly Feminine Lightbody

First I'd like to talk about something that occurred many years ago before you made the decision to create the bipolar sexes. You existed in a physicalized version of your etheric lightbodies that combines elements of both sexes but is predominantly feminine, in your current thought – not physically, but energetically. It is the feminine characteristics that allow one to feel united and in touch with all beings, though men obviously can do this as well.

Before the polarization of the sexes there was a stage where you basically densified your own lightbody. But there was so much feminine energy that the opportunity for growth was limited. [See chapter 6 in *The Explorer Race*.] You might ask, "Zoosh, are you saying that feminine beings do not in their own right grow?" Yes, that *is* what I'm saying. You might say, "That's a pretty shocking statement." But in reality, even now on Earth consciousness-raising groups of feminine beings (women) are simply reinstituting their natural way of being. The natural feminine way *is* the way of most of your universe's beings.

This might be perceived as mostly benevolent. On the other hand, since your Creator wants to stimulate a growth cycle, it was noticeably necessary to bring up the qualities that have been ascribed to the masculine being (which I would call, for the sake of simplicity, the physical and the mental) and to make the mind more of a daily experience and physical action more of a direct relationship to the mind.

In this way [chuckles] it is possible to make mistakes more easily. This is not done because Creator is ruthless, but because Creator realizes that the veil of ignorance, which is upon you even today, is basically inadequate to stimulate mistakes that help you learn if you are still primarily polarized to the feminine. If you are, the tendency is to live through **feeling** and **inspiration** rather than **thought** and **action**.

This is why it was necessary to increase the masculine energy. When the mind is confronted with ignorance, it will restlessly seek the "truth." And when the mind pursues a quest toward the "truth," it will inevitably latch onto something from time to time that feels wonderful. That wonder will not last, because the limits of that truth will be revealed when physical actions are taken to support and sustain it.

The Masculine Raises the Learning Curve

I'm going into all this now because the primary root race of the Explorer Race as represented in the human being is basically a combination of the sexes in the physicalized version of the etheric lightbody. It is necessary to appreciate the learning curve that the masculine can place in your daily experience and to absolutely understand that being a woman or a man is unnatural to your normal state of being. Normally you are *one*. You are normally neither a woman nor a man, although you have these lives now in this current time of learning. Your normal state of being in your lightbody is a lightbeing characterized by its feminine qualities, and your immortal personality is engaged within this vessel. I want you to understand that.

I also want you to understand that the Creator fully recognizes that the potential for violence springs out of ignorance. However, violence will be uncreated to a large extent during that evolution of time-space you are going through now. Creator was not entirely comfortable with taking the risk, but It needed to do *something*. Creator intended to use only a small amount of negative energy, but that energy ballooned much larger.

The true root race of you all is, on an individual level, primarily a *feminine humanoid lightbeing*. We could take it a step back to where it was simply a point of light, but I want to make it reasonably humanized for you.

Can you place it before or after Sirius?

It was before the feminine prototype human being was created on Sirius. When it didn't work to bring about the desired growth curve for the Explorer Race, it was requested that the Sirians create the sufficiently polarized beings that you are now.

When did our brains get split? They're coming back together now.

Oh, I think it was always designed that there would be different hemispheres and different levels of the brain, such as the primitive brain . . .

The reptilian, the limbic and the . . .

Keeping the Mental Body in Check

Yes. They were always designed to reduce the effect of the mental body, to keep it in check. A natural brain that is the physical representation of your lightbody would be so powerful and would have such capacities, mentally speaking (the Zeta Reticulans have a brain like this), that you would all be brilliant beyond the measurement of IQ tests, making growth *again* less likely. The chances of such individuals making the same mistake that other individuals had already made would be nil. If one person made a mistake, that individual might learn from it, but that information would be spread telepathically to all human beings and no one else would learn from it. You might learn in a larger sense, but not *physically*. If one person stubs a toe you learn to watch where you're walking. The pain of the lesson would not relate to all other beings, however; everyone would simply get the idea "watch where you're walking." So the brain had to be reduced in complexity, thereby in efficiency.

Since we were so unconscious then, the reptilian brain would run the body and the limbic brain would allow all that emotion to run havoc and to create lessons?

You must remember that it is primarily the emotional self that keeps you connected to Creator. Neurologically speaking, the word "emotions" is basically a mental interpretation of feelings. Without feelings you couldn't survive; you'd be dead, because you wouldn't know where to go to get what you needed. And not everybody is led around. The feelings relate to the instinctual self, and you cannot survive without it. For example, if you didn't have your instincts you wouldn't get the message that something wasn't safe, even if you didn't mentally know it. So we must understand that the *word* "emotions" is just a way of degrading the actual meaning of feelings.

But we've let emotions get . . .

The Instinctual Body

When a group of people (to anticipate your question) is not trained in the spiritual uses as well as sacred physical uses of the feeling self (the instinctual self), then the emotions will be an accessory, basically an unknown and soon-to-be distrusted accessory. But when you are trained in the instinctual body (which, as I said before, is made up of the **physical**, the **spiritual** and the **emotional** bodies), your life becomes infinitely more simple. Your decisions become quite easily accommodated and you make significantly fewer mistakes, especially painful ones. You make enough mistakes so you can learn, but not so many that you have to suffer.

Creator never intended for you to suffer. Creator always intended that everybody be trained to use their instinctual body. I will not say Creator made a mistake, but I will say that Creator might have underestimated the capacity of the mental body to become isolated from natural life and create arbitrary rules of living ofttimes based on compensation for not having something you want and at other times based on control. That is why most present societies do not teach how to utilize the instinctual bodies. If the instinctual body were a portion of an automobile, then it would represent the engine *and* the brakes. If you don't know what to do with that part of you, you just sort of bounce through life. You spend perhaps forty or fifty years learning how *not* to do things and wind up being very much stifled, perhaps, because you do not have the proper means to make decisions. *Mistakes are perpetuated ad infinitum by learning mentally*. But that's another matter.

We're soon to go back to the instinctual body and we can train it. Is that ahead of us?

Yes, I've actually given some of that training. I might do it again at some point if you wish.

As we move up the dimensions, won't . . .

You don't have to move up the dimensions to learn how to use the instinctual body.

Even if you don't do this training, won't we grow into or change into that?

Well it doesn't happen by magic. You learn just like you learn how to drive a car. Someone doesn't come along with a magic wand and say, "*Pouf!* Now you can drive a car." You learn the rules, how to look down the road while you're steering, and to press the gas or brake pedal. You learn how to do many things at once, and once you learn how to do it, it's second nature. But you do have to learn. It is the same with the instinctual body. It doesn't take much more than a day to learn how to do it, but you have to apply yourself.

So you have to learn. It's not something that just comes.

That's right. It doesn't happen like *pouf!* You learn—another human being teaches you.

The splitting, the inserting of the corpus collosum—was that before the Sirian prototype, or was it before Orion? At some point there was a whole brain in density, wasn't there?

When the being that was basically a humanoid lightbeing self was solidified into something that looked very much like a present human being, that brain was a single mass. It is, in terms of your nearest neighbors, most closely associated with Zeta Reticuli. These people are brilliant.

I had the idea that we used such control wrongly, that we did something that caused someone to come and change the setup. But it is just that we didn't learn.

It's that the brain was too efficient, not that something was taken away from you because you messed up. The brain was so efficient that the chances of your learning in a reasonable amount of time were greatly reduced. Look at the Zeta Reticuli civilization. They have not had the growth curve available to you because they tend to learn en masse, and this makes their growth curve *very* slow. It's sure, but it's very slow. In terms of their capacity to grow, it is *way* behind yours. In terms of their ability to adapt quickly, that also is way behind yours. But in terms of their actual intelligence quotient, that's off your test scale.

You were talking about the individual body, and the emotions and the intuitiveness of the physical body in the lightbody. Do the emotionally attached senses keep a person from understanding and reaching that intuitive body?

Are you saying the emotions do that?

The emotions of the physical realm, as we are taught in 1996, yes.

Well, the emotions are actually the avenue to the instinctual body, and this can be felt. What is physical attention? Is it a visualization? No, it is a feeling. Here is foundational homework on the instinctual body.

Instinctual-Body Training Exercise

I'd like you to bring your physical attention, your awareness, down inside your chest to the solar plexus area. Now generate or feel (either one) a physical warmth that is physically discernible.

Go ahead. Let me know when you have done so.

Okay.

I want you to practice that tonight before you go to sleep. Tomorrow or the next day, whenever it's appropriate, feel this warmth as much as you can. Go into it. *This is the physical evidence of love*, that feeling that holds all matter together. I would like you to go into the feeling and feel it as much as you can.

Then tomorrow or the next day, whenever it's convenient, go out on the land somewhere and utilize these fingers like this. The best way to do it is with your left hand because that hand is less likely to be willful than your right hand. (I'll explain the reason someday, but we're cutting to the chase right now.)

Walk around to various rocks or trees and notice where you feel the warmth strongest (pointing toward a tree or rock). If you notice the warmth come up *very* strongly, then say thank you to that tree or rock and approach it. Don't go over and touch it; just approach it anywhere from ten feet to two or three feet. If you get too close, you might feel it less because the tree, not knowing what human beings are going to do, might become concerned.

Continue to feel the warmth as you interact with the tree. Try not to think; devote your full attention to being inside this warm part of your body. If you notice the warmth beginning to fade, see if you are thinking anything. If you are, don't give yourself a hard time, just say, "Oh," and go back to being inside your chest or solar plexus. Maintain the warmth with that which welcomes you (which, in my example, is the tree). That's your second level of homework.

Practice one thing a day. The third level of the homework might take a couple days or a week – take as long as you want. Go someplace again on the land, preferably someplace with a clear space around you. A high point would be nice, but it's not necessary. Please do not stand on the edge of a cliff, but be someplace where you can see around. Look where you are standing – we don't want you trudging through any anthills. We want you to be able to walk freely out on the unpaved land.

Then use your hand in this gesture. I'm going to call this *the wand.* Put your left palm upward and just move around very slowly in a circle, imagining that you are the second hand of a watch. Move around counterclockwise and stop when you have completed the circle. Then put your hand out like this, with the fingers open or closed, and ask, "Is this a good, safe direction for me to go?"

Notice how you feel the warmth. The warmth reacts to your question. If the warmth is present, it's a good direction. If it's only slightly present, that means it's an okay direction. Put a marker there: a pebble, a button – anything. Go around, again moving to the left. Starting at, say, 12:00 o'clock, go to 11:00, 10:00, and so on around the watch face and ask your question every time. Notice where the warmth is strongest.

When you are done, try to get the general idea of which compass direction is good. Ask the question, "Is this a good way for me to go?" If you choose, walk a few steps that way when you are done, or just take note that, say, northeast or southwest is a good direction to go. If you have the opportunity sometime, travel in that direction for a short drive (not miles and miles) and do the same thing again, directionally speaking. Get out, stand someplace and again move around the dial, pointing, "Is this a good direction for me to go?"

One might say that this correlates to triangulation with mapping a place, but it's instinctual-body triangulation. One can find a good place to go, though you might not know *why* it's good. You can in time get more specific with your question: "Is this a good place for me to go to find work? Is this a good place for me to find pleasant companionship? Is this a good place for me to go to find a good relationship?" – whatever question comes to mind. The question must be simple and it must be about what you want, never about what you *don't* want. For instance, ask, "Would this be a good direction for me to go to be safe?" not

"Would this be a good direction for me to go to avoid danger?" Always ask for what you *want.* Take note of the direction, and after you travel that way for a time, stop and check again.

This can be utilized, once it is learned, to keep you safe and lead you to things that are worthwhile. This is primarily instinctual-body homework. Eventually you won't have to point with your hand; you'll just be able to stand or if you can't get out of the car (it's raining or whatever), just use your left hand and sweep. "Which is the best way for me to go?" says you (like this, with the sweep).

Let's keep it simple. Eventually you'll be able to get out of the car, stand in a circle and move around without pointing anything but your physical body in that direction, asking the question out loud. You *must* ask it out loud because you need to do something physical. This has to do with the instinctual body, which is physical, emotional, spiritual – not mental at all. But the mind can take note. This exercise presents the mind with what it always seeks – *it wants to know*, which is the physical evidence the mind is trained to seek; it is not because the mind is cynical. That is your primary instinctual-body training.

Thank you.

Creation of Souls

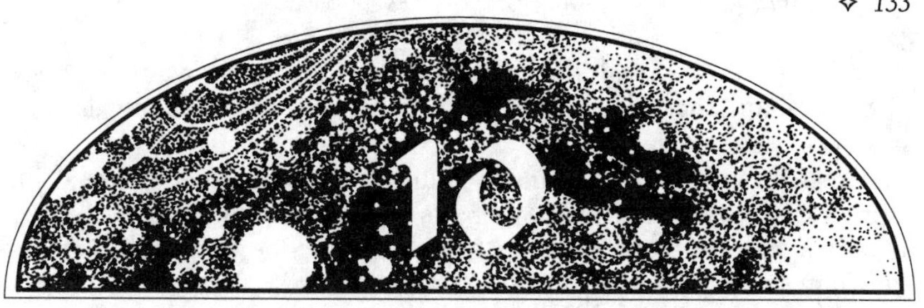

The Origin of Souls

May 21, 1996

Progenitor: What Zoosh was before, I am; what Zoosh will be after he's Zoosh, I am; and what Zoosh is, I am his guide.

It is stated frequently that you are one with the Creator, but this creates an artificial illusion that you were with Creator before It created this universe. But you were not. On the journey here Creator went to different places to see and absorb what has been created in other universes and to acquire light energy and light acquiescence. (This word "acquiescence" is not quite the English meaning; here it means "to be a portion of and included within.")

Alpha, Contributor of Individual Expression, or Fractionalization

In gathering the light energy that wished to come along with your Creator, tonal signs and cosigns were exchanged. As Creator moved along (in some cases physical motion, in other cases dimensional shifting) It was signaling a tone. From time to time different beings would join Creator. One of these beings was a multiple-density, substrate being. This being was multiply condensed – beneath the layer of consciousness but above reproach in terms of spiritual advancement. Creator, sending out (signing) a tone, was cosigned by this being in the same tone, which meant, "I will accompany you."

Creator asked this being what it had to offer and what it felt it might gain. This being said, "I am made up of many, many portions; nevertheless I am one. Yet these portions have never had the opportunity to experience true expression in individuality." The being went on to say, "I understand that it is your choice to express individuality in your creation."

Creator signed, "Yes, it is my intention to explore individuality." This being responded, "Might I, in my parts and in the sum of my parts, offer material that you can invest with your light, tone and intention and utilize in some form to support the idea of individual expression?"

The Creator said, "You will be the material from which I carve individual-identity souls." So this being got onboard.

You might ask, "Did this being have a name?" I will give the being a name—Alpha. Alpha, as you know, represents (at least symbolically) the beginning, and that is appropriate.

You previously told us what his name was, so you don't want us to say who this is?

I'm not talking about somebody who was ever mentioned before.

Zoosh once said that our souls were made up of his body.

This was *before* that. I have my point of view, in any event. I don't need to align my opinion with Zoosh's. It is my job to guide him, not to boss him around.

What do we call you?

I am the Progenitor. Now, the matter of this being was intended to provide generally a veil for an impersonal variety. It's true that Zoosh came along later and personalized that variety, but this being came before that and needed no recognition beyond that.

If you understand that your souls were then with Creator, this being provided the ability to fractionalize. Zoosh provided the ability to personalize and Creator provided the means of expression. Yet were your souls not occupied before this?

What *are* souls, in any event? Souls are basically branches of a tree. They are cells in the body of something larger. They are truly expressions of an organism that knows no boundaries. If you wish to do many jobs in one day, you might choose to split yourself in two so you could do much more. (Some of you here would like to do that.) This is the foundation of individuality for you, but it is not much different for a creator. We're talking about multiples, so it is basically mathematical.

The origin and purpose of the Explorer Race has always been to cocreate with Creator while assimilating as much experience as possible to prepare you for creator duties, as Zoosh has said. And yet there are certain things that Creator cannot pass on to you, such as on-the-job experience. I come through tonight just to announce my presence, because it is possible I will speak more in the next book. So I will say good night and let Zoosh continue if desired.

All right, Zoosh speaking. Well, there is my Old One there for you. His outlook and my outlook do not necessarily coincide in all ways, but that is perhaps appropriate, since my "individuality" is not dependent upon his point of view but is cognizant of it. What shall we talk about?

Can we stay on the origin of the soul?

Do you have anything more to ask about it?

Start with the spark that splits off or comes out of, is birthed from the Creator – how does it work from there to souls to humans? What is the line of light that makes all this work?

The line of light is not actually to humans; it's to whatever form the soul is gravitating toward. If we can assume that nothing happens by accident but only by design, the soul is asked, "What would you like to do with your life?" What a soul would like to do breaks down into a certain number of categories. It might want to experience life as a man, as a woman or as a nonpolarized lightbeing. These choices will assign certain levels of existence.

The Soul before Birth

In terms of material creation, the soul's expression from the point of light is condensed personality liquefied somewhat by the intention of Creator. The mass of the physical self here on Earth, for example, is not actually ensouled at the moment of conception, though that can happen. This conception is blessed by Creator, but the ensoulment rarely takes place at this moment. It will take place only if the soul wishes to experience the full range of growth starting from the point of conception.

Generally the soul begins to move in and out of the fetus sometime between conception and six weeks after. It will not usually stay in residence in the physical self all that time, but will move in and out to consult with its teachers, who will guide it about what to look for and about what life has to offer on the physical plane. The soul is usually present before birth on a full-time basis, but not necessarily in the body all the time until perhaps the age of three. The requirement to be present full time comes right around three years old.

If you are asking, "Is the lightbody formed first and then the physical body?" the answer is yes. First comes the lightbody, but it does not look like a human-being body. It is basically a point of light – a circle, since that is the universally accepted vision of wholeness in this part of creation. That form of the lightbody is quite adaptable. It will step in and out of the baby's body, which is forming within its mother.

You might ask, "When it steps out, does it always go immediately to its teachers?" No. Very often it will go around the physical world and experience things. For instance, sometimes the soul might hang around the environment of the mother and father just to get a feeling of what's going on. Occasionally a perceptive person might suddenly see a pinpoint of light. That could be a soul (and very often is), but sometimes it is a spirit ("spirit" meaning someone passing through). It can be many different things. It is certainly a lightbody in transit.

The soul lightbody can also experience physical life in other forms. Let's say the parents are going someplace for a leisurely walk. The soul in the baby might then step out (float, fly) and go into a tree to get familiar with the physical world, especially if that soul has not incarnated on Earth before or at least not for a long time. There might be a significant amount of interaction with other physical life.

However, the soul will not go into the father, although it might come near him. It will not go into the mother either, except for going through her body into the baby's body. It will not experience the mother's or the father's bodies because it's too easy for it to become programmed by their agendas and become confused about its own identity or (more likely) subordinate its own agenda to that of the mother or father. That could happen, so that's avoided.

Thus there is a certain amount of sequestering of the soul – it can go here but not there. It takes a while for that pinpoint of light to become what I call the human-shaped auric field. The auric field begins inside the body and radiates outward; it doesn't just surround the body. When the child is not yet fully formed as a recognizable human being, there is no need to develop that radionic space effect.

The Shapes of the Soul/Lightbody

But as the child comes closer to term, the lightbody, when inside, generally finds the center of the body (somewhere around the solar plexus) and begins to practice radiating out in the form of a human being, extending beyond the skin of the baby's body. Then the soul begins to identify itself as a human being. By the time the baby is born, if the soul should exit to talk to its teachers or experience something else, it is tethered to the baby (it gets corded right around six months) and can go and do things, remaining a pinpoint of light.

Throughout your physical existence the soul will retain that ability. It will be in the general shape of a human being when it's hanging around you, but once it gets even five or six feet away it becomes a pinpoint of light, because even though it's corded to you, it does not identify itself as a humanoid but as the symbol of wholeness, the circle – or in the third dimension, a sphere, though you will tend to see it in its two-dimensional appearance, rarely a sphere. It will never be ovoid, only circular.

When the body sleeps and you're traveling about in your soul, you are this pinpoint of light. Even though you might see or identify yourself as a physical shape in human form, you are still a pinpoint of light. In the case of dreams, as you get older and your dreams get more complex and human-oriented, you will sometimes experience a dream state where you feel as if you are someone ("someone is chasing me" or "I'm falling" —you know, traditional dreams) and you actually *feel* it.

Now, even though you see other beings, you are *actually in* that dream state (an actual reality) and *are* that pinpoint of light. You can feel physical experiences because you are tethered to your physical body; it is your physical body at the other end of the tether that actually feels these physical sensations. That's why in waking up after such a dream you will sometimes remember certain physical experiences. You know that it was not just a dream, but reality. (There is no such thing as "just a dream," but that is a term you have intellectually applied and use.)

Understand that it is the nature of the soul lightbeing to be a small point of light. I could actually give you a measurement: If you were to see it in its actual size, it would rarely be more than a quarter to a half an inch wide. It usually doesn't get any bigger —it doesn't have to. This is not to say that the soul cannot take other shapes. I've seen souls take other forms, sometimes an odd sort.

When the soul leaves the body and the tether is no longer there, that's what you call death. Of course, there is really no death, because the soul is the immortal personality. The body returns to Earth like a library book and you, the soul, go on. When this happens sometimes the soul takes a different form. Because the soul was in the body in the form of the lightbody and is leaving entirely, taking the tether with it and moving through the veils, you might get a different shape. I can draw you an example. Now, this [see illustration] is just a vague idea of a form it might take, and this form has to do largely with the collapse of the human-being form as it's moving out of body and re-forming into a sphere. (I mention this only because it was once photographed. You can find this photograph, which was real, by the way.)

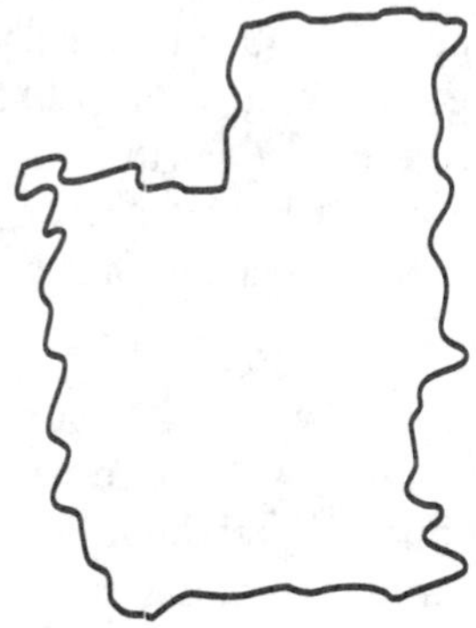

Your lightbody is this pinpoint of light. Sometimes it will be seen as a kind of flat white light, but most often as a pinpoint of very bright light. Sometimes they're spirits, but more often they're souls.

Soul Caravan Routes

Those of you who are living on a soul caravan route might see this sight quite a bit. A soul caravan route is a course that souls sometimes take because the energy along that route is more benevolent. It doesn't happen with everybody, but there are different routes that affect different areas. Some routes used to be under the sea, but they're not there as much anymore because there's too much undersea electronic shenanigans going on. The soul trails are generally over land now because even though you are suffering from an overwhelming amount of electronic pollution, there is actually less electronic pollution over some land areas (especially between mountains, where it's somewhat protected) than under the sea, where there's not as much protection from mountain ranges at shallow depths. Thus some of you will sometimes be in a position to see lots of these pinpoints of light.

You might ask, "How can I tell the difference between a pinpoint of light that is a soul being from Earth and a pinpoint of light that is a spirit?" Well, most of the time you can't, but occasionally you'll see a pinpoint of light or a mass of light moving through that is purple. This is usually a spirit, or even a group of spirit beings. Sometimes you might see it in gold or some other color. If it is any color other than what you identify as white, then it is probably a spirit being that has not coalesced into a form you would at other times recognize as a human being.

What's the step above the soul? One of our teachers says you must go beyond the soul, so what's the next step up?

You are striving toward cocreating with Creator, so there's not a huge step above the soul. People use certain terms such as "higher selves." The idea of the higher self has been worked a little bit too hard. If we're talking about steps between individuality, then the soul is an individual, your immortal personality. I often call it that because "soul" has too much religious connotation, so I don't like to use it much. The steps between the immortal personality and Creator level are very short.

I know others have said, "Well there's this and this and this and this and this," but there really isn't. There *might* have been once upon a time, because you used to need a lot of steps to get to this level of the Ph.D. program on Earth. It was a struggle to get here and you needed rest stops and breathers. Your immortal personality has a significant amount of endurance and can travel in different guises.

After "Death"

After you go through the veils and become your natural combined light-self, you become a condensed version, along with many other immortal personalities, within which you still have your individual identity. If someone were to stand back at a distance and look at you, you would no longer be a pinpoint of light, but more of a cloud of light. Passing through the veils after "death," you still maintain your identity even in your lightbody, even in that cloud of similar others as a world. You're walking and talking, you're being a human being, but at a higher-level dimension. But as I say, the observer would stand back and simply see a cloud of light.

You might say that this is a perceptual experience. If you're inside the cloud of light you see it one way; outside it, you see it a different way. This is not unknown to your lives now; where you stand makes a lot of difference in how you see something. That's one level. I'm talking about general levels, not positions you can take, choices, venues. The next level is right around Creator – not quite Creator, but as close as you can get to Creator without experiencing the full range of everything happening at once.

Now, if you are not going to reincarnate as an individual again, you have the option of joining a slowly developing, larger cloud of light, which is the accumulator for all lightbeings who will replace Creator – kind of like this. This illustration [see drawing] is a rough diagram. Creator is a cloud of light; underneath that is another cloud of light (Assistant Creator).

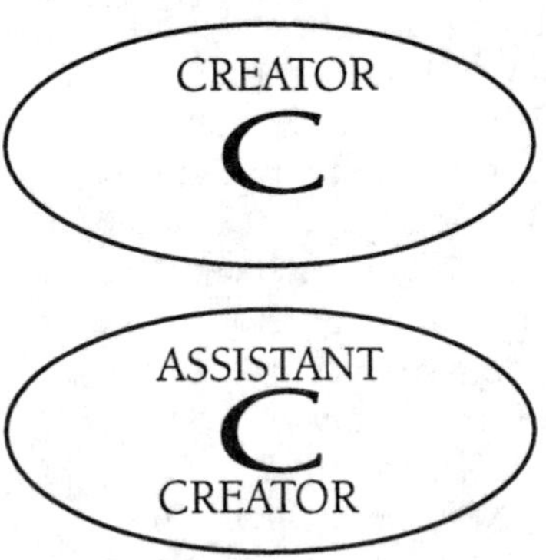

Your choice is either to join up with the cloud of light that's accumulating to replace Creator so It can go on and up, or to return to Creator Itself. This is an individual, personal choice. Now, you'll be interested in the percentages. You might say, "Well, we've come from the Creator, so we're going to want to go home and be loved by the Creator." While that seems to be true and obvious (although once you go beyond the veils you're basking in unconditional love anyway, assuming you're not immediately reincarnating someplace else), you'd be surprised *how* many choose to go to what I'm dubbing the Assistant Creator cloud of light, because there you are surrounded by others who've had the powerful experience of individuality.

Your immortal personalities (individual souls, if you like) actually experience individuality. You might say that the Creator experiences individuality through you, but you might also say that a mother or father experiences a child's life through the child. Having gone through the process (some of you more than others, being mothers and fathers), you know that even though your mother and father might experience life somewhat through you, they don't really know what you're feeling. For those who actually return to the fold of Creator, Creator might not have the full range of experience on the *personal* level of individuality.

Personal and Impersonal Unconditional Love

Here's the key: On Earth you get used to experiencing personal love. Your mother loves *you*. Your mother doesn't love *everybody* unconditionally, but specifically *you* (speaking idealistically here). And your husband or wife loves *you*. Your brother or sister (when they're not mad at you) love *you*. It's personal love compared to impersonal, unconditional love. (Of course, on Earth you will often experience personal *conditional* love. That's the reality of the way things are here now.)

When you go through the veils you immediately experience unconditional love. Although everything is Creator, there is a point at which you approach the center point, the point of emergence of Creator into the expression of some form of physical life or some form of identifiable individual life. As you're approaching closer to that point of emergence and get closer and closer to what Creator is, you begin to experience love unconditionally, but very impersonally. You're not exactly welcomed back into the fold by hugs, though you do get that experience when you go through the veils on the way back to Creator. When you go to the cloud of light that is the Assistant Creator in my illustration, forming up underneath the Creator (only for the sake of the illustration), you will experience personal unconditional love, because you'll be surrounded by others who have experienced individuality in all of its various ranges and forms—ETs, Earth people, everything. You'd be surprised how many individual souls would *prefer* to experience personal unconditional love rather than impersonal unconditional love.

Now, by saying "impersonal unconditional love" I'm not talking about something that is indifferent, but you do not feel your sense of self-identity when you're exposed to impersonal unconditional love. You feel a part of something wonderful ("light—we're all moving together; it's totally unconditional love"), but you are not reminded of your individual personality. However, when you're exposed to personal unconditional love, you feel cherished for being yourself. From 90% to 92% of souls go to that Assistant Creator cloud so they can experience *personal* unconditional love!

That's one of the things that has been developed here. When Creator came in and started this creation, the experience of unconditional love was basically impersonal. There wasn't an attachment or any sense of identification of individuality. But Creator has discovered that when a soul (an immortal personality) experiences one life after another of individualism, after a while that soul takes on a coloration of self-identity, which is a good thing. It helps to create themes; it helps to create a body of knowledge and wisdom and so on. But it also creates a need to be personally cherished as some one person rather than a mass of some greater thing.

A cell in your bloodstream is basically impersonally loving, but if a cell developed the need to be personally loving and to be loved personally, it would most likely express itself by drawing your attention to that part of your body. I'll give you an example: Often people cross their arms and legs when they're uncomfortable, nervous or worried – body language. But the points of contact, especially in the arms (which are more likely to be uncovered than the legs – cross your arms and see where the skin contacts), are where cells go that need to be personally loved. Because physical matter experiences personal love through touch, they will migrate to this part of your body to be touched and feel warmth and, being fairly close to your heart, feel a certain amount of love radiation. And when your arms are crossed over your solar plexus there is also an immediate interaction with your creation energy. So this body language is not accidental.

The expression and the need for personal unconditional love has been greatly amplified in this creation that your Creator has brought about. Once you've "excused" Creator, allowing Creator to go onward and upward to whatever is waiting there for It (I don't want to tantalize Creator, but you understand), the universe that *you* create will have, as its predominating theme, personal unconditional love compared to Creator's all-inclusive unconditional love.

Now, personal unconditional love is not exclusive, but it cherishes the individual creation, so that even though you become some mass of light in which you experience total unconditional love, you still experience your individual personality. Thus there is no loss or blending of the individual personal identity. This suggests that what you will create will tend to follow that theme, at least in your initial creation. So your version of individuals you create in your universe will most likely be highly personified. It will be interesting to see just what you make that into. This will probably mean that the experience of the withdrawn personality will become unknown, psychologically speaking.

Now, the withdrawn personality is very often created because of some conflict or unpleasantness suffered in early years. But because you are not going to desire personality to be suppressed, you are most likely going to create a more benevolent universe (which is a given, anyway) and have individuals in it who are highly personable. Thus an individual in one of your societies who might be considered eccentric or even egocentric here would be commonplace there – for example, the artistic individual who might have a significant countenance, perhaps an extraordinary way to dress, an unusual way to act, a more broad-minded attitude – or perhaps not so broad-minded. The idea of significant identifiable personality will be a foundation on which you will build the initial stages of your created universe.

I'd still like to understand the fragmentation of souls from what we call the monad. Where does the fragmentation take place? You said there were 144,000 souls who came here to become the Explorer Race, and every human who ever lived on this planet is part of one of those souls. How does that splitting or branching happen?

Soul-Seeding, a Way to Expand the Range of Options

That's a good question. When groups of individual immortal personalities come, there is a range of what they would like to do. What they can imagine doing in any given incarnation might be compounded by the fact that they're going from one planet to another, but it would still be limited by the limits of their imagination.

Since Creator wanted the Explorer Race to have the most complete level of experience, greater than Creator's own, the gaps and spaces of potential must be filled by the splitting or the seeding of souls. For example, let's arbitrarily pick two souls from that 144,000. Regardless of how many incarnations they are going to create in this creation, there is space between them. (I'm lining them up based on the range of their accepted and potential experience, assuming that zero is nothing and a hundred is everything.) If we break down those digits into say, 100,000,000 marks between digits, then obviously those 144,000 souls individually experiencing their range of incarnations are not going to cover all those marks.

Let's say that these two souls are close in the range of their expression, their development and their potential expression, but there's a big gap between them. What will happen is soul-seeding, not unlike the way a cell divides in the case of identical twins. This happens between lives. We have here one soul, then another soul [illustrates with a drawing] – two souls, yes? The division takes place and the two equal pieces of the sphere break off and form new spheres. One portion will go on and continue the incarnations originally planned by the original circle; the other will tend to fill in the gaps of experience on either side that weren't being filled in by the differences between those two souls' original personal agendas.

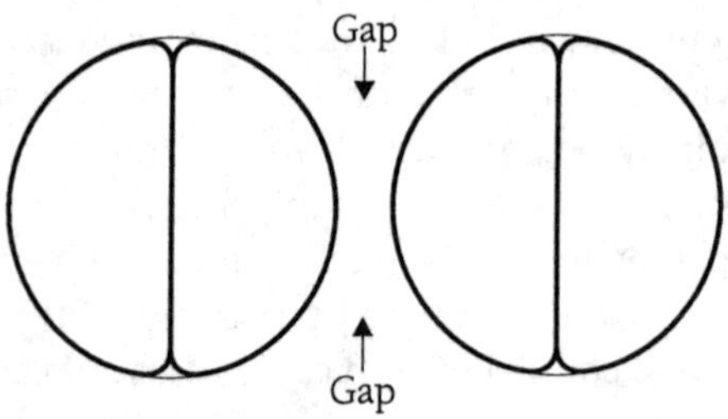

The split is very much like that of the cell [the zygote]. When the split takes place the twins are identical, but one will choose to continue the original agenda and the other the original agenda *plus*. The latter has a greater range of options to explore. It's as if someone came along and said, "Well, I know you're an editor, but now you can be both an editor and also a symphony conductor on the side." Or you might be told, "All right, this is your IQ; you now have another fifty points." You are offered a tremendous range of experience, but this greater potential means that you can choose to create extra experiences, and you will do that simply because you have more tools at your disposal.

Who gives you this greater potential?

Thank you. It is that which creates the impetus for souls, or Creator. In this case it is the nucleus of Creator's personality as carried on by Creator's "managers." (If Creator is the boss, then there are those who manage—the executive council.) This executive council falls largely under the heading of what you would call guides or teachers, not individual guides but a larger sense of a guide. I suppose we can use the term "monads" since you like it so much, but my understanding of monad is a little different from yours. Those who carry out Creator's instructions will supervise this to make sure it is done properly, but there will not be a program that a soul must follow; they are just given an option.

Light Generations (Creator Extrusions, Monads)

What's the percentage of monads in relationship to souls?

I'd say there might be five or six monads, period. They don't need many monads because their job is a fairly narrow range of what they have to do. And because they can do more than one thing at once and can be in more than one place at once . . .

Our perception that we're going toward a merging with our individual monad is wrong?

Well, the same word is given two different meanings. That's why I'm reluctant to use the term "monad." I will give them a title that doesn't sound so magnificent that you will create a hierarchy here. How about supervisors? No, that sounds too authoritarian.

These aren't the friends of the Creator; these are extrusions of the Creator?

Yeah, these are extrusions of the Creator. Let's just call them that. Sounds a bit technical, though. I'd rather call them *light generations*—I like the pun.

The Creator created those and then they *created the souls?*

No, they don't create the souls.

So what's the top of our line of light called that we have called monads? What do you call the spark of light that is the true individuality we are going back to become?

What you're talking about is basically what I'm talking about when I refer to the *point of creation* that is the center of the Creator. You're talking about this as applied to the individual. You can call it whatever you want. You can refer to it as a monad; the Latin word works well.

How many monads are there in relationship to souls? Does each monad have one soul?

You could say there are 144,000, but souls split and sometimes come back together—you know, split off, do all they can do, then eventually come back together. But sometimes they go on and split some more. Let's just say, for the sake of simplicity, 144,000.

Each of us, then, as an extrusion of the Creator has to merge with that 144,000 to become one again?

You make it sound like work, but it is natural. You don't *have* to do it. You're stretched to come here like a rubber band, and when you release the bonds on your physical life, you snap back. You don't have to do anything; there's no work involved at all. That is what you naturally are. The work involves *becoming the individual* – the individual expression – and then holding that. Holding that can be challenging. That's part of the reason you sleep, because it is not natural for you to be, even in this incarnation, something limited; it is natural for you to be significantly unlimited. So to experience yourself even for moments of time as something limited is a stretch. You can't hold it, so while your body sleeps you become what you truly are. Then you forget that again to be yourself here so that you can grow and change and all the other stuff we covered in the first Explorer Race book [*The Explorer Race*].

The 144,000 and the 2% of Creator

As an embodied soul here, we have alternate selves, possible-reality selves, probable-reality selves, higher-dimensional selves – they're all ensouled and each of those is part of that 144,000?

Yes. And of course, going back further (that story is a simplified one), the original creation of personality would be similar to an exponential increase, but on a larger level – one becomes two, two becomes four etc. The term "144,000" has been used largely because Creator (the story is true, but I'm going beyond it) initially wanted you to have within your grasp the means to extrapolate certain levels of *intellectual magic.* And numbers have the capacity to form the foundation of symbolic language and connect you to each other. (Your languages might have different symbols, but numbers are basically numbers even though different civilizations might create different symbols for the numbers.)

Creator wanted you to have an intellectual universal language and apply individual traits to numbers. Without getting into numerology, Creator wanted to give you, regardless where you come from, something by which you could symbolically interact with numbers. Certain portions of symbolic numbers (also known as symbols, such as the circle) fill that duty very nicely.

Creator did not want you to be lost, dangling at the end of the string without any hints. Creator wanted to guide you but not lead. It is Creator's wish that those who guide individual souls also guide them and not lead them – though it is very seductive indeed for a guide, even an old hand, to want to lead sometimes and say, "This way, not that way." But you know, that's how guides learn, because there are always consequences to that. Because you get to live with those consequences, you do learn.

You said only 2% of the Creator was expressing in individuality.

Yes.

Are the fragments of the 144,000 related to that? What about all the other parts that are not differentiated individually? How does that work?

Yes, because the experience of creation requires a lot of energy. It requires an unlimited capacity for constant generation and cogeneration and regeneration. Just think about the individual cells in an individual body (to say nothing of the atomic structure and so on) multiplied not only by everybody and everything living on the Earth and by Earth itself as a planet, but multiplied by all the planets, all the souls and all the beings ad infinitum. That 2% that jumps out of the Creator to become souls is actually the outer boundary of what Creator could afford to birth out of Itself. "Okay, I'm going to clunk along with 98% of myself."

You might say, "Well, Creator is still utilizing that 2% of matter that becomes individuals through our individual creations." Yes, that's true, and you could extrapolate and say that because of necessary limits imposed upon you so you can learn, it is still not a total use of your capacity even when you're out of the body. So Creator is, in a sense, really clunking along with 98% of Itself.

But that 98% is doing many other things besides just watching the 2%?

Oh, of course.

They're maintaining the creation.

That's right.

So how do we get all of the pieces of ourselves together and know everything that all of the little parts of ourselves have done and experienced?

Because you are here in still somewhat linear Earth, the idea of assembling, disassembling and reassembling appears to be work. But remember that you're stretched out on a rubber band to be here, and all of the other individual parts of you are somewhat stretched out, too. There's a lot of tension when you come to a place like this where you're constantly tested and going through creation exercises, consequences, responsibility, flexibility and so on. But the fact of the matter is, you don't have to gather up all your parts. Once you are no longer here in the Ph.D. program but are just having lives at higher dimensions where everything is nicey-nicey, there is no gathering. You don't have to go hunting; it's right there and you've got access to it all. You don't have access to it here—that is the veil of ignorance. But in the core of your being you do have absolute access to it (when you're asleep, go out of the body and wander around doing things). You don't remember that when you come here, because in your wakeful state and even, to some extent, in your subconscious state you must have the veil of ignorance so you can re-create your reality. But you don't have to do any *work* to reacquire all that is yourself once you're beyond a working re-creation place like this. (There are a few other places that are variations like this, but this is the extreme—"this is where the action is, man!"

Where does that happen? When you go beyond the veil and become fourth-dimensional beings?

No, you're still attached to the idea of progression. The assumption here is that once you go beyond the veil after death, you go to the fourth dimension. But it isn't like that. The whole concept of numerically progressive dimensions is only a concept. Once you go beyond the veil, you are not so much in the fourth or in the fifth dimension; you go *beyond* the concept of dimensionality.

Are you talking about death, or when we actually come into the fourth dimension in this body?

No, I'm talking about beyond the body. When you die and step out and go through the veils, you come into this whole thing. You don't go to the fourth dimension; that is a concept.

We're at 3.47 now; when we get to the fourth dimension consciously in this body, then we're not beyond? We're still within the veil? Or what?

Understand that "this body" that exists right now is going to have to change cellularly to function in the fourth dimension. So "this body" – the body that exists as of this date in measured, quantified time – does not go to the fourth dimension. But the fourth-dimensional changed body will be in the fourth dimension, yes.

Will we still be ignorant then, or will we know everything?

You will have greater access, but you will not have total access because your lessons continue in a milder, less dramatic way in what is being defined as the fourth dimension. A lot of you will just go on to other places, but for those of you who go to reside there, negativity (otherwise known as discomfort) drops to 2% of your experience and everything becomes comparatively nicey-nicey.

It's going to feel shocking to experience the body changing form, isn't it?

You know, it's changing right now, and to some extent you will have experiences. Some of you who are especially sensitive are having seemingly bizarre experiences. You might sometimes see lights or catch a flash of something that you know was there, but when you look at it again it's gone. You will experience thin spots in the veil; you will sometimes have a wave of energy rush over you. You are having the experience right now.

If the experience were to happen like that [snaps fingers] you wouldn't learn a thing. But remember: You are creators in training and thus you need to experience things slowly so you can understand them, incorporate them into your body of experience and, most important, have individual reactions to that body of experience. If your friend experiences this and another friend experiences that and you hear about both experiences, even if you didn't experience it yourself you have reactions to it. It's important for people to read and talk about these things so you can react individually to them, and in that moment assimilate to some degree other people's experience through the interpretation of your immortal personality.

I think we'll draw it to a close. I'll say good night.

The Next 50 Years

Garden Magic

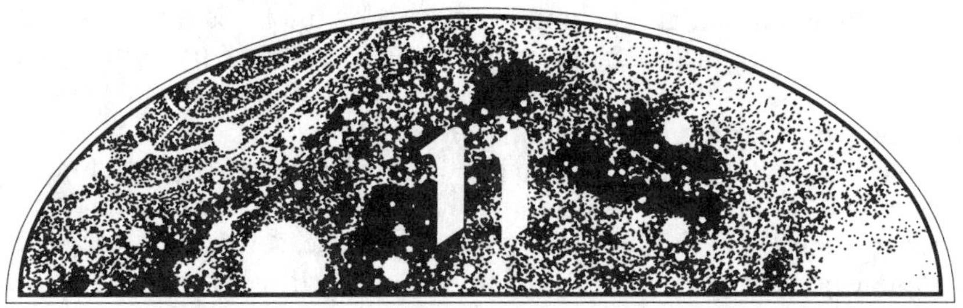

The New Corporate Model

July 2, 1996

In *The Explorer Race* we discussed the core of your being – who you are, where you're from, why you're here, what you're doing, where you're going. In the first part of this volume we went a little deeper into the *who* – who God is, who you are, where you're from and so on. In this part we have a theme also – where you're going and what is the ultimate purpose of existence. I will make some effort to give you a workable answer.

It's important to focus on *what will be*, because this allows us to guide the future. Perhaps the more difficult thing for most people these days is that the future is so nebulous that they are stuck or attached to the past. When people become frightened they invariably cling to something. Sometimes it's some person; more often it's some idea or dogma.

Our job here in this section is to buoy people up so they will not have to cling desperately to outmoded ideas and ideals. This is a serious factor. The dogmas of the past, even the recent past, are no longer appropriate for your present and certainly not for your future. The idea of the people coming together to form something has always been the ideal of the New Age community and of many previous civilizations, even core religions on this planet.

In the future it won't be so much that people will come together to do things, but that individuals will get together and become well aware on an intimate level of the sacredness of life, from the tiniest ant to the most gigantic solar system. It is usually easier to do on a small level –

that is, by going out and talking to the ants or the birds or interacting with the trees. It must be done so that people can move past one of the biggest addictions that you've had a hard time shaking in recent years – moving past the drug of following leaders.

We're going to wander around a little, but mostly we're going to talk a bit about the future so you can have a model to work toward. From time to time I will talk to corporations, because corporations are the people who are creating the initial model for Earth's system. I need to remind you that the corporate model is prevalent amongst many people you will know, although they will have a more benign idea and practice of the corporation. So don't dump the idea of the corporation. It's a good idea, but it needs to become less hierarchical and have more equality amongst people. That will require heart-centering, a sacred mentality and a true vision that what you want for yourself and your family, you must expand to include the family of all people on Earth, even the plants and the animals. When you do that you'll be ready to take the next step, which is to include all life everywhere.

Claiming Your Creatorship and Using Real Magic

The future is not going to grow out of the past. Rather, the past is coming to an end and the future that has been waiting for you all this time is there for you. But in order to claim that future, which is benevolent beyond your dreams, you must claim your creatorship within yourself as well as the core of your loving being, which is more than allowance – it truly desires the best for all beings. This does not mean that you control others and try to make things better for them, but that you allow them to pursue their own interests.

This is particularly important with the animals and plants. How many people, even enlightened ones, pull what they call weeds out of the garden and lawn and spray unwelcome plants? You must remember that plants exhale oxygen, and since Mother Earth's attitude right now is that there are too many people here, she is removing those plants and animals that produce oxygen.

Plankton, we know, produces oxygen; most chlorophyll-based plants produce it. But as you can see, Earth's electrical body starts forest fires and creates droughts. She does not have it in for the trees; she knows that they have volunteered to move into their lightbodies and return at some future time in some other way. Mother Earth is saying to you, "If you are going to create a worthwhile way of life for yourself on *me,* then you will have to create from beyond the material world."

She is challenging you, throwing the gauntlet in your face, as if to say, "Now *you* must be creators." If you must be creators, that means you must begin to use *real magic*, and real magic is not pulling a rabbit out of a hat but actually creating something out of thin air, as it were. Instead of saying, "Hocus-pocus, dominocus" to create *pouf!* something neat in the air, you need to learn how to attract things to you on the basis of your needs and simply because they are worth having and it will not harm others for you to have them.

You see, that's the key—knowing that all life is sacred. When you attract something to you, for that magic to work it must bring beauty, joy, comfort and nourishment into your life without harming others. You must put your mind on that and simply state out loud, "This is what I need."

You will have to go past the drugs of the past. I'm not going to quote Lenin here and refer to religion as the opiate of the people. I would rather say that those who have been in religious teaching positions (Buddha, Jesus, Mary and others) are all rooting for you to ask for what you want and need and (through spiritual means that I will discuss or that you already know) let that need go and let what you asked for come to you without questioning its form.

Basically it is a time for true magic, and Mother Earth will no longer sustain life for you human beings on this Earth without your doing something in return. She will not necessarily welcome all plants or life forms that exhale oxygen, which you must have to breathe. So you will be in a quandary. Some of you will move to places where Mother Earth is more benevolent—rainforests or tropical zones where plant life is more plentiful and encouraged by Mother Earth.

More Oxygenated Water

Some of you will get the far-reaching idea that it might be possible to transform the molecular structure of your body so that it does not need as much oxygen and can, for example, live on nitrogen. That is not realistic, so don't work on that one. Work rather on the creation of oxygen-bearing liquids; in other words, learn to be essentially like the fish and assimilate oxygen from living waters.

There will be people amongst you over the next five to ten years who will be doing a lot of experimental work with water. Ways will be found to create water that is more oxygenated than the water you have now. And there will be spiritual teachers who will teach you how to drink that water and assimilate that extra oxygen. I'm not talking about heavy water here, but about water that is denser in oxygen than the water you have now.

In order for water to maintain more oxygen, it must have a significant amount of life in it, and one of the best life forms for this purpose are sea plants such as seaweed and kelp. This type of living plant life encourages water to become oxygenated and retain a greater level of oxygen than is normal. Although most of these types of plants grow in saline water (I do not expect you to drink saltwater), you will be able to hybridize some sea plants that will grow in fresh water. As we speak, some individuals and corporations in Japan are working on this right now, so be assured that this is in the works.

Don't be discouraged, you inventors out there, because we need to have you come together and share some of your ideas. If human beings are to survive here as a race, you must learn new techniques to assimilate the oxygen in water – which will not go away, by the way, regardless of how many droughts you have. You can always desalinate the ocean water, but please *do not* use water as fuel by extracting hydrogen from it to burn in a combustion engine. That's an interesting idea, but that engine is useful only on a planet that has more water than it needs. *You do not* have more water than you need!

How about drinking hydrogen peroxide water? It is something we used on the farm to get the soil to open up after it had been chemically fertilized.

It's certainly something to consider, and it might be possible to create solid hydrogen peroxide to use as a filter, not unlike sipping through a straw, wherein minute quantities of oxygen would affect the water. I think it would be better to use it more as a filter, in which case it would affect the taste of water far less. It's a good idea, and it is something that people will work on.

Spring water and ozone machines and ozonation – is that what you're talking about?

Perhaps, but I think you'll find that people will utilize methods that allow the water to become slightly more oxygenated – hydrogen peroxide transforms into oxygen inside your body. I think that the more oxygenated water might take on certain appearances. For example, it might become noticeably thicker, actually viscous. A thoroughly oxygenated water, when rubbed between your fingers, might feel like highly saline water, which has a slightly viscous quality. It will be different from what you've become accustomed to, and it's going to have an effect on your physical bodies. If you can get enough "liquid" oxygen in this form running through your body, it will give you more energy and transform certain conditions in your body to something more healthy.

For example, many parasites in your body cannot live in an oxygen-enriched environment, so they would simply be transformed in your body; that alone would give you more energy. Your physical bodies as they exist today function optimally with from 30 to 35% oxygen in your atmosphere, but it's been a long time since most of you have been exposed to such air. There are still some places on Earth where one can experience 20 to 22% atmospheric oxygen; even though those places tend to be hot, people there have greater physical vitality.

I want to steer chemists and the health food and pharmaceutical communities in this direction because there is a tremendous benefit for human beings to become more oxygenated. There is a significant support for most corporations to go in this direction. I do not want to support or encourage the idea of masks, however, or even small oxygen tanks, because these are artificial and will not work very well. You need something that you can consume, because most people eat once, twice, three times a day and have snacks in between.

Nutrition, Unchemicalized Food

In the future there will be fewer hungry people. I would say that within 20 or 25 years there won't be any hungry people; by then you will have resolved the problem of nutrients (in terms of food) and you won't be trying to radiate food to give it an indefinite shelf life. There will be a revolution in the next five years, and soon food sold in the regular markets will be so healthy that there will be no need for health food stores. That will happen certainly within the next fifteen years, and I expect great strides in that direction in the next five years, because people are not going to stand for the present quality of food anymore. Smart people in major supermarkets are going to get the idea – "Hey, we can sell the best stuff and make good money, because many people will come to us."

You're going to resolve the entire food dilemma because people will actually be able to eat less food yet feel fully satisfied when they are fully supported by that food. The reason people must eat so much now is because the food does not support you; you have to eat ten times more dead food to receive the amount of energy you can get from eating a little bit of food full of life. "Full of life" means not only nutrients, but that which can turn to oxygen within your system. As much as possible I want to steer corporations in this direction because it is the wave of the future.

You were saying that some people might be moving to places where there is still plenty of plant life, implying that plant life in many places will greatly diminish. You also spoke about magic. Will people successfully attempt to work with the devic-angelic energies, as you have called them, to encourage plants in places they are needed?

The problem, you see, is that there are too many people "unwelcoming" plants. People have a tendency to divide plants into categories: plants you like and plants you don't like. The plants you don't like tend to fall under the heading of weeds. Plants you do like tend to be food or decorative plants or simply plants you enjoy.

You need to get past that fixation, because *there are no weeds*; there are only plants whose purposes, benefits and treasures you haven't figured out yet, such as the dandelion, whose potential is great. From various parts of itself (the seeds, flowers, stem, root, leaves) the dandelion can cure probably 90% of the diseases you don't seem to be able to cure. Don't synthesize it chemically – it won't work. I might add that dandelions growing near swiftly moving water such as a river or a creek would be the most powerful.

There are people now working with plants and their devic energies and oversouls to welcome them and sing them out of the ground and so on. But there are so many people destroying plants and making them unwelcome that it's overwhelming all these good efforts. Even people who've been singing plants out of the ground for a long time have had trouble getting the plants to come up, because the plant kingdom in general feels overwhelmed by being unwelcome. So when Mother Earth comes along and says, "Well, there'll be a drought here" (she doesn't talk, but it just happens), "and then I'm going to have some fires and so on," the plants say okay because they don't feel very welcome. Not many people go into the forest and say, "Wonderful trees" and so on – at least not enough to compensate for the people who are spraying chemicals on plants to kill them.

Welcome All Plants, Study What Is Natural

I'd have to say that you have to change your point of view (I'm talking to people in general) to one that welcomes all plants. I'd like to see people *immediately* stop spraying anything on plants to kill them – the sooner, the better.

You could say to me, "But Zoosh, some plants will take over the world." Well, the plants that grow like crazy, that you can't seem to get rid of, are not trying to take over the world. They're trying to get your attention! And the reason they're trying to get your attention is because they have something to offer you! Work has been done by biologists, naturopaths, medical people, mystical people and even pharmaceutical people on these plants (especially those you can't seem to get rid of – kudzu is one), and they have a great deal to offer. There have even been some inventive people who have on their own discovered things that plants are good for, but they have not received the respect, the support and the financial assistance they need to get their projects off the ground.

I'm going to tell you right now that medicine will be a lot cheaper when you welcome plants and the people who are working with them (dandelions, kudzu or tropical plants); when you discover (as has a Filipino researcher in Los Angeles) that the water hyacinth, for example, has many properties that are good for human beings; and when the pharmaceutical companies discover they can ease into naturopathic medicine. You won't have to have huge chemical plants; you can steer your chemists toward studying what is natural.

The main thing is that you're off track because you got onto the track that says, "We can do it better than nature." But in point of fact you *cannot* do it better than nature. You need to learn more at the corporate level about nature and the natural world and the way it works so that you will stop trying to control nature and let nature support you.

You pointed out the corporate level because that really is the big financial driving force behind all the changes?

That's right. Look at the situation now. Your first "world order," which is actually in place now, is based entirely on the corporate model; this is what I've said all along. International corporations are gradually consolidating, and the first world order is hierarchically related to the corporate model. So it is essential that corporations get the word, because they have the ability to change rapidly. If the world's 100 biggest corporations shifted their point of view tomorrow onto a path more natural, within *six months(!)* your world would change dramatically, hunger would disappear within a year and production would triple within three years!

People would feel better. You could kiss off sick leave; you wouldn't need it anymore. There's a lot to be gained by the corporate community. There's a huge market to be tapped, and that market is basically the people's unfulfilled potential. You can't continue to have more and more people buy products, because the Earth can't sustain it right now, and she's going to say, "Okay, you're going to have to fix your population soon." You're going to have more potential within the individual.

Corporate Survival and a Benign World Order

I'm speaking to corporations now because it's up to you. You've established the world order for the moment, but it is not going to last if you continue trying to create an artificial environment. You'll just destroy the world. Aside from all of the misery that would create, it sure isn't going to look good on the bottom line. If you want your corporation to live into the year 3000 or even 2050, start functioning (not merely planning ahead) as if 2050 were today. The corporations that do that will survive and thrive. The corporations that cling to the past and attempt to control nature will be gone entirely in fifteen years!

What would they actually do if they functioned as if it were 2050?

You would have to live in harmony with nature. You would have to start thinking about what's missing. You've got only *x* amount of oxygen to go around, and the oxygen-creating forces are being depleted while oxygen demands are increasing. Most animals and all human beings need oxygen, yet that which produces oxygen is rapidly being depleted. You have to start encouraging oxygen and tapping it from a lot of sources. It won't be too long before you start getting oxygen from stone. As we speak there are a few Japanese corporations doing a lot of advanced research in that field.

It doesn't hurt the stone?

Sure it hurts the stone; it transforms it into something it wasn't. In the future if you can work with stone that volunteers, that would be better. But I'm not expecting you to be at 2100 right away.

What does stone transform into?

If you take any chemical that contains oxygen and extract that oxygen, it's going to return to the element that remains. I'm not trying to get you on that track. What I want to say is that corporations need to have the long view because a more benign world order *will evolve* within the next 25 years, but only those corporations that actually encourage a higher quality of life for all people (including people who do not work within their corporation) will survive. I am not anticorporation; I *want* them to thrive, because they can evolve into something very similar to the unity of planets that exists on other dimensions, to say nothing of your own universe.

If you broke down the Network of Planets and looked at those societies, you could very easily fit the organizational system there into the corporate model and see that it works very well. The reason I nurture corporations is that I *don't* want them to go away; I want them to evolve into something that has a higher purpose.

When it is 2050 you will have a more benign world order, which is important because people need predictability; they need to know that they are not only welcome but that society is happy to have them here. So continuity is important. But they will also need to be encouraged to make the most out of their lives and acquire what will support and sustain them and help them produce what others need for support and sustenance.

As you get more into magic, some corporations will try to patent it! Guess again. They'll call it a formula or an "application," but it's not going to work. You're going to have to go through that one. You'll discover that Earth's raw materials are not enough for 5.5 billion people —you don't have enough for everybody *if* you stick to the idea that that's all there is.

If you're going to become a spiritual magician (to say nothing of a corporate magician), you are going to have to say, "There is only this much material, and we have this many people and animals and plants and so on—but there is also the potential within the energy of space" (meaning the space between one thing and another).

Space: A New Energy Resource

Put your two hands together right now [he claps his hands]. What's between your hands? *Space*. Is there something there? You bet there is. There is *energy*. You are beings living in a vast sea of energy, and this energy is more than just space—it is a cosmic pulse of life, which is the energy that exists in the mass of all being everywhere. Cosmic energy is literally the matrix that supports and sustains all life as you know it (and, for that matter, all life as you *don't* know it.) Thus it provides the stuff of magic if you create something magically—say, for example, loaves and fishes. What did Jesus produce those loaves and fishes from? Was it just *pouf!* and there were enough loaves and fishes for everybody?

No. Jesus knew how to utilize the cosmic energy that you call space the stuff between your hands as you bring them together. He realized ("realized": *think, feel, do*) how to create the cornucopia.

Now, you budding magicians out there need to recognize that space is not nothing; it is merely *something unidentified,* which means that it's a renewable resource you haven't identified as such. I need to steer you budding magicians out there onto this source, because that source is the matrix of life. The space that exists between your hands is exactly the same space that exists between the hands of a Pleiadian or an Andromedan or an Orion. This stuff, regardless of your atmosphere, whether it be oxygen, nitrogen, methane, whatever – the stuff that is the matrix of what you call space – is the stuff from which all the loaves and fishes you might ever need can be manifested.

If you understand that, you will appreciate not only how to create but where to create from. I want to steer you (that's my hint here) *where to create from.* You corporations that live consciously in the year 2050: Start thinking in terms of magic. Move past the idea of fixed resources. Economists have been telling you for years that resources are fixed, and that this is the terrible danger of having too large a population. I will tell you this: Resources are indeed fixed *on the 3.0-dimensional world*, but you are moving past that. And since you will move further during your lifetimes, you need to get past the idea of a limit.

The loaves and fishes didn't suddenly become 400 basketfuls; it wasn't like that at all. (This story is real, by the way.) Jesus was a cosmic magician, amongst other things. As people took loaves and fishes out of the basket, the basket was never emptied. You could, as a mathematician, be boggled by that: *The basket never became empty!* It was a fairly small basket; one person could carry it easily.

As a cosmic magician, Jesus was able to flow in the energy of space around him and re-form it into loaves and fishes. Cosmic energy is not just something out in the cosmos; it's right here and now, between your hands! I want you to understand that.

Those corporations that exist in the philosophical space of 2050 will survive if they can get past the idea of continuing to tear things out of Mother Earth's body by mining and drilling and scraping, and understand that they *must* begin to move past technology as you have known it into twenty-first-century technology, which is creationism.

Heart-Centered Creation Technology

The next step to becoming creators for yourselves is the technology of creationism, which will initially start with utilizing crystals to amplify the magical abilities of human beings. However, you will soon move past the crystal level, because crystals are very much like training wheels on a child's bicycle. They get you started, but if you stay with them, they will keep you where you are. If you keep those training wheels on the bicycle, the child will never learn balance. At some point you have to move past crystals and do it yourself. A crystal will raise your energy a little, and as you focus on what you can do through it energetically, you raise it a little more. But eventually you'll have to do it on your own without the crystal. I'm just going to steer you, not tell you exactly what it is.

Know that creation technology is entirely heart-centered, which tells you that it is need-centered. This means that a corporation's staff, if it's going to serve the human population (to say nothing of animals and plants), cannot just say from the mind, "What's a good product that we can make and sell?" You must say in your heart, "What is a good product that people actually need?" As a corporate individual you'll need to be able to move past the sole idea of selling products toward creating a product that people actually need. And when people actually need a product, you will feel it in your *heart*—you will *know* they need it.

For instance, in the next few years you will see that many people need to be fed proper nutrients, and you will say, "We must be able to create a food that is compact and does not drain resources to produce, and see that all the people involved in its creation are heart-centered and will focus through love (yes, *love*) to create this product."

It is a truism that the food Mother prepared tasted better and was more nurturing because she put love into it. The corporations that have employees who lovingly create something and are treated lovingly by their corporation will as a result treat everyone else in the corporation lovingly. This is a function of 2050, and even though right now you might think it is impractical ("That is impossible! We can't do it!"), you *can* do it.

To the Japanese

I will speak for a moment to Japan, since you are on the leading edge of many technologies and are working toward things that other countries haven't even begun to think about yet, and certainly haven't prioritized. Here you sit on an island that is, at the very least, creaky. Your island does not have the capacity to survive into the year 2050 *unless you begin to use heart-centered spiritual magic.*

Here is my recommendation: Those of you who are focused in these fields, begin to restructure the actual chemical structures underneath your island. Pack them and build them up with "frozen" oxygen. Do not pull it from anyplace other than cosmic energy. Begin to see your island rising and becoming strong without harming the surface civilization, easing up perhaps an inch a year. In fifty years the whole island will have risen fifty inches. That will help.

There are things to do, but you must begin. For the future of corporate structures on your planet, *Japan must survive!* The Japanese civilization must survive, even with all of their problems and societal anomalies and even though they do not understand women as a true resource. *All future magicians will be women.* Yes, heart-centered magic must be done by a woman—or a man who is totally and completely focused in his feminine energy (and there are not many). It will be much easier for a woman.

When the Japanese culture balances that one little factor—the feminine energy—it will become the most important culture on Earth. I'm not trying to say the Japanese people are better than others, but that they have a culture that looks at a problem and devotes its full resources to its solution.

Other cultures will begin to emerge that way, and certainly there are other cultures that have that capacity—Sweden, Germany, Finland, Norway, Switzerland and others. (I have noticeably not mentioned the United States because it is a conglomerate of cultures.) I'm not trying to say that one race is better than another, but that cultural influences cause this. There's no reason that those cultural influences cannot be shared with the world in a benevolent way, not "be like us or die" as in the past, but rather "look at what we have done, what we've created; don't you want to be like us?" Everybody will be saying, "Yes, show me how!" "See, that's how to do it." As they say, if you build a better mousetrap, everybody in the world will want to know how to do it (although I don't support mousetraps—but that's another story).

The corporate model is important because of its influence. I don't have to tell you; just look at the influence of television. Some of you in this room can remember the days before television. People were a lot different then, and you might not fully approve of the change that took place A.T. (after television).

In the coming years it will be possible for corporations to influence programming to make it benevolent, so that instead of being irradiated by various electrons and so on, you will be irradiated by something beneficial, and watching television will actually make you feel *more* healthy, not less! That *is* possible. In the year 2050 people will be sitting in front of television for its health benefits! I'm not saying that this will be an advertising campaign ("Eat this and you'll feel better"), but that it *is* possible to change that radiation.

Look at the future based not on a negative past or a cynical idea of what it might be like, but see it from the core being of all human beings —*what you would like it to be.* When you get past all the unhappiness that most human beings have had to put up with, they *all* want the future to be something wonderful, where everybody is welcomed for being who they are and happy to see everybody else, enjoying others for being who they are and appreciating the wonderful diversity that your microcosm of life offers here on Earth.

That's what life will be in the year 2050. Any corporation that expects to survive will have to begin right now working toward those goals. If you're going to be here in 2050, you must start acting now as if it were already 2050. I know you can't do it in all levels of your corporation, but you *can* do it in some levels, so just begin.

It's the first of July 1996. Are we still at 3.47 between the dimensions?

You are now at 3.4756, easing toward 3.48.

In 2050 where will we be?

At 4.0.

So this planet will be fourth-dimensional? It used to be that we were going to pop up into two levels.

You might, but you'll still be in 4.0. Things will look different because beings will feel welcome.

The book where I first said howdy [*Allies*] had a cover that shows the planet with a somewhat greenish atmosphere, which I refer to as Terra. It is green because the plant life feels so welcome. For plant life to grow happily amongst other plant life and for animal life to live amongst them and people too, everybody must be absolutely *certain*, without any doubt whatsoever, that they are welcome. So yes, 4.0-dimensional Earth will look very much like the cover of the book *Allies*.

All that talk before about popping out of something and into something else . . .

It's really true. It's very much like this . . .

Now you're saying we're going to create it as we go?

What is the whole purpose of creationism?

To learn to be a creator.

And how do you function as a creator? Understand that magic is divinity in itself. So if we're going to do the *pop!* (which Van Tassell also referred to) . . . it's true that true magic often simply turns something inside out. So the *pop!* is still there; we're simply examining it, slowing it down a little to examine what happens in that moment [snaps fingers]. We're slowing it down here to examine it.

But we're still creating it moment by moment?

That's right.

In fifty-five years we will create it.

As far as I know. Maybe more so by the time we get there, but by that time I'll probably be talking through other people.

Organic Farming, Off-Planet Colonies

You talk about growing. What about the insect kingdom? I live in a farming area where they've tried to speed plants up by fertilizing chemically, which I see now is totally wrong. They also treat crops chemically to keep the insects from eating their produce. How does that fit in? If the soil is balanced and healthy, usually the insects stay away.

If you read *Findhorn Garden* you will get a pretty good idea of a different approach. Insects are there first and you have to plant something for them. Understand that plants become less of their own nature when they are rushed. Do *you* like to be rushed? Nobody does.

Natural fertilizer is fine; you can use all you want, but chemical fertilizer rushes the plant and it becomes anxious. Anxious plants actually broadcast a signal into their world to attract what they understand and know in nature – natural beings. This is why insects come flocking. The plants signal, "There's something wrong with us; come help, *touch us*." When they are touched by other natural beings that they know and understand from years and years of exposure and contact, they feel soothed and comforted, you see. You might say, "But Zoosh, many of these insects might eat the plant." But the plant says, "But this is something I know and understand."

My best recommendation is to read *Findhorn Garden* and apply it to the best of your ability, and while you're at it try to use natural methods. Organic farms *really do work*, and it's reasonably profitable. I urge all farms and farmers to go for organic. I know it takes time to establish it, but it is absolutely the wave of the future. By the year 2050 there will not be a single chemical used on plants other than that which is created by natural beings (any farmer knows what I'm talking about).

How many people are you seeing on the planet in 2050?

There won't be as many as now. But by that time you will have begun sending people to other planets. You'll have thriving colonies on Mars, in terms of projects that are already started. Even now as I speak, people are planning and working on the techniques to create thriving cities. There will probably be domed cities initially, but eventually they'll be able to oxygenate the atmosphere. By 2050 you'll probably have on Mars at least half a billion people who were born on Earth, and you'll probably have a pretty significant Moon colony going then, maybe 100,000.

You might say, "But Zoosh, that's a pretty rough place to be." These people will probably be involved in some form of manufacturing. Even so, you will be reaching out to other planets and solar systems, because by then you will be very much plugged in with the Network of Planets. By that time your Earth population will probably be down to somewhere around 3.1 billion, and it will continue to reduce gradually until it reaches around 2.7 billion. This is not by unnatural means – fire, famine, flood or any of that stuff – but mainly through mostly natural means (maybe a little technological). It's going to stay at that level for a while.

How would that work on the Moon? It's sort of under the control of off-planet beings right now.

That's going to change. You will become more spiritually evolved and more advanced in working with machines (crystals and so on) in a way that is loving rather than forceful ("If it won't work, I'll *make* it work"). When that idea disappears from technology and you get more into creationism technology, other planets will recognize that you are ready and they will support and even encourage your outposts on the Moon. They will come and teach you many things. It will be a revolutionary time, a *great* time to be a student! Instead of studying the past, you'll be studying present cultures and civilizations on other planets – oh, it will be fabulous!

Yes, yes. Can we colonize and reuse the moons around some of the larger planets in the system?

It's possible, but even with your own Moon you can remove only a certain amount of material, because it needs to have a certain mass to do what it does. Even now it's being mined by certain extraterrestrials. I know you're talking about other moons in your system, but I'm talking about moons in general. There's only *x* amount of material you can ever remove from any moon if that moon has a benevolent effect on the planet it's orbiting. It's possible that you'll be able to mine other moons, but by that time you will have discovered enough about magic that you won't have to do *any* mining.

I meant, can we colonize them to live on because they are beautiful places?

Well, if you're going to do that, you'll probably colonize planets, not moons. You'll make an exception with your own Moon because it's near. But generally speaking, I think you'll be colonizing planets where you're welcomed. And the Council of Planets will let you know where it's okay for you to go – and where it's *safe* to go, for that matter.

So we'll go beyond this solar system . . .

Initially it will be your own solar system. By the year 2050 you won't be past your own solar system in terms of colonization.

Where else can we colonize besides Mars?

Probably not much else, but there would be some potential for creating underground civilizations in planets that have vast underground caverns. If the technology used is correct, if it works with the planet's own choices and is thus benevolent, then it would be possible to create an underground city on, say, Jupiter – someplace like that.

Are any of the ETs who are on the Moon and elsewhere using creationism, or are they all technological?

No, they're using technology. You see, when I'm talking to you about creationism technology, I'm talking about something that isn't particularly widespread. You normally don't find creationism technology at the third dimension – or the fourth dimension, for that matter. Sometimes you find it in certain advanced cultures in the fifth or sixth dimension, but you don't usually find it until you get up around the seventh dimension, where it's common. So we're going to help you develop this benign technology in a place where it's never really been developed and utilized and certainly not *applied* much before. It was applied initially by certain beings, but it certainly hasn't been applied by the average born-and-raised-on-Earth being.

Now we'll call it a night.

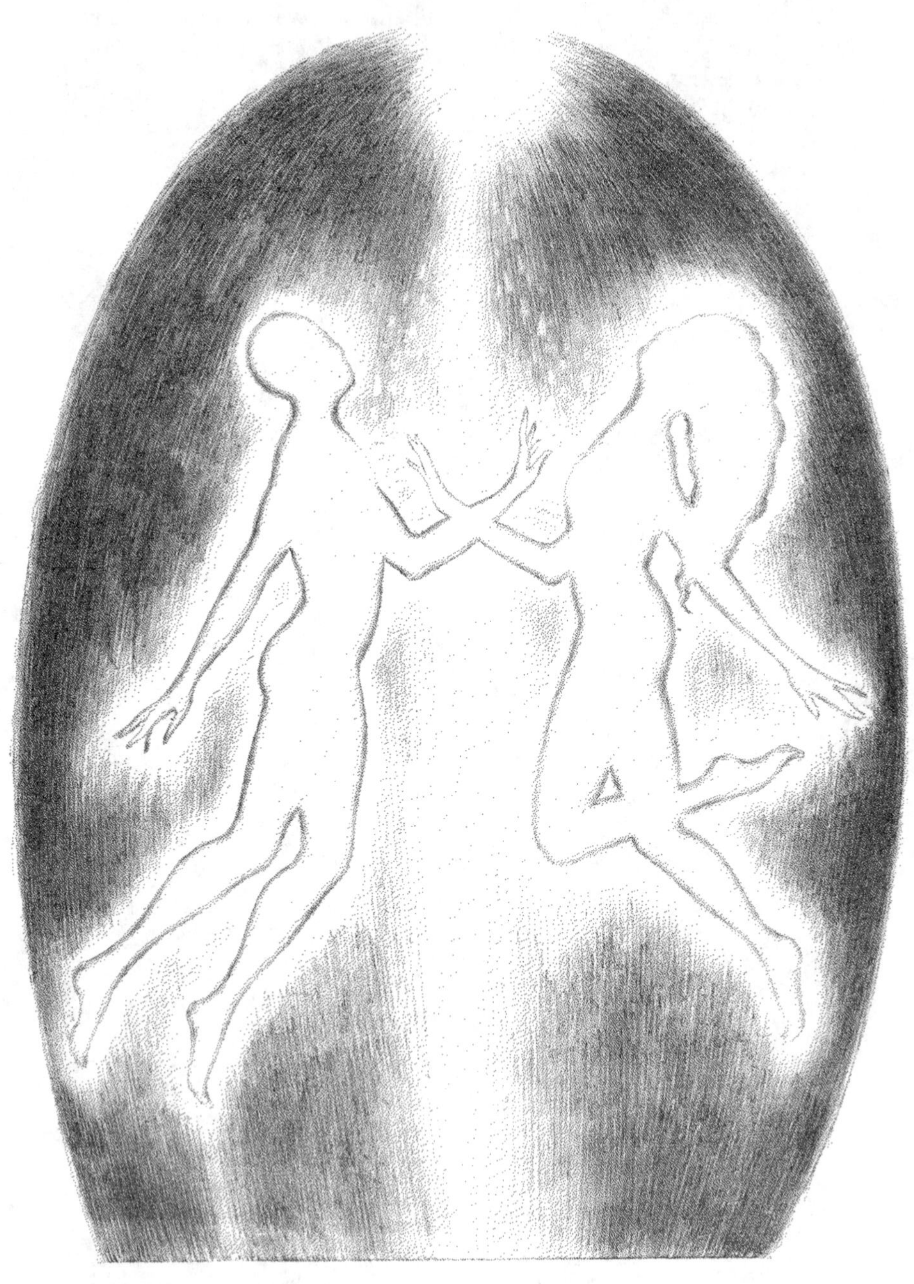

Feeling

The Practice of Feeling

July 9, 1996

Now I will speak of truth and love. It is most important to understand that love contains no element of truth. It has become the biggest challenge for all beings to understand that your society suffers from being saturated by the concept and practice of truth. For you to truly understand this, we must begin with the feeling of love. Love that you feel in your body is truly the name for Creator. That is why in ancient texts it was often referred to as "that which is not named." Even in some of today's religions that remains the case: one does not speak the name. This has passed down to certain individuals and they have put it into the records of the religion. Although it is not clear, I will make it so.

The function of Creator as love is intended to be ultimately manifested by all people here on Earth, and to the extent that you manifest it, the animals and plants and elements will reflect it. Let us understand briefly what has happened—how we came to have gods practicing religions who speak words that are to be followed and if not, punishment comes. This is not an insidious evil force, but reflects mankind's need for authority.

Words vs. Feeling

The true nature of Creator I have explained. When we call Creator a word instead of a feeling that is felt, we immediately give rise to the potential for creating dogmas, even karmas to some extent. Once we have created the word *God*, we immediately reduce what Creator can do for you to approximately 0.0000001 of its potential. We have a situation where God becomes defined by a word. The moment that happens, then God must speak words, and the words that God speaks are to be considered sacred. You understand, of course, that God does not pick up a pen and write. No, no, God is a concept created by man. *True* God, *true* Creator, is *a feeling of love.*

Because man has created the concept of an intellectual God, then man necessarily writes down through inspirations (even though from Creator Itself) words that describe God and what God might say. As long as God is described by *any* word, the mischief available for such a concept . . . look at your history. How often has one god or another been misused against people, eh? What we have, then, is a need to return to the true nature of Creator, which is a feeling of love.

In order to truly ordinate oneself, one must move beyond words and experience Creator as feeling. This is the true nature of the feminine. You see, the feminine, in an attempt to legitimize itself, has utilized the word "goddess." But truly the feminine of Creator is *love as it is felt.* If this is so, and if the masculine of Creator is also love as felt by whatever being or form you are, then that tells you that the *meeting* of all beings is love. You see, the biggest and most difficult task confounding you now is to separate truth from love, because mankind's pursuit of truth has led him to reduce Creator to God.

There has been a lot of talk about how it would be possible in fifty years' experiential time to make great strides, although mankind has actually made great strides backward over the past few centuries. What you can do is reintegrate the natural language, which is felt. Picture this: A child is ready to be initiated into the family's philosophy and undergo an awareness of Creator. Several individuals surround the child and take him up to a special altar. He is told that on that altar he is going to meet Creator, so he's very excited. At the altar all three of these individuals bring the feeling of love into their chests and bodies. They all then touch the child in his chest, maybe in his back, and suddenly the child is filled with a warm glow of love. The adults (the initiators, as it were) then say to the child, "What you feel within you is Creator. This feeling of love – this wonderful, euphoric blessing – is Creator."

Creator cannot be described by a word so much as the feeling within oneself that is then described somewhat. But to describe Creator Itself, which is love – well, we know that the greatest poets have been stumbling over that for a long time. That is because love is a feeling unique to each individual, even though there are certain similar physical properties.

Your society needs to move beyond spoken language – even the written word or idiogram or, for that matter, data. Anyone who writes programs for computers must speak the computer's language, which is a series of symbols input by man. All these things are forms of written language that keep you from understanding the true nature of Creator.

In order for you to fully grasp the simplicity of Creator, in order to feel Creator's editorial policy, as it were, focus on feeling this love. I want all of you who are reading this now to focus into your chests, your hearts, to see if you can feel love. For some of you it will be tingling, for others it will be warmth or heat; to feel it best it will feel like warmth. For those of you who are simply tingling, do not fret. Keep focusing physically. It is not a mental exercise. Do not be deluded by the mental body's efforts to put words into your mind, be they definitions, mantras or whatever. In point of fact, true love is the ongoing absolute of Creator.

Now, you can define unconditional love. You can even define how you are feeling physiologically during the feeling of love. And yet you cannot simply take that aside and say, "This is Creator," because you are surrounded by all of Creator's great and various possibilities. Certainly, the animals sometimes provoke, do they not, a feeling of love or affection – or amusement at the very least?

You might ask, "How can we achieve Creator and creation technology in so short a time? Well, as you begin to feel and communicate your feelings, you shall achieve this. How do you communicate your feelings? Do you tell people what you *think?* How often you have spoken to each other, argued amongst each other about what you are feeling – otherwise known as what you are *thinking* about what you are feeling? But how about the true communication of feelings, as in our hypothetical example of the child who becomes aware of the feeling of Creator?

The Practice of Communicating Feeling

If you feel something and cannot describe it to another, the best way to impart it is to reach your hand (right or left), hold it over that person's heart or solar plexus (with their permission, obviously) from 4 to 12 or even 18 inches away and tell the person, "This is how I am feeling." Then it is up to the party receiving that feeling to note how she feels in her body. It will probably not be a definable emotion, but a physiological feeling. If yours is a pleasant feeling, then that is how she is feeling. If it is unpleasant, then she will briefly feel unpleasant or uncomfortable, or she might be able to define some emotion. The moment she has that feeling, she says, "You can take your hand away now. I have the feeling." Then you are silent while she explores the feeling. You don't have to constantly radiate that feeling; she simply becomes aware of it and then examines it.

This is not so terribly different from those who do long-distance viewing of objects or people. For a split second (no more than one second) they view these objects or people at a distance and then stop and examine what they have viewed, in this way not interfering with what they are viewing. Anything more than three seconds is interference, because people are feeling beings. Even rock, stone, animals and plants are feeling beings; they will become aware of your presence and energy and it will interfere.

The cycle must be learned now. You have tried, many of you, for years and years (certainly those of you with political ideals, aspirations, goals, agendas, whatever) to show the relevance of your cause to those who just cannot agree with you. Now it is time for you to get beyond words and negative (as you call it, or self-destructive, as I call it) actions with a demonstration of your political agenda, and move to the true language.

Here is a hypothetical meeting. Let us put in one corner of the room a member of the Palestinian Liberation Organization and a member of the Israeli cabinet. No matter how much they talk to each other, they can never convince the other person of the rightness of their cause or why the other individual should agree. They are politically polarized. How do you get them to communicate?

For one thing, they must understand their commonality. All human beings feel emotions in certain physiological ways. When you are happy (or joyful, amused, euphoric) certain physiological things happen in your body. You laugh, you feel a sense of lightness; it is a pleasurable feeling. When you grieve, your body tightens up in places and you must cry or take some action to release it. Everyone is alike in that way.

Let's say that the PLO representative speaks to the Israeli cabinet member and says, "My friend, I need you to understand what I am feeling, because my feelings truly express what I have been unable to put in words you can *feel*." The person is saying that no matter what he has said to the Israeli representative over the years, no matter how long they have negotiated, the feeling of compassion between the two has not grown. He continues, "Do not be upset, but I will now hold my hand briefly about 14 inches above your solar plexus. See? My hand is empty," he says, "I will not harm you. We have been talking for years; now it's time to move past talk."

He holds his hand over the Israeli cabinet member and asks him, "Can you not feel in your physical body what I am feeling now?" And the Israeli cabinet member, trusting his associate in communication and dialogue, allows himself to feel. In a moment he feels the passion and compassion of his fellow advocate, and he says, "Oh yes, now I feel. I understand." This does not mean he has *mental* understanding, but that he understands the *physical* feelings, the emotions, the passion of the other individual.

As you know, it is always difficult if you are an artist or a police officer or a schoolteacher to pass on the true definition of the passion you are feeling to those to whom you are communicating. Yet if you can pass it on this way, the person has a means to feel what *you* are feeling.

The Israeli cabinet member says, "This is magnificent. I now understand what you are feeling, and as a result I can understand your words better. Do you mind if I do that with you?" "No, no," says the PLO representative. "Please, it is fine for you to do that with me. I don't mind at all."

So the Israeli cabinet member then holds his hand a similar distance away from the solar plexus of our PLO representative, and the representative suddenly says, "I can feel what you are feeling. No wonder you have not been able to understand my words!"

I must paint it for you this way because for you to truly move to the next level of communication, you must get away from words. No matter what culture one comes from, words (even those defined clearly in a dictionary or in a group's mythology) do not always mean what is being stated. For example, men and women use the same words but do not necessarily have the same meanings. Or a child might use an easily understood word (*happy, glad, sad*), but you might not know how the child is *feeling* in that moment.

For instance, child might have a nightmare and run into the parents' room upset. "I dreamed there was a car crash. You were both hurt and I was frightened." The parents might wake up and sleepily say, "It's okay, dear. It was just a nightmare. Forget about it; go back to bed." If one of the parents says, "Come here," and gives the child a hug, in that hug the actual feelings of the child are transferred and the parent understands *immediately* how upset the child truly is. When this happens, more is exchanged; there is love and openness, enabling the parent to understand.

The Next Step

You now need to move beyond emotional exchange between intimates (as between a child and parent) to an exchange with each other. The bus driver tells you, "The fare is a dollar in coins; you cannot give me a dollar in paper." If he is speaking to a recent immigrant who does not understand, the bus driver can whip out a picture of coins that the immigrant immediately understands. In order to improve your communication, there must be a greater sense of actual exchange. In this example it is a visual reference.

As you all know, there has been an attempt in recent years to convert to universal traffic signs, the idea being to go past language and use a universal visual reference. This universal visual reference is associated with your true communication.

Let's take our emotional communicants to the next step, using our PLO representative and Israeli cabinet member. Now that they each know what the other is feeling physiologically, emotionally, this has created another level of connection. They feel more like comrades because each can identify with the other's physical feelings even though they had totally disagreed before. There is a common language. They might say, "If only we can utilize this language."

Our PLO communicant (who has been initiated in these techniques for the sake of our example) says, "Now we must take this to the next level. I want to hold my hand over your solar plexus and feel what I am feeling. I will focus on something I want you to experience as I have, the whole experience of it with the smells and passions and feelings."

The Israeli cabinet member says, "Yes, yes, let's do this. We must expand our communication." Then our PLO individual holds his hand over the Israeli cabinet member's solar plexus and feels the tragedy that has befallen his people. He relives it visually, and as he does this some pictures pop into the head of the Israeli cabinet member, perhaps pictures of similar things he has seen with his people.

This new level of exchange might take practice, but it can be done. To move toward creation technology (which is feminine technology), we must get past the separation of words. Words have always sought to promote truth, but truth for one person can oppose truth for another. However, each person feels love very similarly. Everyone knows the difference between feeling love and feeling anger – it is crystal clear and without doubt. They know that the passions and compassions that rise within them are universal.

We must move to the universal language, and I encourage you to begin. Please take this as permission from me to begin practicing these techniques.

What you do is similar to what an actor does. An actor knows the role, but a profound actor who has the capacity to emote can *feel* the role. By feeling the role and being in the experience, the actor *becomes* the role and radiates the feelings to the audience. The audience is not simply listening to that actor mouth lines (no matter how dramatically), but it feels the feelings of the actor who is communicating the truth and sharing the tears and fears, the passion and the compassion of the experience. We must move beyond the fallacy of "truth," which seems so seductive and important in spite of the fact that it is feelings that are physiological and can be readily identified from person to person.

You might ask, "But Zoosh, how do complex concepts get communicated?" You'd be surprised. Complex concepts do not have to be communicated when one is not pursuing "the truth." There is no need to take the roundabout way from point A to point B with complex philosophies when one can *feel* and *picture* and *relive* what one is picturing, as in the case of our actor. I would like all of you to begin to communicate in this way.

I'm not asking you to throw out your languages yet, but it is essential that you begin this now. You have seen some things in the past few years that tell you that you must have a better means to communicate. Parents, imagine the struggles you have in communicating with your children. Employees, communicating with the boss is so difficult that words fail you most of the time, and all you can do is talk about trite things – sports, the weather, a needed raise. But there needs to be an absolute unity in a company to produce a product in the best way so that you feel it is a contribution and that any flaws can be communicated by picture and feeling as well as words.

You have often asked, "What is the next level?" *This* is the next level. You are surrounded by teachers: the stone, the animals, the plants – all communicate this way right now. Recently you have seen and heard things that cry out to you, "We must communicate with each other in a way *that cannot be misunderstood*."

The main problem these days as you are reading this is to understand, period. It matters not whether you are an Oxford scholar or a graduate of MIT or whether you have grown up on the streets and know street talk. Even though you speak exactly the same language, too often even your best friend does not understand how you truly feel or what you are trying to say. Husbands and wives know this, lovers know this, certainly children and parents know this. You have so much more in common than you have differences. You will find out when you practice this. You can think about it, you can theorize about it, but the *actual experience* of feeling the feelings of someone else is profound.

You do not have to stay connected to that other person. Just get the feelings and then pull back and say, "That's enough, I have it." Then feel it for a moment. If it is uncomfortable you can say, "I have it" and offer it back to them – or give it to Creator to transform.

In order to truly know, to truly be able to reproduce and re-create, you must have an absolutely unimpeachable means of communication. We know from your history that it is not words or even ideas. It is *experience*. Those of you who have been in wars come home and try to tell your loved ones what it was like, but you just cannot convey the whole-bodied experience. Words are a pale substitute. This is why you don't have to explain yourself when you talk to a fellow veteran.

Only a machinist knows how it feels when he has machined the parts perfectly within the closest tolerances and can say to himself, "This is right" and feel those feelings. You have a unique opportunity now to feel each other's true feelings and move into the future with a means *to know your common bonds*.

Can this technique be done now?

Certainly, begin practicing now with those who want to do it. Don't just run up to people on the street and say, "Have I got a new way of communication for you!" Pick a friend who says, "Yes, I would love to learn this." Remember, true telepathy is in pictures and feelings interpreted into your cultural frame of reference, otherwise known as language. Why not cut out the middle man, the interpreter, and *know* what the other person is feeling? Here's how to practice.

Exercise in Communicating Feeling: Training for Telepathy

Pick your partner and say, "On this day we will attempt to communicate to each other something about which we each feel strongly." It can be anything. Perhaps that day you have seen a magnificent painting that has stimulated you and filled you with awe. Perhaps you have seen something that has upset you or warmed your heart – something profound. Take a week, and if you don't see anything, then seek it out. Go to galleries or places of amusement; go be amongst the people. And do not deny your feelings when you see something.

Then when you are with your friend or partner say, "I will impart this to you." Do not tell him what to expect. Reexperience what you have seen; see and feel again what you felt then. Remember your feelings, and when you are fully visualizing that event, place your hand (preferably the palm of your right hand; if using the left hand, use its back), holding it anywhere from 14 to 16 inches from the solar plexus or the heart area of the other person (avoid the heart area if it is something tragic or upsetting). If the person feels nothing and they ask you to do so, move your hand closer. Touching is unnecessary and can be a distraction.

Practice with your partner as often as is convenient, first one, then the other. If you wish to test yourself, after you get the feeling say, "Okay, I have it now." Stay with the feelings or pictures and write down what you are feeling in your body, such as, "My stomach feels tight, my gut is churning" or "I feel total relaxation, almost like I'm ready to sleep." If you got pictures, don't question them, just write it down. And don't compare notes until you are done for the evening.

This is training for telepathy?

It is training for true telepathy. Don't name or describe the emotion you were feeling, but what you were *physiologically* feeling: "I felt warmth" or "I felt tingling." And if you saw something, say, "I had a flash of a mountaintop" or "I saw myself looking at a four-way cloverleaf from above." Don't question it. Whatever it is, write it down.

The idea is to discover that you all feel emotions and visualize in much the same way. The mechanics are the same. And that is the universal language. Granted, there are colors, tones, even math in the universal language, but this goes beyond a language between people.

Let us say that you are a farmer and you discover that there is something wrong with your best milk cow that the vet can't identify. The cow appears to be unsettled and skittish, and it is affecting her milk production. You can't just say, "The heck with it!" You have to do something.

So you walk over to your cow (she is used to having you around, so she does not mind) and reach your hand out, filling yourself with love and concern (not worry, but compassion). With your palm perhaps ten inches away from the side of the cow or near her shoulder, you impart this feeling. This requires total concentration. If you don't know what to picture, imagine a white wall or stare at the ceiling of the barn – something familiar to the cow.

You will know that it's working because the cow will seem to relax or she will look at you, giving you a feeling of connection. After a moment take your hand and reach over (you can do this with your left palm toward her). Don't touch her, because you'd be surprised how susceptible you are. You can come within eight inches and hold it near her. Move away from that spot, perhaps up along her spine, and just see if you can feel what she is feeling. Your bodies are not the same, but you will get some sense of feeling. Because it is foreign, you might feel strange. What you are attempting to feel is not her physiology, but her feelings, her emotions.

In this case she is anxious because she has seen a truck near the barn that takes cows away. (From a cow's perspective, her friends go away and are never seen again.) So you get that picture or a sense of anxiety in your gut, mortal fear. As any good rancher or farmer knows, cows don't give milk when they are terrified. So you make a point never to park the truck anywhere that she or any other cows can see it. It's a simple thing, but it's possible to do.

I tell you this because even domesticated animals can communicate to you with feelings and pictures if you are prepared.

To relax the cow, couldn't you give it a picture of its being okay, whatever that picture would be, to relax her and let her know that the truck is not going to take her away?

What would be better, perhaps, is to give the cow a picture of a place the cow enjoys – her normal stall, her favorite pasture, luscious green grass that she likes to eat. Don't give her a picture of something technological unless it's where she spends her time enjoyably, something that will relax her. Good question.

Farther down the road and already achieved is a technology that can record those feelings and pictures and sensations, right?

There's technology right now that can do that. But the problem is, this has been developed for military purposes and is for the most part incorrect because it is based upon a technical (masculine technological) means of measuring or quantifying physiological, measurable symptoms. Her blood pressure goes up, her heart beats faster, her skin sweats more – this kind of stuff. No instrument is better for measuring human feelings than another human being. That machine (in the case of the technology of today) does not want to be a machine but instead, its original components in a mountain or wherever.

You see, the machine itself was manufactured from uncooperative materials dragged out of Mother Earth. Almost all machines are like that. Understand that the machine itself rebels, it has resistance, and the resistance of the native materials in the machine tends to distort even what you are getting now.

But the technology of tomorrow and that of other civilizations has this ability to record, to make libraries of what people see and feel and . . .

Yes, often stored on crystals.

They call them ROMs.

Yes, they're often stored in various technologies, but even if you store on any material the entire knowledge or myths of your culture or even a benevolent dream —whether a crystal the size of a grain of sand or in a book called an encyclopedia —you are *still* using masculine technology.

In feminine technology something *wants* to do what it is doing, it *wants* to participate. It uses almost no naturally occurring products — and certainly doesn't use anything that's been forced to be something that it would not normally choose to be.

But in the true technology of tomorrow aren't there things that want to be used that way and records where you can experience a culture by immersing yourself in it?

Yes, but do you not understand that I want you to go *past* that. Why do what you have already done in lives on other planets? Even though it might be fascinating to explore the thoughts, the feelings, the records, the joys and pleasures of other cultures, why not experience them directly? For example, here you are living on Earth and I'm giving you this training. You can begin with each other or your dog or cat. Isn't there anyone who wants to go beyond *that?*

Telepathy beyond the Planet

How about feeling the feelings and getting the pictures from somebody who's on the other side of the galaxy, to say nothing of the other side of the universe? We know that you can feel the feeling that is eternally felt by Creator when you feel love in your *physiological physical body*. (I want the redundancy because people are so used to defining their feelings as their thoughts.)

What if you want to feel what someone is feeling on the Pleiades? "What's love like on the Pleiades?" you ask. Well, you can read books by Sasha (channeled by Lyssa Royal) or dredge up a copy of *Allies* and read about life on the Pleiades. The advancement of the Pleiadians came about largely because they didn't have any contradiction or polarity; they are not much more spiritually advanced than you.

On the other hand, you could connect with a Pleiadian. You could ask, "Is there someone on the Pleiades with whom I can exchange feelings appropriate for them and for me along with pictures of what their life is like?" As you sit on a rock staring at a moonlit sky, you begin to relax and suddenly begin to feel something. Someone will get your message.

Practice this with Earth beings first so you can get good at it, then try it with someone from another planet. Make sure you ask first that it be the right person for you, that it be a loving experience. And do not send your friend on the Pleiades anything other than love or joy or picture anything that does not give you a pleasurable feeling. If you build up a trust with this person, maybe connect with her many times, she can become open and more vulnerable. It would not be appropriate to suddenly picture an accident you saw where a car went off the road and people were injured and an ambulance came and took them away. It was frightening to see it, and if it comes up for you for a moment, you must immediately break the connection by doing something physical – clap your hands together, feel a rock or your leg, anything. Then go back into those good feelings and continue. You don't have to explain; just continue. It's better to stop and continue later. These things will happen.

The more you practice with other human beings who have a higher tolerance for discomfort, the more you can do this long distance. It will begin to create a *felt alliance* between you and Pleiades and, in the larger sense, between Earth and Pleiades.

People could also start doing this on this planet; for instance, is there anybody in Australia who would like to be a pen pal?

Not until you do it *physically* with another person. It will take you fully a year practicing this with a partner once a week or every other week before you can begin to do it long distance with anybody else. (If the person in Australia happens to be your husband or wife, you might be able to do it a little faster, but it would still take six months.)

Here's the thing: Let's say you're working with your partner to learn to get the feelings and the emotions. Eventually the person will be able to do with you what the PLO person did when he showed the Israeli something he'd seen that he was passionate about. You feel it briefly, then you immediately withdraw to examine the feeling of it.

Practice for Accuracy by Learning How You Feel Different Feelings

We want you to have the full gamut of feelings sent and received, but we want you to begin with benevolent things and work up. When you feel you are doing it very well, then we want you to begin doing it at a distance. You will still hold your hand out, but instead of doing it up close, the other person will sit across the room. Practice that for a while until you can do it five or six times out of ten. You see, we want accuracy here, and accuracy has to do with your physiological feelings and your ability to get pictures.

Later you might be able to see that it relates somehow. The partner of the person who saw and reexperiences the magnificent painting might not see a painting, but for some reason sees children playing and feels amusement or happiness or even calm, knowing there is continuity and that life goes on. You bring your notes to your partner, who says, "Well, gosh, there was a beautiful picture of Mount Everest and I just loved it."

The main thing is, you find out how the person was feeling. When you are transmitting, you want to be filled with the feeling (the compassion, the joy and all of that), but you also want to notice how you are feeling physiologically. Therefore you must know *how* you feel joy and love and compassion, all these.

Eventually you and your partner are able to communicate successfully 50 to 70% of the time. Then you separate yourselves a hundred yards apart, then farther and farther. Eventually you might be standing on one side of a valley and the other person on the other side ten miles away. Then you can begin to practice even longer distances. You want to build this up slowly. This ability is your natural state of being when you're out of the body at higher levels; you communicate in this way on the master level. So we're actually asking you to master what you have already mastered at that level. It is here in the physical that you are learning to be responsible for your actions based upon their consequences.

The person who's receiving the image, the feeling: What if they're so sensitive or empathic that they not only pick up the image but also the physical state of the person?

Good question. If you are empathic or tend to be highly receptive, you will need to experience it only briefly. If you need to break the contact because you already have the feeling, then look away from the person. Break the contact when you have enough of the feeling to examine it so that it doesn't build up anymore.

If you're going to have a partner, it works better if both women and men work with women, because most women have a natural ability to feel their feelings in a more identifiable way than men do, because men have been raised in an unnatural way in your society.

Can you come up with a name for this? For instance, we have the term "remote viewing." Is this going to be "remote" something – a new science?

A New Term: Compassionate Knowing

It's not a science. Science basically attempts to re-create and improve on perfection. I will call this *compassionate knowing*, known in future vocabulary as C.K. The main thing I want you to understand is that you now have the ability to do these things. To truly change your world you must change the way you *function* in your world. You cannot simply change what you think, because thought's primary function is to examine, and its ability to take action is extremely limited. We must deal with the physiology of the body because that is something that people can understand.

If people eat too much, they feel it. If they have a craving for popcorn, that first mouthful of popcorn feels like, "Oh boy, *this is it!*" Anybody can identify with that feeling. You see, your common ground must be something that you *experience* so you can move past the fallacy and the seduction of the "TRUTH."

But my favorite thing is looking for the truth!

You know how often spiritual teachers have told you that you will find truth within yourself. *This is what I mean!* Last week I told you I would give you teachings that go beyond the simple statement of "I'm seeking for the truth" and "seek the truth from within." That has been sufficient before, but we must go beyond that.

Now is the time of absolute urgency and application. You cannot simply go along with things as they are. You cannot have twenty billion or fifty billion people on Earth. You cannot say there will always be hungry people. That is irresponsible, and no Creator is irresponsible. You are Creators in training and you must relearn feminine technology at this level, which is creationism. You have accumulated sufficient knowledge intellectually. Now it is time to take action, and this is an action I recommend. I will certainly count on it and recommend other actions.

The purpose of this book, you understand, is just as much to be a workbook as it is to confound you with my capricious ideas.

How does the idea of magic fit in? You were using that term with creationism the last time.

Yes. True magic is the ability, with complete cooperation, to invite, incorporate and manifest. This homework I'm giving you on the new communication – compassionate knowing – is designed to help you to *know.* Feelings are essential to knowing who is cooperating with you and who is not. If you want to become a spiritual magician, you will need to *know.* If you want to produce a rose from the bottom of your cane, as it were, you must know the energy forming the rose there before you can give it to the child in the front row. You to need to know which energy wishes to do that and which energy is doing something else. And you will know that through your feelings. Therefore I will always give you steps that will lead you ultimately to your destiny. Good question.

You are doing well, those of you who are reading these books and attempting to expand your consciousness. Perhaps I do not throw the gauntlet in your face, but I do challenge you now to *do*, and do this work in ways that benefit you quite immediately, and they will inevitably benefit others as well. Good night.

Songs of Magic

Benevolent Magic

July 23, 1996

Who will our gods be in 2050?

Aaah, *there's* a good question! You know, you can well ask that. At that stage of the game there will still be some belief in a greater force, a creative force, a benevolent energy, but right around 3.7, 3.71, you begin to shuck the dependency on an exclusively externalized God force. By that time the average man and woman will be reasonably aware and alert politically so they can realize how religion has created the stage for history, rather than history having unfolded, as it were. I'm not blaming religion, because it has in the past certainly attempted to improve mankind's way of being, but too many voices speaking insistently tend to create history, not improve it.

Right around 3.71 we begin to see an awakening within the human consciousness that creation takes place just as much within. This comes about to some extent through consciousness-raising groups and through the manifestation of benevolent prayer.

True magic has just as much in common with benevolent prayer as it does with *pouf!* For true magic to function it must be benevolent – for the greater good of all beings (which includes more than human beings, of course) – and be directed and channeled through a benevolent energy. As more and more people begin to do this, they will discover the limits they've been putting on true benevolent magic. Benevolence is a force to be reckoned with because it is fairly inflexible; it must *be* benevolent. When you begin to realize that your wishes, hopes and dreams can be limited by your concepts and you begin to pray in a more benevolent way, there will be an important effect.

A Future Prayer

Here is an example of a future prayer:

We ask that the almighty and benevolent being that encompasses and cherishes us all will guide us toward our loving, benevolent future and give us landmarks along the way to know and recognize that we are on our true path toward this end.

A prayer like that is inclusive. It's no longer "I."

We're taking prayer beyond the singular, because that emphasizes separation. Just look around the churches and religions and you can see lots of "I's," lots of singulars. But when prayer is "we," it immediately incorporates. (That's why the corporate model as your first world order is not all bad; corporations do incorporate.) A prayer like that assumes by its very words that the Almighty is benevolent. Even if one were to say that the Almighty also contains malevolence (It does not, though It allows it to some extent), this prayer would program Creator to be benevolent or program what you expect, want, desire and wish to participate in from Creator.

A whole generation, because of its experience with computers, now realizes that what you put in is what you get out: WYSIWYG – "what you see is what you get." This whole generation is getting prepared now to understand that how you frame or mold something is the way you will live with it. In the larger sense, this is creator training. As your problems become more and more overwhelming (social problems, for example) and you get to the point where the usual means of resolution are not working, you will need to incorporate true benevolent magic as well as benevolent creative energies to transform things. You will also have to become a we, no longer an I. So in the future, prayer and meditation will become an incorporated effort. On many other planets (higher dimensions, you might say) meditations usually unite a body of people or a single being with the mass of all beings, rather than simply be for an individual. The need for singularity in religion is on the wane.

Religions

In some religions membership is dropping off, so this is probably not the best time to begin a new one. As people begin to explore creation technology, which is feminine technology – the willing and joyful participation of all parts – they will be less likely to join some outfit that will attempt to explain their world in terms of us/them. Any religion or philosophy that approaches life on the basis of separatism is not going to last much longer. Religion in general must be modified. The more that religious empires can incorporate *we* in their religion and let go of competition (even the friendly kind), the more likely they are to live on.

I've spoken before of certain religions that are likely to exist in the future. Of all the Christian religions, the LDS [Latter Day Saints] religion is the most likely to survive, *providing* it incorporates, embraces and fully integrates women and the existence of the Goddess into its religion. If it doesn't do that, fifty years from now it will be gone. What's the big scare? If half of the people in your religion are women, why keep them out? Don't use religion as a means to control people. Religions like that won't last; they are doomed to destroy themselves. I see no reason to put a time bomb in your religion.

Religions say, "You have to come through me to go to God." But at 3.70 and beyond we will be looking for God within, and doing it within a group.

Yes, and also within yourself. Of course, religions that cling to such dogma will not be around for long, only long enough to serve those who still choose such practice. Religions are not going to be forcibly closed down—don't worry about that. But as the parishioners become older, the young ones are not likely to be drafted into such a limited belief system.

I've said in the past that Shinto might survive because it involves certain aspects of the natural world. It has not been a religion of the common people but of a more intellectually enlightened group. Even so, it has the mechanism of the natural world within it. But in the long run the religion that is most likely to survive is what I call the natural religion, meaning a religion that takes place all the time, not just when you're in church or with your fellow brethren and sisters—a "walking-about" religion. It is founded on the principle that everything is alive, and you interact with everything as if it were alive and experience the life of other beings in the most benevolent way. It is a way to experience life.

For example, when you are out in the forest, you might notice a beautiful tree. If you feel welcome, approach the tree; if you don't, say thank you and go on to another tree. When you don't feel welcome it means the tree is doing something else. Trees have lives; they do things. When you find a tree where you feel welcome, just sit with the tree for a while. If you are a spiritual person, you can attempt to blend with the tree. (Don't give the tree *your* energy, just ask the tree to give you some of its own.) You must go to a bigger, older tree to do this. You might feel the difference in density of the tree compared to that of your body. You might feel the leaves, feel the sunlight or the rain on the leaves.

The more you can feel emotionally and physically the experience of life in other beings, the more likely you are to identify the true, endearing and unending qualities of life in yourself and your fellow men and women.

Last night you said that at 3.70, the secret government couldn't touch us anymore. I see this as the point where duality will end. When we see that nothing out there can get

us, we can really open up, let the barriers down and feel more. Life is going to change dramatically.

Yes, when 3.70 to 3.71 comes around, you will be less approachable by those who would manipulate you and also more likely to see your common ground. You might ask, "What about the midway mark at 3.50, 3.51?" You begin to shed things at that point, but you don't really begin to gain a lot until you're around 3.70. In this way the fourth dimension does not overinfluence your experience shifting between dimensions, and by the time you are at 3.70, you will have shed enough that those already in the fourth dimension can welcome you. (A soldier coming home from the war doesn't bring his weapons, but leaves them someplace.)

Since no human being has ever done this before, you're looking ahead and talking to us from the future. You're not looking at someone else's experience and extrapolating to us, because no one has ever done this before.

That's right. No body of people has ever done this while they were still living in their physical bodies, that's true. Of course, everybody does it at the end of their natural cycle, but that's another thing.

You're sort of looking at it from the front end.

I'm sort of standing up in front looking back down the pipe and talking to you, true. I'm giving you sketchy details because nothing is cut in stone. I can give you generalities, but I can't say exactly what will happen, because if I did that I would tend to program you. I'm only trying to encourage you, sort of cheer you on.

I see this as goals, models, possibilities, something to go for.

Yes, quite.

The other thing is that at 3.70 or so, we're going to be so different because we're going to have much more heart energy. What percentage of our heart energy will we have then?

Oh, I should think at least 40%.

Forty percent? At 4.0 it will be 100% – is this in 2050? Does it vary for different people?

Yes, it will vary for different people; there is a range. Some people might get it before, some after.

If you're present there and realize that you create your reality and that nothing can get you because there's nothing that you haven't put there, I can feel my feet in a place that's totally different, where there's so much joy and possibilities. It's a whole different feeling.

You begin to slowly acquire the tools. Spiritual tools always come to you slowly so you can play with them and get to know them. At that point you will have shed enough so that you can acquire them in a benevolent way. In the material world one functions within the same general parameters as one does in the spiritual world. In the material world there is often a desire to acquire things, yet this is simply a reflection of how one tends to acquire tools, abilities, means and methods in the spiritual world.

When we realize that we can deal directly, that there are no interpreters between us and the Creator, how can religions exist at all?

I think they will be there because people will enjoy the camaraderie of churches; but there will be fewer church buildings because people will be less caught up in the separation of religion and there will be less glorification of symbols. There will be some symbolizing of spiritual growth, but they will not be considered as holy, such as the cross is now by many. People will look at a forest and think how beautiful it is; then it will dawn on them that it would have to be destroyed to build a church or any building. Right about that time people will start thinking about what really needs to be constructed, what kind of cities they want, how to use materials, what materials to use. Floods of questions will arise associated with needs that must be filled. Then teachers will be amongst you to teach you true benevolent magic.

A wholly different way than we've done it in the past – cooperation with nature.

Crystal Cities and Sprouting Buildings

Yes. You might have crystal cities that come together, although not formed out of individual crystals. The chemical composition of quartz is silicon dioxide, and you have a lot of it on this planet. Those who are trained by teachers of true benevolent magic will know a special way of singing. They will form small groups to sing into form hollow crystals with doors that are open all the time or with an energy that one passes through to get inside. These crystals will grow from the Earth's own stone. These hollow crystals will begin small, but eventually, as needs increase, they will become much larger and be underground.

I am often asked from time to time, "Zoosh, how do they get sunlight underground?" Aside from the artificial suns, one way is to have an underground crystal city, with crystal spires coming through the surface that act not unlike skylights on a roof. They gather the sunlight and bring it down into the city to where it's needed through various reflectors. It doesn't take anything from anything else, and the crystal that emerges through the surface is not overwhelmingly large. Eventually some of them might become large enough that observation decks or platforms could be arranged around them and people could be there.

The crystal cities will have sufficient air conditioning, so don't worry. Technology will become more benevolent, and it will be possible to live in places where you have not lived before. You will begin to realize that your needs directly correlate with what is provided for you. As you begin to remove steps from your path, instead of saying, "I have to do this and this and this and this in order to have *this*, my goal," you'll begin to realize that your needs alone can produce the goal if you do not limit the means by which true benevolent magic can deliver it to you.

Instead of having to tear down a forest, dig up the earth, lay foundations and do all of the work done now to put up a building, a building will basically sprout, with the assistance of true benevolent magicians singing and striking various harmonic chords while they hold pictures of what *could* be that they envision or, more appropriately, that people need. People need food, shelter and so on, and these will be provided. It was always intended that you reunify with your higher-dimensional abilities at the "lower" dimensions by simply relearning and joyfully rediscovering the skills you already have at the higher dimensions.

I'll put my order in right now for a crystal bathtub! Noted?

Noted.

We're talking about the needs of humans; are we talking about 3.70 coming at a point past the time the Earth stops spinning and starts spinning in the other direction, which knocks down all the buildings and everything?

I'm not going to support that idea. I can see why the idea of the Earth stopping and then starting again is popular.

Doesn't it change its direction of revolution?

I'm not going to support that. It *might* happen, but if the Earth stops spinning, it's going to create a big mess.

A hell of a big mess.

You don't need to be a nuclear physicist to figure that out. You would have at least three days of serious chaos. Even though that is a very interesting idea and I can see where it might be very attractive, I do not think that has to happen.

What can happen?

Well, again we're back to true benevolent magic. Traffic signals aren't built until it's well past the time of needing one. When people are overwhelmed by social problems, nuclear waste (shooting it to the Sun is not realistic), armaments everywhere—at some point these things are going to be in people's faces.

How can we stop someone from using these dangerous weapons destructively, capriciously? The technology exists to prevent ignition, and if you can prevent it, a lot of weapons systems would simply not work. (That's part of the reason some armament makers have been trying to create ignitionless ammunition—but that's another story.) The technology exists to stop all of these things. Every extraterrestrial who is allowed to interact in any way (study your people and so on) knows that *everyone* is immortal, so they are disinclined to interfere because they know that if something globally destructive happens, your bodies will stay here but you will go on. They are waiting for you to create true benevolent magic. When you do that you will be signaling to them that you are prepared to meet them on an equal footing and resolve your own problems.

Teachers of Magic

The initial practitioners of true benevolent magic are practicing today and have been doing so about ten years. They haven't trained many people. They've been trained through vision, insight, spiritual teachers, spirit beings and so on. Their number is small. I tell you this because it's important for you to know that the teachers of true benevolent magic are with you now, even though there will be many more of these teachers when you get to 3.70. For those of you who are truly interested in benevolent magic, ask around, especially in spiritual communities. Put out the call; say your prayer. Put it out there that you need to learn true benevolent magic so you can help create the *we*.

You're saying if we can clean up the mess and stop the pollution, the Earth won't have to do this stopping?

It's not so much that we need to stop the violence. People have been trying to stop it for years, but it hasn't worked because it's very difficult to stop violence.

Without being violent?

Nope. It's infinitely easier to *start* something else. Starting things is much, much easier—that's why projects don't get done. [Chuckles.] Stopping something is significantly more difficult. Here's homework for everyone: I want you to consider the meaning of *benevolence*. Look it up in a good dictionary or define it yourself. It's going to come out to mean something positive. And when you define it, I want you to begin to integrate it as a feeling in your physical self.

Homework

Imagine yourself being benevolent or receiving benevolence—anything. The main thing is that once you can experience how you might *physically* react to the feeling of benevolence and *being* the feeling of benevolence, I'd like you to do at least one act of benevolence toward yourself and one act for others. How will you know it's truly benevolent? Well, I'll tell you.

When you can imagine living in a benevolent world or being a benevolent being, you will begin to notice certain physical feelings in your body. It will feel good. It might feel like love, it might feel like peace or calm or even grace. And when you are doing this benevolent deed for yourself, those feelings ought to be present. If those feelings are not present, relax and refocus and bring those feelings to you. Create those feelings and then while you are feeling them, do something benevolent for yourself.

This is not easy homework, because you must use your feelings. You must be aware of the physical evidence of your physical self. It will also help you to become more conscious of something that is incredibly influential on your planet—the difference between what you feel and what you say.

What Is Said vs. What Is Felt

People, with the best of intentions, often say one thing, yet their feelings are totally different. For example, a child falls down and bumps her knee and she cries out, "I hurt my knee." You might be Mom, Dad, Uncle, Aunt, Grandma, Grandpa, a teacher, a neighbor or anybody. Of course, you turn around and say (if you have feelings of love and caring in that moment), "Here, let me help you," and you take her in and wash off her knee, perhaps putting a little disinfectant on it, the kind that doesn't hurt, then a fancy little bandage with flowers on it. Or you might sing her a song or something. You cheer her up a little bit and let her know that you care about her, and all the while you're feeling these feelings of caring.

That's a true message. It will go into that child's heart, and someday when she grows up and her own child falls down and hurts herself, she'll be able to do the same thing because she was taught by how you, the adult, *felt*, not what you said because it makes no difference what you say. There are many languages in the world, but feelings are the same wherever you go. All people feel in the same ways, although they might feel different emotions in different places. Some people might feel love in their gut, others in their heart and still others in their head. The same emotion might not be exactly the same physiologically, but the way it *works* is the same. But love is love. Regardless how you feel it physiologically, it will be an identifiable feeling within your physical self.

Here's another example. The child comes up to Daddy, who is busy, perhaps washing his car or talking to his friend, and she says, "Daddy, I fell down and hurt myself. Help me." Daddy says, "Oh, I'm sorry, honey. Run in quickly and see your Mom. She'll take care of you."

Now, on the face of it, that sounds pretty good. (Of course, Daddy could have carried her in, but it's okay. Sometimes you can't do it.) The main thing is, what was Daddy feeling in that moment? Was he talking to his friend? In his inner feelings, was he annoyed?

How did the little girl receive what he said, which sounds okay on the surface? Whatever Daddy was feeling is what the little girl will hear/feel. If Daddy was feeling loving and caring, the little girl will understand that adults feel loving and caring for her when she's hurt and needs help. That is what she will remember—physically, not mentally. It's what you remember *physically* that affects your life.

Think about it. Maybe a child (and you've all been children) will need love at some time and maybe the parent has a perfectly good reason for not stopping what they're doing to pick you up and hug and hold you and kiss you and tell you they love you—and mean it when they're feeling it. But maybe Mom is cooking or Dad is working or Uncle Charlie is talking to his friends and for some reason they can't or don't stop, but just look over their shoulder and say, "I love you, honey." But they go on.

I just want you to understand *how you learn*. Once might not do it, but if it happens most of the time, the child learns to identify love with what the people were *feeling* when they said, "I love you." I cannot underline the importance of this enough, because what you are learning here – the true responsibility of a creator, and you are all creator students – means that you are aware *all* of the time. I know that is a lot, but when you have fewer distractions (as you will at 3.70), you will be able to be aware most of the time. By the time you get to 4.0, you'll be aware almost all the time. It is not just a conscious, *mental* awareness (for thought is much denser than feeling). The important thing to remember here is that true awareness means *how you feel*.

You can't just say to the child, "Oh, I love you, honey," and then go on back to what you were doing. If it happens regularly, that's exactly where she stops growing emotionally. She begins to identify love as impatience or "someone doesn't have the time for me," or "I'm not important." She begins to lose her sense of self-value. When she grows up she unconsciously identifies those feelings as love, so she seeks someone as a lover or marriage partner who, when they speak of love, has those same feelings.

Magic As a Form of Prayer

Therefore, if you are creators in training you must be conscious, aware of what you are feeling. *That* is the core of true benevolent magic – not what you are thinking. I want you to be thinking about this because your tool these days is thought. When you begin to take action (I'll teach you more about true benevolent magic in the future), *what you feel in the moment of asking* for that magic to take place is very important. You don't *do* the magic; it's not a spell, but a form of prayer.

Maybe the Earth doesn't have to slow down, stop and spin the other way. Maybe true benevolent magic, as you were saying, can head that off by creating other possibilities. Let's not say what they are. The minute I say what they are I immediately limit those possibilities.

The root of true benevolent magic is that solutions for anything are limited only by your concepts of what those solutions might be. That's why I say thought is dense. You might think how something might be solved even though you do not have the means to apply it in that moment. For example, there's a radioactive waste dump that is affecting people's health. It's percolating into the water no matter what people do. We don't know what to do about it, but we have to stop it. What can we do? Obviously, the physical solution is overwhelming, and likely the water of the whole planet would be at least marginally polluted before you could correct the situation.

Now, let's practice for a minute. I don't usually do meditations, but I'll make an exception.

An Exercise in Feeling-Prayer

I would like you to center into your heart space or wherever you can. If you can, feel the warmth of love or the loving gold light. Go ahead.

All right, you can relax. That was your first level of the homework, to just feel that. When you get around to asking for something, it might take the form of a dance or a prayer or some positions of the fingers. It'll be different for different people. You will then ask, but ask for something for yourself or for your group, not for the world at large. I appreciate your desire to change the world at large, but I want you to begin small. Decide on some problem to be resolved; maybe you need a larger house or a newer car or better health.

If health is your focus, fill yourself with that benevolent feeling and ask out loud, "I would like to feel better in my physical body now. I ask those who can help to do so."

Continue to feel that loving energy until the feeling either goes away or you are tired or fall asleep. If you ask for something like that, it's simple. If you're experiencing something uncomfortable from another person (maybe you're a youngster at school and there's a bully giving you a hard time), don't say, "Get rid of that bully." Say something like, "I would ask that going to school be a wonderful experience for me in every way, and that I will look forward to it and enjoy it because I will be safe." In that way you do not limit the experience. If someone is seriously threatening you, then ask that they be moved far away from you, not that they be harmed because that would not work.

Now, if you take your understanding of everything we said about the little girl and her family and project that as humans interacting with the planet, speaking to the planet from that place of heart feeling, that alone would make an awesome difference. The planet is a living being, and she speaks to *you* that way all the time. Just because you don't speak to her that way doesn't mean that you can't learn how to do it. It's the same thing with communicating with the animals. It tends to work better with wild animals because they are living as part of natural life, but by all means practice.

If everyone on the planet would do that, it would change the planet's awareness. That would take care of some of her needs.

It might very well. Imagine many people on a given day at a given time all using true benevolent magic, all asking that Mother Earth's wounds be healed (not saying how) and that she be restored to a whole and complete being in the most comfortable way for her. *It would then happen in some way.*

There is a great deal you can do with this, but I want you to start small. You can ask for benevolent things for others, but sometimes you will have an uncomfortable feeling. If you do, it means that the person or persons you're asking for have a different agenda, that there's something else they need to do – that's all. Just stop what you're saying, go back and begin the prayer again. Don't assume the worst; just know that there is something else that being or group needs to do before they can experience the benevolence you wish to shower on them.

Let's go to 3.70. Now, we know that we create our reality and that there are no dark forces after us. We've cleared, opened our hearts. Everything's going to change with that attitude – politics, commerce and so on. Who will our role models be? Astronauts?

No. It's an interesting thing. At that time people will begin to see the common ground they have, and they are less likely to need role models. Role models, while seeming benevolent, serve separation – you are here and they are there. Besides, a role model might be totally inappropriate for another culture; one culture's role model might be another culture's anathema. I don't think that role models will really factor in much at that point.

So are you saying that we will continue to have all these cultures on the planet and their different ways of looking at reality?

Yes, we'll have cultures, nationalistic or even racial groups, but people won't be captivated by the concept of the "chosen people" – chosen to rule, chosen to suffer, chosen to learn, whatever. That concept is a separatist one that creates an elite or a permanently downtrodden culture.

So we'll feel ourselves as the same, but just different flavors?

Yes, that's right. Why not grape or lemon today? They are very refreshing.

In the sense of a little girl or boy saying, "That's what I want to be when I grow up" – we're talking about the mythology of the future.

At that stage the child begins to identify what they want to be with other children, and peer groups will become more benevolent because children begin to see each other's true potential. By the time 3.70 comes around, that generation of kids will really wow a lot of people. They'll learn fast, they'll say amazing things, they'll be bright, cheerful, endearing, healthy – all those wonderful things that parents hope for. In the past the role model was your mother or your father, but by that time Mother and Father are also going to be more conscious.

Wait, I am asking the questions from where I am now, but if I look from a 3.70 viewpoint where the soul wants to express itself, it won't be looking outward, but inward, right?

Do you mean, will the individual look to the soul and model themselves after it?

Well, there will be more a desire to express what's within instead of looking outside for something.

That will begin to be more prevalent. When I say that children will look toward each other, I mean the eyes are the mirrors of the soul, and children will be more benevolent with each other. Children do not have to be violent or hateful or mean-spirited; about 90% of that kind of behavior is learned. When that stops being taught, society will inevitably become more benevolent. And as people become more conscious, they will be willing to make certain sacrifices, not the least of which is to have fewer children. Children will then become the precious commodity they truly are. If a couple has only one child, the parents will value that child.

In some Polynesian communities, you know, children are a shared resource, not just a responsibility. If a woman doesn't have a child, then she is able to participate in being a mother with somebody else's child. This is also going to evolve. Many of the finer qualities of Polynesian society will gradually emerge for the average person, because many Polynesian ways of being have a great deal to do with the feminine being (the Goddess, one might say) — the model of love and life based on the nurturing feminine.

Earth Avatars

Every 2000 years there's an avatar. Are we expecting one or are we it?

Where on Earth do you get that number? Why every 2000 years? Why not every year?

All right, every so often an avatar comes to Earth.

You'd be surprised; they come regularly. It's just that at different times they are more appreciated than at others. At any given moment at least two or three people would easily qualify for that role. It's just that if society is pointed in directions other than desiring a true avatar, avatars will go unnoticed, ignored or, as happens all too often, treated as if they are crazy. That is why in the past in more sacred societies people like this were cherished, protected and insulated from society. You might object, "That seems to go against the whole point." But when that group protecting them felt that the larger world was ready for the teachings this being could offer, then that being would be released to society.

With the cynicism that prevails now, an avatar might appear as a ball of light in the sky, glide down and land in the middle of an athletic field during a soccer match, and although some people would be impressed initially, because so much goes on now (TV and movies and people disbelieving their own hearts), there would be a strong resistance to accepting the beauty and love of such a being even if everybody in the stadium was overwhelmed by the loving energy. They might enjoy it for a day or two, but that would be it.

That doesn't mean it's not worth doing. But the reason it doesn't happen much nowadays is that avatars and those who hang around with them recognize that now it must come from within every individual. The light has to be turned on from within the individual, because the externalized light has simply not been working, though it certainly has been tried.

In order to understand your future, it is helpful to understand your present. It is also helpful to understand that you can influence the outcome of your future. Those who predict disasters are not evil beings or corrupted in any way. Many of them have had true spiritual visions of these things and are giving these warnings because they have been guided to give them. But I must tell you that these warnings are often designed for another purpose entirely: to stimulate your unconscious magical abilities to change them. I cannot tell you how many disasters that have been predicted have not occurred *because* the word got out that they were going to happen. From time to time foreseen disasters that were intended to be told about didn't get spread for one reason or another, and they happened.

When disasters are predicted, understand that it is usually to stimulate you to unconsciously re-create the future so it *doesn't* happen. Now, however, you are given the opportunity to re-create the future consciously through true benevolent magic – through prayer that is centered in love and through the understanding of true creationism. We will talk more about that in coming chapters. Good night.

Nine Faces of Creation

Future Politics

August 6, 1996

In the future, politics will be redefined not so much by people's dogmatic instincts, but because you will be more unified than ever before by the time you get to the arbitrary year of 2050. There will also be significantly more knowledge of how everything works together. Politics will become feminized and will be geared more to what I'd call brotherhood and sisterhood, wherein the common ground of people is accentuated and cultural differences are shared and absorbed. The more rigid modus operandi is either dropped or simply not catered to.

By the time you are in fourth-dimensional politics you will have realized that what is constructive for people in general must be applied to people in particular. Some cultures do this now and have managed to maintain themselves. By the time our fourth dimension is locked in you will find that this benevolent politic is a reality. Let's broaden this a bit. Instead of having a titular head of government (even the United Nations is somewhat hierarchical, but it is certainly demonstrated in the politics of sovereign nations) you will have something more akin to a politics based on feeling.

Your most significant common ground, one that you can feel, is your emotions as your body expresses them. Even women to some extent are raised to ignore these feelings, but your experience of feelings will be heightened significantly by who you are in the fourth dimension and by the desire to join in a commonality. By the time you are in the fourth dimension you will begin to seriously express your common Explorer

Race roots and you will want to do things together.

For example, team sports, where one team tries to defeat the other, will be modified. There will still be teams, but they will find ways to help each other rather than compete. So competitive sports will change its face a bit. You will find something more akin to co-ops, wherein groups of individuals come together for common purpose. You might have one team helping another team, without awards or medals or even public recognition—just support.

This kind of support is something that almost all of you appreciate, although some of you have been trained otherwise. It is natural for a child, especially a baby, to want and need support. An adult who does not want or need support has been trained or propagandized against that natural human trait. You are made in such a way by Creator to *require* support. For example, it is not possible for you to live without eating, and you usually need shelter. This is basic support.

Thus politics will take a form that is more universal and benevolent. Any specific questions?

Could we look at how we will organize ourselves? I would love to have you talk about the Federation. You've mentioned only the Network of Planets. How do we come together as we expand levels of organization – star systems, galaxies and so on?

The United Nations is a pretty fair model of the last vestiges of a workable body politic. You will come together in the future in a way that is more akin to a fraternal order wherein there is a shared benevolent experience. You will want to come together. People will not be classified as socialists or communists, but by their traits—what they like to do and what they're good at, which will be clearly defined when they're young. And if they show potential in other areas, that will be encouraged. Many adventurous projects will be supported because it is in your nature to be adventurous.

Allowable ET Gifts

By the time you're in the fourth dimension you will surely be having regular interactions with extraterrestrials. However, these extraterrestrials will be very clearly instructed to avoid as much as possible bringing their ideas and their systems to you. This instruction to basically back off went out a few years ago, and most ETs have done so. They are being informed that it is up to you to provide something for *them*. Those who will be allowed to contact you when you're locked into the fourth dimension will be those who can provide things for you that do not greatly degenerate your adventurous nature. As a matter of fact, they will be encouraged to provide things for you that support and stimulate your adventurous nature.

For example, you have one of these things in a rudimentary form now—games. Virtual-reality games are gradually becoming more popular. Of course, they are limited by the fact that so-called virtual reality does not look particularly real. It is virtual reality only in that there is interaction with something that is basically an animation.

When do we get the holodeck?

ETs will not be providing you what you call the holodeck (refer to *Star Trek: The Next Generation*) because that would allow you to experience all the adventure you need right here and you'd never need to go anywhere. It would work against you.

The ETs will bring highly complex communication systems that are basically three-dimensional, but not clear enough to give you a pristine, three-dimensional image of another planet. The system will be purposely degraded so you can see clearly the individual you are talking to but not much of the background. You might find yourself sitting in an audience in a round or crescent-shaped room, and on the stage is a technical gadget like many ships have that people have been on. It will have a screen twelve to eighteen feet wide (larger than a ship would need) for an audience of maybe three or four hundred people. You will be able to see and talk directly to beings on other planets and perhaps even experience a certain amount of feeling and sensory input. It will be designed to attract you to the planet, not to substitute for a trip. It'll encourage you to go there.

The holodeck will likely be kept from you for quite some time, and will be made available only when you are well into the fifth or sixth dimension and voyages in space are like voyages a naval crew makes, where they are at sea for six or eight months and need some form of entertainment. By that time a rudimentary form of the holodeck will be your experience. It is probably through this method that you will be introduced more grandly to the Network of Planets. You know I avoid this use of the term "Federation" because it's been worked to death.

But what do they call themselves? There literally is some organization.

Network, I think, is a closer term. Those of you who are networked through computers understand that you do not have to have an allegiance. You might buy your way into a computer network, or if it's the Web you might simply share it; but you do not need to swear out an oath to protect each other. It isn't a political instrument. It is more like a shared cultural resource whereby something you need can be made available or you can be told where to find it, which in your case is most likely.

Voyaging to Star Systems Related to You

By the time you're at about 4.2 in dimensions, you will not, as a population, have ships that can travel great distances from one end of the universe to the other through time. You might have ships that can travel in some rudimentary form of light speed, which is actually not that fast. When you're traveling in light speed you're still traveling in physical space, which is necessarily slow, considering the vast distances of space. It is like a safety mechanism.

Most likely, however, you will be getting rides on various extraterrestrial ships. For example, you might have a vehicle that comes here from present-day Orion, and contact will be made with your body of governments. By that time you will be Earthians. You might still have some loose-knit identity as Americans if you live in either North or South America. There won't be this United States thing – and that's good. Understand that you are going to have to give up the idea of nationalism. Some of you will be very reluctant to do so because you have been trained from birth to the idea of "we're better than them," otherwise known in sports circles as "number one." This idea has got to go. The younger generation, fortunately, is gradually being trained toward this ideal. By the time you're at 4.2 you're basically going to be Earthians.

Not Terrans?

Well, Terrans, if you like. But it doesn't really identify you right now, does it, because now you're on Earth.

So perhaps you'll board an Orion ship that travels in light. The people who board will not be trained astronauts or cosmonauts, although there will be some of them who will be given a certain amount of exposure to the ship's instruments and might even stay on the ship after everybody else gets off. Most of the people will be passengers representing a broad spectrum of humanity. There would be just as many artists as physicians and just as many teachers as former soldiers. (By that time the military will have evolved into something more akin to explorers.)

When would we get to 4.2? In about 2100?

Right around 2080, probably before 2100. Your voyage would be to Orion, Pleiades, Sirius or someplace that has a direct association with Earth people. Nationalities won't be considered in the passengers – you're Earth people, period. All races will be represented, probably in a fairly equal manner. By that time the idea of interracial marriage will be no big deal, like "so what?" As I have said, the future skin color for this planet has always been considered to be a light brown, a slightly thinned-out chocolate-milk color. This comes about as a result of the blending of the races. It was always intended that the races blend so that you could pull the best qualities of all the races together to create a prototype Earth male and female. It might be hard to believe that you can get there in such a short time, but you've made quite a few radical steps in the past five years. There'll be quite a few more to come.

After you arrive at this place associated with the Network of Planets, someplace reasonably close that you can identify with, when the passengers disembark they will be met by representatives of what they do, their occupation. For example, the physicians will go with the planet's healers to learn and exchange, to find out how each does it where they're from. The artists will go with the artists, the explorers will go with the explorers and so on. The idea is to share cultures.

There won't be any kind of council meetings. It's important to understand that the idea of council meetings is very reassuring to you, because if you are told that this or that council met, you think there are organizations that are somehow overseeing you. But the higher you go in dimensions, the *less* organization you have, because you do not require any kind of regimentation or need to be told what to do. You *know* what to do based upon your common ground, otherwise known as your feelings. You do what feels right; you do what feels good – and you honor that. That is the feminine politic, as I said.

You will continue to hear from various channels about this council and that council, and I'm not going to say it's baloney. Some of that is to reassure you. It's important for you to begin to know these things, because I can assure you that creators get together, but they don't sit around a big table and have any kind of pecking order. It's more of a pickle-barrel experience where you just sit around and chat.

The whole thing is geared to be an adventure. At that point in time you will be encouraged to *want* to go other places, to *want* to find out about other cultures and, most importantly, bring your unique personalities there for their sake. At this level – say 4.2, maybe 4.3 – when you start interacting with Pleiadians face to face, you might need to have a certain amount of cleansing before you impact them.

Those of us who look after these things cannot allow the Pleiadians to meet Earthians who are totally devoid of any kind of negative energy. Although Pleiadians *want* you to be 100% pure, we're not going to allow it because that defeats the purpose of their growth. They're going to choose people to meet you who are trained to handle 1 to 1.5% discomfort, which seems like nothing to you now, but to them would be very noticeable. Of course, the ones who volunteer to meet you will be adventurers. The whole idea is to get the adventurers together.

I don't want to dash your idea of a council of planets or a council of galaxies. It is so loose-knit that it borders more on being a coffee klatch (if I can use that term) than something for which your presence is required.

We get telepathic information from Wes Bateman about the Federation. When they make decisions it's based on what they call the "light of divine direction," wherein they get a light or a clearance to do something – from somebody or from their own self, perhaps.

Let's convert Wes' terminology. When Wes says "light," does this mean some light external to you? No. Let's convert that light to feeling. Some of you know how to see auric fields, and ofttimes it is for the purpose of healing or assisting someone. But sometimes you simply perceive light around the body. This generated light of an individual has everything to do with what they're feeling. It's not that the light and the feeling are somehow stratified, meaning hierarchical. I've said before that light and tone (sound vibration) are equal. Light and feelings are also equal.

For example, the feeling of love (since it is so unique to any individual) might be perceived as many beautiful colors. If a person were filled with love you would never see gray or dark colors, but some beautiful color. If a person felt passionate you might see a little red or pink. If it were a universal feeling, you might see white or gold. Directed toward an individual (say, a sister or a parent), it might even be pink or with a little pale green (connected with healing).

The reason I bring this up is that this light is not a vague thing – it is a *felt* thing. You can't reach out and grab light in your present bodies, but you can reach out and feel feelings. Most of you can feel the feelings of others and are constantly reacting to them. So when Wes refers to the consent of the light, this is basically something felt from within.

Some of us have had the idea that when we got to the fourth dimension we'd be free – "We're out of here!" – but it sounds instead like going another grade in school. We're still watched and meted out information and herded in the direction we're supposed to go – all that, right?

My, you certainly have a way of reducing it to its lowest common denominator! Let's just say you're nurtured toward your ultimate intention. You're not herded with cattle prods and treated like nincompoops.

How long will this go on, then?

Applied Unconditional Love, Patience, Curiosity and Passion

As long as it takes you to acquire applied unconditional love. The lightbeing, the light of oneself, is naturally unconditional love. But to have unconditional love on the physical level, especially in the third dimension, you've got to work at it. Or how about something even more challenging – patience! Creators have to be very patient, and they cannot be attached to the outcome. They might *desire* an outcome or have a view toward something in particular occurring, but (allowing for infinite possibilities) one might have to experience one's vision sparingly, because it will happen here or there but not everywhere.

It is necessary for you to go through school. You will continue to be nurtured in the fourth dimension. You will be protected in the sense that you will not be exposed to a broad range of higher-dimensional technology that is available for most other beings. If we simply integrated you into the mass population of ETs, it wouldn't take long for you to cease being the Explorer Race, to say nothing of your ultimate cocreation with Creator.

What kept Creator going to the point where It began to create is curiosity, the *only* thing that can stimulate creation ("I wonder what would will happen if?"). If that question were to be removed from a people, that would be the end, because they become benign, loving subjects of Creator rather than equal, cocreating individuals.

You're saying that if we were allowed to interact with the mass of these benign, loving creatures, we'd be entrained into that vibration and wouldn't keep this passion we've worked so hard to attain, this curiosity?

Yes, this passion. We are culturing that passion in you because we want you to keep moving forward. It's not a manipulation; we have the total permission of your souls. Before you volunteered for this experiment your souls told us: "No matter what happens, don't tell me before it's time." You can identify with that. So we're not going to tell you before it's time.

As so often is the case with any successful venture, it's getting there that's almost all the fun. Whenever you exit your body at the end of a life cycle, you naturally rejoin all and the Creator. You will know all (whatever is necessary), you will certainly *feel* all, and some of you will exit the Explorer Race for a time and go on and have lives elsewhere – benevolent, easy, perhaps comfortable lives. But you will inevitably return to the Explorer Race experience because that is truly where the action is. No one comes here to Earth or joins the Explorer Race trip without having some desire to discover something new.

Souls and Soul Fragments

Are all the members of the Explorer Race fragments of the 144,000 souls that Jehovah brought, or have a lot of others joined up over the course of time?

Others have joined up over the course of time. A lot of the individuals have to do with that 144,000, but others have signed on. Think of all the walk-ins and soul braids. It's not a terribly exclusive club. [Chuckles.]

How many? How do you count them – the souls when they're back together, or the fragments?

I tend to count them in their completion rather than in their individual incarnations. It's just too high a number. In their total completion there are about a thousand. A small number, yes?

A thousand? I thought there were 144,000 plus?

They weren't complete, either.

If we had to trace ourselves back to be one of the 144,000, we'd have to go back a lot further than that, wouldn't we?

Yes, you could say "back," or you could simply say, distilled into your pure essence. It's not a big number.

They were already fragmented before they got to Orion.

Some of them fragmented before they left Creator! You know, there's no set pattern. Some went out relatively complete and fragmented slowly. Others fragmented right away. Individuals do it individually.

So the thousand will be the ones who step in for the Creator?

That's right.

Ah, you never told us that before – that's a new piece.

Yes.

The Council of Nine

Why haven't you connected the 144,000 to the Council of Nine?

How does the Council of Nine fit in with all this? Well, you know, councils, as I said, are so loose-knit in reality that at that level you feel connected with everyone and everything. I cannot draw a straight line (it's more like a dotted line), but there *is* a connection.

Are you saying the Council of Nine has not always been the same nine beings?

I am saying that the Council of Nine is really one being with nine faces. One might take an individual core personality and break it down, not into its ultimate multiple incarnations where it is shadings of different portions of itself, but into eight or nine actual distinct qualities of that individual personality. In the case of the Council of Nine, nine is really one.

You hear in metaphysics "we're all one" and all that jazz; in reality that's what it's all about. You might say that the Creator has fragmented Itself into various facets of Its personality, and that's where you all come in. But without getting unnecessarily complex, let's just say that the nine is really one and that it is nine faces, nine core individual personalities that this individual most easily represents.

Can you say just what the nine qualities . . .

I'll give a few of those personalities, not all. One of them is *ultimate curiosity:* the kind of individual who asks questions but, stimulated with the answer, continues to ask questions – ultimate curiosity. Another one would definitely be the *adventurer;* otherwise it wouldn't be connected with you in any way. If you don't have the adventurer, then you're not involved in the Explorer Race.

Another one would be *the lover*, and this covers not only unconditional love but intimate love as well. Another one would be the *jokester*, or comedian – not the trickster, though; that wouldn't figure in. Another one might be the *dramatic* side, and another might be the *organizational* being. And another one might be the *mentor*. That's enough.

Which one is nine? The guy who calls himself Nine?

I don't think you understand. The nine are one.

I know, but when they separate out and appear as nine, which quality has number nine?

No, no, no, it doesn't work like that. It's not "count off according to qualities." Number nine might be adventure one day and amusement the next. But amusement is really part of the jokester. No, no, it's not broken down like that – that's an organizational need. As a matter of fact, you ought to add, "Ask *organization* that."

It's important to realize that one individual even now, one of you, would perhaps have at any given time many different faces to your personality. Some people feel free to demonstrate that, such as an actor or actress or even a drama coach. But a person might demonstrate different facets of his or her personality, such as an attorney or a politician or even a mother or a father—one moment being loving and nurturing, the next moment being strict and disciplinary.

It is not an accident that you have these different qualities. It is in your very nature—perhaps in the very nature of life itself by this Creator—that variety expressed in great numbers or singly is an absolute in this creation. Even an ant starts out as a different form of itself, and while you might think they have strictly a herd personality, every once in a while you might see one stop and explore a blade of grass or just stand there briefly for a moment. They are individuals; they're simply cooperative individuals. [Chuckles.]

When the nine are one, is there a name attached to them that we've ever heard?

Let me think about that a moment. I think not a name so much, more of a tone. I'd say probably somewhere in the key of G you'd find that tone.

Nonexclusive Groups and Political Parties with Broad Issues

We've talked about the future drawing us forward, so how would politics draw us through this next few years?

You might ask what you could do (politically speaking) to grease the rail toward this kind of future politics. If you can, encourage, join or otherwise support fraternal organizations that are not exclusive. (It's okay if it's all men or all women—I'm willing to accept that for the moment, although it doesn't have to be.) If you can, do that and find causes everybody can agree to. For instance, setting up small-town hospitals, where different kinds of healing practitioners can work side by side as equals without having to compete with each other—now, there's a worthy cause. I cannot tell you how many practitioners of various healing arts would love to work in such a circumstance. But many of them were raised to believe in competition and were also guided toward exclusivity. Anything you can do to support a broad spectrum of individuals coming together for some common cause is worthwhile.

I will make one exception. The computer network is being done now, so you don't have to do that; and it will continue to grow on its own. I think you have to bring people together physically, and organizations like that do exist now. Generally speaking, people involve themselves in something that is urgent, but it is possible to create a group where some general idea is considered worthy enough to work toward. I'm not talking about a political action committee, but something that actually supports everyone to the best of your ability and includes the people who are working toward it.

It's only a very short time before you get political parties that are geared more to broader issues. In Europe and other countries you have environmental parties. I'm talking about cause-oriented groups that everybody can generally agree with. I don't expect you to start the Pure Water Party, but that is a portion of an environmental party's loose-knit dogma. That's probably the best thing you can do politically right now. There would be some resistance from the status quo; that is to be expected. But I do not expect it to be catastrophic resistance.

You could say that up to now we've been tempered by survival issues and by the stress created by the controls and manipulations of the secret government, which you said would end at 3.7.

At 3.7, maybe 3.73, you will no longer feel the secret government's stranglehold on you and you will begin to move more at your own speed. You will begin to pick up the pace because you're not dragging an anchor behind you anymore. You might move, for example, from 3.73 to 3.80 in a week. Time drags longest from about 3.47 (now) to about 3.60; then at 3.65, 3.70 the juggernaut will build. The next few clicks on the dial are slow.

At that point our hearts are going to be more open and we're going to be more conscientious and responsible. First we'll clean up the planet . . .

Wait a minute now, you're jumping to conclusions. When you say you'll clean up the planet, how are you going to do it? Are you going to return all the mined materials to their native rock form and put them back? I don't think so, although once you get up around 4.5 you might be able to do that.

Stop Mining and Polluting and Start Major Recycling

What you will do initially is *stop* doing a lot of stuff like mining and drilling for oil—even drilling for water—because by that time you will have learned how, through the use of various sacred techniques that I've referred to as feminine technology, to bring pure springs up to the surface. This is a bit of magic, but it is also need and thankfulness and a certain amount of honoring.

You will begin to bring to the surface what you need through *invitation*. There will be movements to return atomic wastes to uranium mines (which I've suggested myself, I must admit). Perhaps that's all right because that native rock might tolerate radioactivity better and eventually deal with it in some benevolent way. But I don't think you're going to be dumping aluminum chairs into bauxite mines.

I was trying to get a time line. We'll quit polluting and be more conscientious . . .

You'll stop, yes. You have on Earth now about 5.7 billion people.

Understand what it means to quit polluting. Everybody's basically going to have to go natural. You've got too many people now to do that, but I think you're going to make a serious effort. Basically you're going to have to recycle a whole lot of stuff.

Well, there's a concept called appropriate technology, and computers will continue to be manufactured and used, right? Airplanes?

But you're going to have to recycle to the maximum. You're going to start mining your rubbish heaps. Mother Earth will not mind that too much because rubbish is basically discarded in canyons. You're going to dig out those canyons and recycle virtually all of that rubbish. Even if you didn't start mining it for another fifty years, most of it would still be there with only the smallest amount of degradation.

If anybody wants a worthy long-range investment, you might want to grab former dumps. Aside from their being reasonably decent (if somewhat minor) natural-gas producers, in forty to fifty years they'll be considered treasure troves of glass and metal. There's some pretty good recycling going on with metal, but it's still shocking how much of it gets dumped. I can assure you, you'll be mining the dumps. And you're right, you're going to keep making a certain amount of stuff, but it will be crystal clear to you that you will not be able to do it in the old way. You'll have to make all the new stuff out of the old stuff.

Beyond that we're still in school. So we still have mentors, we're still basically in quarantine, although certain people and concepts are allowed in. But somehow I always had the idea that there would be this blast-out and we would be out of here. But it's just incremental . . .

I think that's more of a personal desire [chuckles], if I may say so.

Choosing Future Lives

It's incremental, it's a constant educational process. So this isn't our last . . . the reincarnating goes on?

Oh yes. An individual at the end of his natural cycle might leave the Explorer Race scene for a time and return later, or continue to reincarnate along the lines of the Explorer Race but perhaps a thousand years into the future. You know, you don't reincarnate one life after another chronologically on the calendar. At the end of this life a person might easily say, "This is where the Explorer Race is now. In my next life I'd like to be with the Explorer Race 3000 experiential years in the future and see what we're doing then." That's not at all unusual. (I'm neither recommending nor advising it.)

But you can go sequentially if you choose, right?

Absolutely. It is highly unusual to do so, though. Most individuals after an Earth life (even a reasonably benevolent one) will choose to go someplace different for a time.

For those who inhabit the physical bodies, as they get lighter and lighter and lighter, you get freer, more loving, a higher frequency . . .

You access more knowledge and inspiration, but your ability to create is never stifled. This means that ultimate knowledge is denied you. If it were given to you, your desire to create would cease to exist. Therefore a certain amount of the gift of ignorance is given to you as long as you're in the Explorer Race.

Say that you wanted your next life to be on Sirius as a water magician. In that case you might have available to you all knowledge of all water beings who have ever lived at any time, without any restrictions whatsoever. But in the life after that you decide to rejoin the Explorer Race, and you immediately don the garment of the level of ignorance that exists within the Explorer Race at that time. You want to know and do it better, then you want to know more.

That's a beautiful phrase. So when you incarnate during the middle of an age, you have to deal with the level of ignorance at that time. No matter what you know as a soul, your level of ignorance is always determined by where the mass consciousness is and what is desired from souls at the time you choose to incarnate.

If you go way back in time, sometimes all that was desired was that people get along reasonably well, find ways to work together and survive the best they could – not high expectations. Nowadays more is expected from most people, although not everyone. To this day not much is expected from some people. But those circumstances are exceptions.

Becoming Creator

Let's get back to the thousand souls when the Creator leaves. What is their status? They can't have the gift of ignorance if they're going to be standing in for the Creator.

No, at the level when you're getting ready to give Creator a coffee break, as I like to say, you discard *all* ignorance. You have Creator's full knowledge because you started out from Creator, plus anything you learned along the way that was more than Creator might have personally experienced, which is the intention of the veil of ignorance – so that you can re-create.

Anything that's anybody has ever done.

Yes. So you then have all of that and you also are open to re-creation. From almost the moment you take Creator's place, you immediately start thinking about giving birth to another race of beings that will replace you. You will have become so enamored with the experiment that you will want to pass it on.

And do it better.

That's right. Because the idea of the Explorer Race – the whole concept of creating some portion of yourself that goes on to become more than you (as any beloved parent would want for their child) – was really something this Creator invented, as far as I know. And it's not going to be something that you're going to want to deny yourself. Now you're experiencing it as the child of Creator, but when you become

Creator you will want to experience it as the parent.

So if He's going off for a coffee break, that means He's coming back?

Well, let's just say, if He's going to the next level, He's expanding. Creator's basically becoming more. You might ask, "Does that mean He actually leaves, or that He simply expands and becomes more?"

To encompass more.

Yes. If I answered that question I would essentially defuse Creator's desire to go on, so of course I won't. You know, as a teacher one must always be thinking about how to motivate the student. And Creator is always thinking about how to motivate you but not rush you.

But in all of creation only these beings who become individualized as the Explorer Race need to be motivated. The rest of creation just is. How does that work?

Oh, I wouldn't say that that's really true. On the Pleiades (I like to use them as an example because they look so much like some of you) you would certainly want to motivate your children to become expanded individuals, perhaps to become more than your parents. No, I think that the idea of motivation exists in quite a few other places. There are some places where it is not necessary, but I would say . . .

But they're individualized. I was counting them as part of all individualized beings. You said that 99% of Creator did not become individualized, did not give up . . .

Correct.

Do all these beings who run creation and do all these things get motivated?

Creator and the Explorer Race

I don't think you understand. Ninety-nine percent of Creator is still Creator's body. Creator arrived here ("here" being where Creator is) as a being, let's say, and all of the individuals that exist in all of this Creator's creation have come from about 2% of Creator, plus the other 1% that creates the Explorer Race. So 97% of Creator is part of Creator's self and is not individualized at all.

But it does things. Do you count the Time Master [a being who is part of The Explorer Race: Friends of the Creator*] as individualized, or as part of the Creator?*

I'd say the Time Master is part of the Creator.

So are these parts of the Creator who do all these other things? Do they get motivated?

Let's just say, to use the brain analogy, that they're cells in the brain of God, and each cell has its own function. But unlike a cell in your own brain, it goes on indefinitely. It does not die out.

You said we were cells of his curiosity? How did you say it?

Yes. You are the more curious part of Creator—that is essential.

Would you say of his brain?

I'm using brain only as an analogy. Creator doesn't have a brain (on the other hand, Creator is not brainless!).

But basically, what's being motivated here in this great big creation is this little Explorer Race, this 1% of creation.

Well, just think on your life. Perhaps your life has had only 1% wonderful moments; that would perhaps be pretty good. I don't like to say this "little creation," because it's pretty impressive. It hasn't been done before that I know of.

So we are unique?

Special and . . .

We're meant to go out to motivate the other 2% that are individualized but are not part of the Explorer Race?

You know, the thing I like most about the Explorer Race is that for fragile beings, it is very durable. You were saying?

I was saying that we are bred and created to go out and bring this lust of life and curiosity to the other individualized 2% of creation that is not the Explorer Race?

Yes, and also to bring questions. I cannot begin to tell you how many other races have (according to them) evolved past any idea of questioning and are now living totally on faith. While that might seem to be wonderful, at some point there is no growth when there are no questions. The question has to go beyond *why* and utilize *how.*

You see, the Explorer Race has basically stimulated the whole question of *how*. *Why* existed before you, but *how* means, "How can we do it? How does it work?" This means that someone tells you how it works so you can learn to do it yourself. *How* is not something you invented, but you popularized it. You culled it out of mediocre obscurity and turned it into something bold and prominent.

Do many creations focus on a particular thing, try to perfect it and then that's it? Is that a way to say it?

That is one way to say it. Or they might focus on specific things and try to perfect them and then fit them well together, interconnected in complete harmony. But rarely do you find cultures that are constantly seeking something beyond what they can even conceptualize.

Is it fair to say that this Creator is the first one in all of creation that's ever, ever created individualized beings out of Himself? And these others – do they have free will?

No, other creators have created individual beings out of themselves, but I think it's fair to say that this is the first time it has been done with the agenda your Creator has – not only to cocreate with Itself, but to acquire more wisdom – not knowledge – than your Creator has. The only way to acquire wisdom, as you know, is by *living* your knowledge to discover what actually works and what is interesting but not applicable theory.

It has to be experiential.

That's right.

When Creator created these other individuals, was that at the end when He was trying to find a use for negativity?

No, no, no. Creator started out creating all the other individuals.

Creator was interested in variety. And it was only at the point where Creator was beginning to repeat Itself that It said, "Wait a minute. Am I running out of ideas here, or what?" It was right around that time that Creator became aware that there was more to Itself elsewhere.

So the purpose of these benign individuals, who were created and who are out there not going anywhere, was for variety, but it wasn't an attempt to use negativity.

Correct. No, *you're* intended to be the problem solvers, and this means that problems that exist in other parts of this created universe that can't or haven't been resolved elsewhere, get to be resolved by you. Negativity is one of those; you had to resolve it and you did.

The Best of Negativity, Then Full Realization

But did Creator create any beings who were too negative, and were there experiments in trying to use negativity before the Explorer Race, where it all went negative?

Yes, there were a couple of flops, but what you're experiencing now is a success even though sometimes you might not believe it. You have managed to create courage and sacrifice and laying down your lives for your friends, which utilizes the best of negativity. Laying down your lives for your friends is not negative, but it certainly utilizes the best of both negative and positive. You've managed to create that and other good things from negativity while not going totally negative. Others have done so.

The idea of polarity here was to put a safety mechanism on negativity. In a single pole (monopole) society, if you put in negativity, all will go negative. But that was not known in the beginning. I say that as if I am the great and powerful wizard, as it were (if I can borrow from Frank Baum); in reality it was discovered by trial and error. By polarizing your society it meant that negativity would not go beyond 50%, or if it did it would get only a little beyond it and no more.

You previously said that once you obtain total Christ consciousness and your heart is totally open, you allow everything. You wouldn't lay your life down for a friend.

That's right. You wouldn't think of doing that.

So the Explorer Race has got to be nurtured to have an open heart, but somehow not get to that point of total . . .

Not become fully realized, no. We don't want that. We want you to want the best for yourselves and eventually want the best for everybody you meet.

Once we get to that stage we've kind of worked ourselves out of being able to play in the Explorer Race. Is that right?

That's right. Once you become a fully realized being, then you allow people their destiny. And if their destiny happens to be a boulder coming down the hill that hits them, you kind of smile and nod and go on. Nowadays people wouldn't think of doing that. You'd say, "Look out!" at the very least.

The Explorer Race will be useful only within a certain range of frequency, because once we go beyond that into this realized being, then we are not functioning as Explorer Race beings anymore. We've graduated.

That's right. That's why the whole experience is designed to shake up and also stimulate all of this Creator's creation. At some point that will be done, because you don't have to go everywhere in the creation to make that happen. You will be going to certain places that also have emissaries who go to other places. The intention is to create a chain reaction of growth, but you don't have to go to every place that exists.

The future bears more heavily on you now than is readily apparent. But you do not have a fate; you are not evolving to something that is already created. It is more that there are rough parameters, sort of guidelines, and you will follow along these basic guidelines. They will not be thrust upon you; you will simply choose them as you begin to acquire more of your natural selves, more of your heart energy, more of your feeling energy, more of your sensing energy, more of your basic instincts and more of your true mind, which is vertical wisdom.

As you begin to acquire these things, you will naturally want to be more. You will naturally follow these threads into the future. Yet you will still have enough ignorance that you will be able to re-create everything as you go along. That is, after all, the intention of the Explorer Race — not only creation but *re-creation* in its infinite variety. Good night.

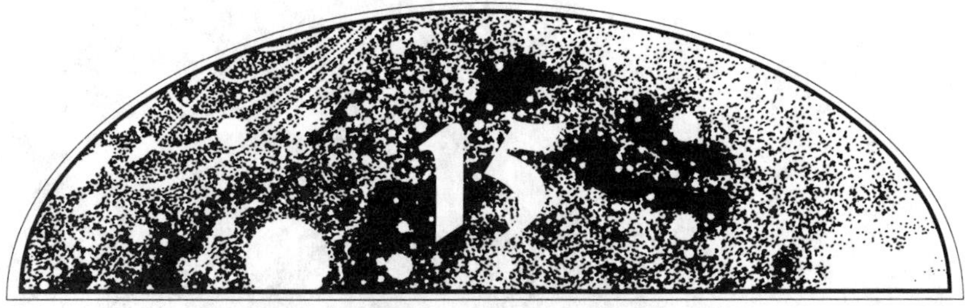

A Visit to the Creator of All Creators

August 13, 1996

Sit back and relax. Tonight I'm going to suggest a scenario. It might not be exactly like this, but it will be something very similar.

Around the year 2051 to 2053 the knowledge of extraterrestrials will have been integrated into society worldwide for at least forty years, and most of what had been kept secret will have long since been revealed. By that time there will have been a selection and training program in existence for almost ten years wherein many thousands of Earth citizens—some highly educated, some just eager, enthusiastic and everything in between—will be selected to take a voyage to Saturn. The vehicle will be flown (or driven, you might say) by several different extraterrestrials, some surely from the Pleiades since they look like you, and some from certain planets on Andromeda and Sirius, for they too look almost exactly like Earth citizens.

The crew will number perhaps fifty to seventy-five and the selected Earth astronauts will number several thousand. With much fanfare and good-bye hugs and "don't forget to write," the ship will take off. Being an extraterrestrial craft of significant proportions, it can travel from Earth to Saturn in a very short time. It would actually be able to get there in less than five minutes, but because it is a voyage designed to show individuals the beauty of what's out in space, they will purposely take several hours, giving you somewhat of a travelogue on the way.

Magnificent Voyage

When you arrive on Saturn, complete arrangements will have been made for you. Everyone will debark, and the dwellers of Saturn who welcome you and integrate with you will look very much like you; they will perhaps have red hair (as you call it here) and freckles. They will invite you inside the planet, where you find your accommodations and spend several days to perhaps a couple of weeks. In your comfortable quarters you will easily find the right foods and all the comforts you need. You will even be able to have pets, but they will not arrive on your ship because they carry certain microorganisms the vehicle would be unable to tolerate. They would already have been transported and would be awaiting your arrival. So there might be a few dogs and cats – maybe even a horse – to help you feel at home.

A Saturnian Meditation

While you are there you will participate in an ancient Saturnian meditation during which the Saturnians unite completely with each other, with those who have gone before and perhaps even with spirits intending to manifest soon. Saturnians are thereby able to keep in touch with their civilization's intentions, provisions and ultimate conclusions. You will join them for portions of the meditation, during which different Saturnians will befriend you and come to know you very well. In these meditations one cannot guard anything as a secret, nor would you want to as a cosmonaut-astronaut, because you want to be amongst people who accept and appreciate your unique qualities.

After a sorting-out process of a week to ten days at most, several smaller ships would arrive, possibly a Pleiadian craft because your needs and their needs are so similar. There would be several other ships. As a result of this meditation different ships would go to different places. Twenty to at most forty individuals from a given like-minded group (in terms of desires, interests or excitement) would go to different planets and star systems. They would be welcomed and feel very much at home there (at least enthusiastic) because what they desire, need, wish for and hope for will be accommodated on this journey. Perhaps thirty might go to Zeta Reticuli and forty to the Pleiades and Orion and Sirius and Andromeda and so on. We will follow a couple of those in our scenario.

Let's say a craft is going to Zeta Reticuli for the sake of those who are particularly fascinated by the sciences and mental arts. They will travel quickly to Zeta Reticuli through time, but also through space, so the voyage is not instantaneous.

By that time the extraterrestrial craft will have been sufficiently adapted to your physiological needs and you wouldn't need to undergo radical methods of sustenance and survival techniques to travel through time. Certain Zeta ships have utilized these radical methods, but Pleiadian ships have a more benevolent means of sustaining life when time-traveling. When you travel through time you are basically reducing the physiological entity to light and "personal impulses," which you might call a direct-dialing code for your DNA and your soul pattern. This can be a radical procedure if not done very carefully. There has been a great deal of Zeta research into how to do this, and they are still working extensively with the Pleiadians to create a method to do so most comfortably for citizens of Earth.

When you arrive on Zeta Reticuli there will be groups of no fewer than three, so you'll always be amongst other Earthians. You will go to different areas. There will be what I would call camps set up to meet all of your needs. Your specific interests will be explored in a multimedia approach —reading, talking and a particular method of awareness wherein you might sit or lie in a very comfortable chair, like a reclining chair in which visual and physical stimuli help you absorb the maximum amount of knowledge and wisdom in your field of interest. This will continue for as long as you choose, but generally no less than three months, to get the maximum wisdom, knowledge and general information.

There might be those who go to Sirius. One ship might go there with as many as forty-seven people, usually to study the original forms of what Earthians consider animals—how these individuals look on their native planets (whales, dolphins, sea turtles and so on). They look significantly different physiologically in their native form, but have a similar emotional and mental makeup. They might also study different species of plants, many of which have cousins on the Earth, to discover how these plant forms communicate and so on. This extraterrestrial study will go on for at least three months, but no more than eight or nine months, after which time the astronauts will have a significant increase in their own particular areas of knowledge and the extraterrestrials will have a significant understanding of who you are as individuals.

Then everyone will rejoin the original ship, this larger Pleiadian spacecraft, and you will say good-bye to your extraterrestrial hosts, making plans to stay in touch in the future. Then you will share your experiences and each other's company. With the accumulation of all of this wisdom, then the *real* voyage begins.

The Voyage to Creator's Grandparents

At this point the choice becomes obvious. You have studied a great many things and, having had perhaps a week to ten days to discuss it amongst each other, the Pleiadians tell you that they have the technology at the leading edge of their knowledge to take you to the point of origin of all life as they know it. Their teachers' teachers by that time will have given them this knowledge for this voyage (I'm telling Pleiadians at the same time I'm telling you now), thus giving them about twenty years to investigate, experience and otherwise enjoy this portal of life.

This voyage will take perhaps eight to ten days even traveling through time. If one traveled in time between here and, say, Sirius, it would take no more than about seven seconds – three seconds to navigate your immediate solar system, two seconds to navigate theirs (measuring from an outer planet) and a couple of seconds traveling between. If it takes eight to ten days to reach the origin of all life, that gives you a pretty fair idea of how far away it is.

During this voyage, since you are traveling through time, you are all basically in a form of suspended animation to protect you and the crew (even the more advanced beings – the Sirians, Andromedans and so on). Even lightbeings would be in a state of relaxation, if not in stasis. It would appear to be automated, but it is largely through magnetic energy, because the place of emergence of all life has invited the ship and its occupants and pulls you there magnetically. Magnetism (which, by the way, exists without electricity) is the source of all life as I know it and, according to my understanding of life, is certainly the source of life, period.

The Creator of All Creators would certainly have the capacity to get you there in the twinkling of an eye, but it does it slowly to make it easy on your physical selves. During this journey this source of all life communicates with certain beings, and it is important for you to understand who they are. You must understand that the Creator of All Creotors is *not* your Creator, but is beyond your Creator, from where your Creator has been offsprung.

This source is in communication with your Creator and some of your Creator's pals. It utilizes the form of creator communication that is most easily understood and appreciated by "people" of creator status, a form of communication that is a low-pitched frequency that I most closely identify with the feeling of love. This loving vibratory communication goes on between the source of all life and your Creator. And because your Creator pulled in several other creators (as discussed in Part 1), there is a shared communication going on. Here we have members of the Explorer Race, who are basically representatives of those who will replace your Creator, coming to be with the Creator of All Creators.

At your arrival you will feel a joy to be amongst that which has created, now creates and will create, and you will feel a love for these grandparents, if you would, of Creator. One loves Creator, yes, but one might also love the grandparent.

As your cosmonaut selves, you find yourself no longer in the ship. You are not returned to your physiological being but are still in stasis, in a slightly condensed state in which all of you blend together to form one blob of feeling. (When you go past color, there is ultimately feeling, vibration.) This feeling is absorbed into the vast feeling of the emergence of all life. A universal prayer or question arises from you astronauts at that point of first contact. You are enjoying this tremendous love, but even so there is the plea, *What is the ultimate purpose of existence?*

This question is answered initially with a feeling, which goes something like this. The Creator's creator blesses you as a cosmonaut, loves you and, most importantly, allows you to love it. (There's nothing quite like being loved and being allowed to love. It is the ultimate fulfillment.) During this moment of ultimate fulfillment you have the explanation for existence. Creator's creator "says" something like this to you: "The greatest love of all life is to reproduce that love in as many flavors and forms and as many feelings and concepts as can exist, the expression of life being equal to infinity. This is felt; one understands totally in that moment timeless infinity, timeless love, timeless creation and timeless fulfillment."

Timelessness and Limitlessness

Relax for a moment. Meditate on the following, then put the book down for a moment.

> *Timelessness is the most critical factor, because if there is no limit to your creations, no limit to your love or your ability to be loved, there is ultimately no limit to consciousness, no limit to anything. Thus you have the capacity to experience total fulfillment in every single super/subatomic particle of your being. Atomic particles are, after all, held together by the cosmic force of life, which is love.*

Sit back and feel that for a moment. This is your ultimate training, you see. In order to replace Creator, you must be able to appreciate not only consequences and applications and other things that I've talked about, but you must have the *motivation* to be a creator. Why create? After a while you run out of ideas, don't you? But no, the Creator of All Creators reminds you that every single subatomic particle within you has its own unique thoughts and visions. And since there is no limit to the infinity of the subatomic particle (meaning smaller and smaller), there is no limit to creation. You create for the sheer love and pleasure of it, because it is an act of love by which you are loved. Thus it is ultimately a totally balanced, benevolent purpose.

Existing within this conceptual experience, you as the Explorer Race, through your representatives of a few thousand lucky ones, bask in this experience for a timeless interlude. This experience is hard to describe in time, but sufficiently long that for the rest of your physical lives you can, at a moment's notice, lie down, relax and immediately remember it, reexperiencing the full range of those feelings and that nurturance.

After this timeless interlude the Creator of All Creators, source of all that is, passes on the invitation that through these meditations you individuals can return at any time to communicate with the Creator of

All Creators, since this being is motivating your Creator's offspring (you) to become creators. (Normally it does not function this way.) The Creator of All Creators is interested in *this* particular experiment of *this* creator.

You who have visited there experience ultimate love. To define it again, *ultimate love is giving and receiving unconditional love in the same moment.* You have certain side benefits from this experience; although you retain your original appearance and individuality, you no longer have certain needs and functions.

Creator Inoculation

You begin your voyage home to Earth. You no longer have to be in stasis because your molecules have entirely changed. For example, after this experience you discover that you can live on love. You receive all the nutrients, all the support and sustenance you need by loving unconditionally or by receiving unconditional love. You have discovered not only the ultimate sustenance, but you have learned how to teach *others* to receive it!

This is what I call the *creator inoculation* for your society. Those individuals will not return as avatars to transform everyone, because, surprisingly, not everyone will want to learn even this kind of ultimate gift. But over a period of about 200 experiential years more voyages will be made. (You see, some people won't want to learn what these people learn because they want to do it themselves; you can understand that.) So after a couple hundred years, maybe a couple thousand tops, everyone will have been inoculated with the lesson of giving and receiving unconditional love (in other words, the motivation to be a creator) either by going on the voyage as these particular astronauts did, or by having been taught by one of these cosmonauts. All people on Earth will have been converted to *creator-motivated status* (CMS, if you like). When this event takes place you then depart from Earth, re-form as a mass soul and replace Creator.

Understand that I'm giving you a precise (within certain limits) scenario of the time when you will actually replace Creator. I've given different time lines in the past, but I've always said that things are not cut in stone, and as time goes on (experiential time) you will often choose to speed up or sometimes slow down the process. I'm giving you this scenario because I want you to understand that being the Explorer Race has ramifications far beyond that of going out and contacting civilizations on other planets (discussed mainly in *The Explorer Race*). As I said in Part I, the ultimate intention is that you not only cocreate with Creator, but that you have consultants, advisors and motivators like Creator through this creator grandparent. I am not aware of any creator or any form of life beyond this one that creates. As far as I know, this is the ultimate creator's creator.

Who told the Pleiadians how to get there?

Zoosh did.

Well, modesty forbids.

The Pleiadians

Why the Pleiadians rather than the Sirians or the Andromedans?

Because the Pleiadians have so much to do with you; your civilization (your intentions, your purposes) and their civilization and physiologies are so similar that among your immediate neighbors, the Pleiadians are certainly the most like you, being able to pass for an Earth person – including the idea of being the true believer, following something blindly. I'm very fond of the Pleiadians, don't get me wrong, but I must admit they have that quality just as you do on Earth. They have certain qualities you have here, not necessarily preordained, but challenges designed to help both of you learn.

If you were in a room talking over things with Pleiadians you would find it to be a lively discussion, because they would have as many attitudes as you do. You would have a lot of common ground, so it wouldn't be a battle royal, but you would also have enough differences to stimulate each other and bring out the ultimate in variety and in creation – which, of course, is the ultimate intention of all life. In the short range it is intended that the Explorer Race stimulate all extraterrestrial races they come in contact with. The idea of parking a couple of Earth people in a room with a couple of Pleiadians is a great idea, I think. It certainly will come to pass.

How does that work? They're exactly like us, but they haven't experienced this deep descent into negativity. Is that ahead of them, then?

Not exactly. They will not have to go as deeply as you did, because you've already done it, but they will have the opportunity to experience 2% negativity. They're not totally like you physiologically; their intestine is at least six to eight feet shorter because they do not have to digest the kind of stuff you do – chimichangas, pizza and barbecued chicken. They don't eat meat. By your standards their diet is bland, so their digestive system need not be so complex. Also, they do not have to deal much with oxidization. Most of their atmospheres don't have oxygen – which would not work for you except in controlled atmospheric situations, of course.

What do they breathe?

Most of them breathe nitrogen or, as a possible meeting ground between the two of you, nitrous oxide. (I don't have to tell you what bizarre effects it would have on human beings, as many dentists have discovered, much to their initial shock and horror.) There are certain atmospheres (helium, for one) that are a lot more gentle on the physical self. Oxygen is designed to oxidize – to rust you, as it were – so you will not have to stay here on Earth too long.

When there was enough room for more people and you could live 700 years, did they pull back a little on the way the body was formed?

Well, they didn't encourage any more births than would sustain a fixed population. This was fairly easy to do because they have to make alterations in their physical selves to be fertile. When they make love both the man and woman are normally infertile. They can make minor alterations that are akin to yogic techniques. By breathing in a certain way and holding one's abdominal muscles in a certain way the male becomes potent. The female can become fertile by utilizing muscles in the buttocks, the backs of the thighs as well as the solar plexus. To create a pregnancy they have to hold these muscles for about five minutes.

Once impregnation takes place, the female no longer has to hold that fairly uncomfortable state of muscular tension that stimulates all of the proper organs. She can then relax and carry the baby to term, which is in their time maybe seven months of experiential time. Birth, I might add, is not painful for them because, for one thing, the baby is smaller (none of these humongous ten-pound children). A birth would take perhaps twenty minutes from start to finish.

As you begin to interact with them more on an open and general basis, they will be more than willing to teach Earth female beings how to do this. They will be allowed to do this only when Earth has decided to fix its population to a more manageable number.

Choosing the Astronauts

How are these beings chosen who go to see the Creator's creator?

By that time the population will be screened from about age three up to seven. Children will be observed for their general personality characteristics. The initial choice will be made from among people who have an adventurous spirit, who are open-minded, who have a reasonably upbeat attitude, who work easily in a group, who enjoy outings and so on. If you match a certain profile, a panel of perhaps eight to ten people (including at least two extraterrestrials) make the choice, and it would have to be unanimous.

What about our Creator? Are they not going to meet our Creator?

This will all be done with your Creator's total approval. After all, your Creator has the great thrill of knowing you're going off to meet Its creator. It's like going to meet Grandpa and Grandma.

They won't actually meet our Creator, then?

I think they won't do that, no. Because when you rejoin with Creator – allowing Creator, as I like to say, to take a coffee break and go on up – that moment will be so wonderful that you would not want to tamper with it in any way.

Will this go on? This is only a few thousand out of trillions of people, so what is . . .

I should think it won't be trillions by then. Trillions only if you're talking about everyone who's *ever* been a member of the Explorer Race.

So will they be merging into bigger pieces by then? How many beings will there . . .

If you're asking if it will be just these few thousands, no. There will be other voyages.

This is so awesome, this fifty years.

It is important for you to understand how crucial this time is that you're going through now. This is really the finishing-up time, the end of finishing school. Now you're at the point of preparing how to be a lady or a gentleman and how to get along with those you don't know or understand. The individual you don't know and don't understand is really *the ultimate core personality of all of you!*

Core Personality Traits

All of you are right now [August 13, 1996] coming back into contact with the immortal traits of your core personality, which sometimes fly in the face of the traits you've picked up in this life. That is an effect of finishing school. You are confronting traits in yourself that don't actually fit, but that you've adopted in order to learn what you needed to learn in this life, or at least finish up what you need to complete in this life. Some of you are going through a lot of confusion, bordering on feeling almost schizophrenic at times when you seem to have more than one facet to your personality competing for full-time expression. I don't want you to worry about that because this is a temporary situation. Many if not all of these qualities are core personality qualities, such as my beloved scribe's.

For example, a few of my scribe's core personality traits might be curiosity and a desire to understand how A came before B and how C follows B—where it came from, how it got here and where it's going. Another core personality trait would be humor, the ability to laugh in the face of the hurricane, as it were. Now, you all have core personality traits, and I've used my scribe's as an example. You also have personality traits that you've picked up along the way in this life, sometimes even carrying over from other lives traits you weren't done with and through which you are utilizing certain methods to achieve a complete understanding of something, whether positive or negative.

But now, as you're coming back into your core personality, certain things that used to be very important will no longer seem so important. These are often traits you don't need anymore. Some things that didn't seem very important at one time or seem utterly new are often core personality traits. I want you to take note of these things. That is what I refer to as the end of finishing school.

This voyage will take place around 2051 or 2053, and that is not very long from now. During that time there will have been for forty years widespread discussion of, contracts with and cosmopolitan interchange with, extraterrestrials plus all of what has been kept secret from you for many years from that rather goofy policy (if I may be judgmental for a moment) of "they can't be; therefore they don't exist. We can't trust the people with this knowledge." (This is one of the larger mistakes ever made in political philosophy.) All will have been revealed and enough time will have gone by so that the people who kept these secrets will have been forgiven.

Now, you'd be very surprised that those who still keep these secrets have an abject horror that this knowledge will get out, mainly because they knew about it but didn't tell you. Please be prepared to forgive them, because they honestly felt overwhelmed. It won't serve you to be angry at them, nor will it serve your ultimate purpose. They simply didn't know what to do.

I mention these things because you're coming down the home stretch. With that in mind, you need a lot of your core personality traits back so that you can integrate your strength and begin to live it. (Your core personality traits are your strength.) You also need to thank and release the manners of expression you have adopted to learn certain things, because these are not your strengths.

As you come back to your core strength, many interesting things will happen, not the least of which is that you will begin to feel more confident—and I cannot tell you the importance of feeling real confidence based on something solid that you feel, have physical evidence for and know is absolutely true and that you cannot be shaken out of by mere words. Your core personality traits feel right and good to you; you can't be talked out of them. Your body gives you physical evidence that they are real, and they come easily and feel comfortable—they're *you*.

Confidence Replaces the Immune System

Confidence based on that is ultimately what will replace the immune system. I told you a long time ago that the immune system was a temporary thing set up so that you could have a buffer between your temporary confused state, which is a veiled place, and the full knowledge of who you are. The veils give you the gift of ignorance, thus the opportunity to create and re-create here in this reality.

The immune system is not the ultimate means by which you avoid discomfort, to say nothing of deadly diseases. What allows you to avoid these is absolute confidence in who and what you are—confidence based on the absolute knowing through feeling, fully possessing your core personality traits, feeling good about them and *being* those core personality traits. With this kind of confidence you do not attract negative circumstances because you feel drawn to this magnetic experience of what is compatible with and associated with your true self.

Being involved in things not associated with your core self often precedes disease, because you're either attempting to do something that is not you or has nothing to do with you—and is thus connected to confusion, so you must therefore constantly reenergize yourself. As I say, confidence based on these core traits is what keeps you free of disease and gives you a healthy, basically trouble-free life wherein you experience the best of your core personality expressed through your physical self. You no longer need substitutions for your fulfillment, such as overeating or drinking too much alcohol. These are substitutions for what you really need, which are things that will give you the true strength and experience of your self. But once your life and your core personality are compatible, you don't need substitutions. The weight falls away; the craving for cigarettes, drugs, alcohol, food or whatever falls away because your life experience is compatible with your ultimate, immortal personality.

So when do we lose the immune system completely?

Right now the immune system of most beings (not all) is about four-fifths to one-third of what it once was. I'm not talking about beings who are sick, but people in general. You might say that pollution is the cause, because it is reducing the oxygen content of your air, and your bodies were intended to live on at least 25% to 35% oxygen; 40% oxygen would be optimal for some of you. But because oxygen is significantly more limited now, especially in big cities, you experience a deterioration. When the latent characteristics of the cells are not being fed what they need, they degenerate.

The immune system is going to fall away and, in an overlapping scenario, your core personality traits will be revealed to you. Those who can accept, include, embrace and practice those core personality traits will ultimately be able to live and *thrive* as your core true selves. Then no matter what you're exposed to—even a significant (not lethal) dose of some toxic substance—it will not affect you. I'm not telling you to test this by drinking cyanide; I am saying that significant doses of some pollutant, or even a common disease-causing organism such as salmonella, will not affect you.

This is another incredibly vital thing, core strengths. Can you say more about what people love that makes them feel powerful? Some of them we can't feel right now because we have a facade in this lifetime, right?

Do you mean as a substitute for the true power of your core personality's confidence? Control, manipulation, dominance—people can be addicted to these traits too, which give them a temporary illusion of superiority or can even sort out a complicated world and superimpose a simplistic schematic over the true complexities of life.

Over what can be a very subtle and complex and uniquely intuitive field?

Yes.

So day by day we need to really look at our life and feel our way into understanding our true core, right?

Take a Poll of Your Core Traits

Yes, and notice what traits you have that actually feel very good. Sometimes it's difficult to notice your own, so help each other. As homework you might hand out pieces of paper (a variation of the homework assignment I used to give out) that have at the top: "What have you observed are my (fill in your name) core personality traits?" Hand this to at least three people, maybe five. Don't tell them you're giving it to the others; try to have them as dispersed as possible so it's less likely they'll compare notes.

When you get the lists back, see if you can see the similarities. The points of agreement, or the things that resonate well within you, are 60% to 90% likely to be core personality traits. At least it'll give you a clue. Do that every year or two. After a while you'll have a pretty fair idea of what your core personality traits are if you have not already discovered them on your own.

This does not mean you should discard other traits, because the traits of your core personality will not come in all at once. The idea is not to dump the personality you have now, which is based somewhat on adaptation to your circumstances of life, but to support, sustain and strengthen yourself. So bring them in slowly, like putting training wheels on a child's bicycle to help him learn, then taking them off after the child has learned to balance.

Well, they're always there; we've just never looked for them or become aware of them.

No, in some cases the traits were so latent that they were not obvious.

Especially if you're working on a particular lesson and you developed this part of you to survive something.

Yes. For example, you might be working on a lesson that will give you compassion about something with which you've had no previous experience but that you'll need in a future lifetime. You might be working on something far from your actual personality so that you can understand, sympathize and appreciate somebody else's predicament in the future. In this case you might utilize traits or behaviors that would not normally be present.

So people reading this book, no matter what their age, once we get past 2010 will likely keep the same bodies?

Oh, I wouldn't say that. It is certainly true that by the year 2009 there will be significant ability in medical science to perpetuate physical life to at least 150 years (granted, through a lot of genetics). By that time they will be able, whether it's desirable or not, to pretreat the mother and father so that the child could easily live to be 150. Even people who have health conditions or people who are reasonably healthy and are already 70 or 80 years old might be able to extend their life to 130. But this is going to be initially available to the few; the bulk of the population will live out their life to the end of its natural cycle.

Diet

Certainly by 2030, when there is ongoing extraterrestrial contact and the revelations of secrets will have been integrated, a significant change will have taken place in your culture. You will probably be eating very little meat, if any. People will know that the mood of the being they are eating – the feelings that they had when they died – has everything to do with what processes through their bodies.

If animal husbandry and agriculture have gotten to the point by 2015 (no later than 2030) where animals raised for consumption are being treated wonderfully well, their death is painless and respectful, and farmers and ranchers are known as sacred farmers or sacred ranchers, then it will be possible to eat meat that will give you a high, and you will be improved by the experience. Once that takes place you will be able to live longer and enjoy it more, but certainly not indefinitely. You do not want to live indefinitely in the same body because at some point in time you will invariably get tired of everything that body can do and you will want to experience life differently. Maybe you want to be taller or shorter; maybe you've done everything a man can do and you want to experience it as a woman. By that time the third sex will begin to appear, so maybe you'd want to try that.

In any event, you will, certainly after 150 or 175 years, want to change bodies, so family planning will take on an entirely new perspective. Philosophy will have become highly involved in the healing arts by that time, to say nothing of genetics. You can literally plan your child physiologically to avoid disease and to be as healthy as possible.

By that time complete physicians will have a philosophy that will cause them to be ten times more ethical than the most ethical physician or healer of today, largely because of interaction with deeper levels of philosophies developed here on Earth by various people and interaction with their extraterrestrial roots as well.

People reading this book now will probably not be the ones who go to see the grandfather; and the children born right now will be the ones who really start seeking their core strengths, who they really are.

Yes, and that is why I'm getting this material out now. We want the knowledge, the philosophy, the new myth for mankind, if you like, to be available on a wider basis so that when these youngsters are growing up they'll have some glimmer of their future. Of course, I do not wish to sound immodest; obviously many others are putting out such knowledge as well. We need to put it out now so that the next generation can have it.

So if we as humans want to have this incredible experience, we need to either lengthen this lifetime or be very specific about being reborn in this time?

I think you need to be specific; it would be easier to be specific about wishing to have a life during this time than lengthen this life.

That is one of the most awesome things I've ever heard you say – the chance to go not just to this Creator but to the ultimate creator.

I'm glad to hear that.

And you're saying there is a time and a physical location of this experience?

There is an experiential location that is a significant distance from here at a higher dimension. Without the invitation the proper technology won't get you there, but with the invitation *and* the proper technology, it is certainly a reasonable trip that one could make, you know. But you sure wouldn't be the same when you came back!

That's my question: How will this person's life be when they come back? They'll be so totally changed.

Yes.

They will not be – I don't want to say limited humans, because at that time they're still going to be encapsulated humans . . .

Yes, they'll be encapsulated, but they will have had the experience of going beyond their encapsulation, of being totally joined during the experience of the ultimate creator's creator. And they will be able to literally live on love. They will have the choice, from that experience, of remaining in their bodies for up to a thousand years. Most likely they will not live past 700 because they will want to experience something else, but when they step out of their bodies, they'll simply lay their bodies down, stroke their bodies good-bye, step out into light and that's it – they'll be gone. They will not be living anything like what they did before and they will, of course, all become teachers of great wisdom.

It sounds like the ultimate motivation.

Let's just say that it won't take too much recruiting to get volunteers, and it will not be necessary to offer pay.

I wanted to give this scenario so I could fulfill what I've talked about for this book. On that note I will simply say good voyage, good night.

Loving Embrace

Approaching the One

May 28, 1996

Counting down the moments toward the end of this volume, I'd like to say that I have enjoyed this series. To our readers, there is more to come.

Understand that the true nature of the human being – all you are, all you might be and all you might become – is not fully encoded into these books, but you will find a code (a trace, you might call it) when you read The Explorer Race, ETs and the Explorer Race, this volume and the various articles sprinkled in the Sedona Journal. This trace is designed to be a launching pad that takes you from your current level of understanding to that alternate universe where wisdom is the way of life by way of your desire to fathom all that you see, hear and know and by your intent to move beyond your present position. With that thought in mind, I bring this volume to a close. I have complete faith in your ability, knowledge and wisdom to grasp the remaining few words that close this book.

You need others in order to experience fully your light and sound, your feeling capacity and also your endurance. Truly you are not in this alone. These days especially you need others to define who you are by bouncing off of each other and to fully grasp the meaning behind what is One. You have been told for a long time that you are all from the One and that as you return to higher dimensions you will approach that once again. Tonight I will talk about that approach, some of its experiences, feelings and perhaps even consequences.

As you begin to notice how similar you are to each other—your common ground and your common bonds—you will also notice that the things that bother and annoy you, the things that make you feel uncomfortably different (not your uniqueness, but what you are afraid to tell others) are felt by a great many others, and you will begin to relax about yourself and feel a greater sense of confidence. The final hurdle to embracing your oneness is simply embracing one another.

You do not have to go out and put flowers in the hands of people (but you can if you choose). What you can do, however, is try to make an effort to notice how you are alike and point that out to people, especially when they are grumbling or complaining about some behavior in another person. You might get to be known as "a cockeyed optimist" [courtesy of the song "A Cockeyed Optimist," sung by the Nellie Forbush character in *South Pacific* by Rodgers and Hammerstein].

I want you to be willing to accept such "criticism," understanding that you could be criticized for worse things. It is of vital importance that you have both the willingness and the ability to see your common ground. It is all right to advocate for yourself, your family and those you care about, but it is also important not to make judgments that keep people far away from each other.

A Time of Union, of Re-forming Soul Groups

This is a time of union, and you can come together in either disagreement or concordance. Wouldn't you rather have concordance? You see, you really have no choice. If you have the foreknowledge that you are going to come together and be one again, wouldn't it be well to prepare yourselves—see how you are all alike, notice what is similar and begin to pay attention to the traits you all have in common?

In the coming days and weeks you will begin to notice more and more how similar you are. For example, right now as the vibration increases, the motion toward higher dimension is felt strongly, and as you and your whole world are moving inexorably closer to your higher-dimensional expression of yourself, you will begin to literally experience aspects of yourself you didn't know you had. You will begin to see lights, hear sounds, receive messages. You will begin to feel the telepathy of the feelings and mind but also of the heart. You will have precognition and know the answers to things you don't know how you could know. The answer might be something profound or something mundane or seemingly meaningless—yet there it is. You can almost hear it in your mind like a voice. The knowledge is right there for you.

This is an aspect of oneness. While you might not have known the answer or read or been told anything or been exposed to that particular piece of knowledge, some human being living in your time knows it or a portion of it (if you receive a partial answer). *Someone* knows it, and because you are becoming more of the one being, that knowledge will be yours as well.

Even now as I speak [May 28, 1996, 7:25 pm MST], something has occurred in the last three weeks that is beginning to heighten: precognition based upon soul-group awareness. You have soul groups, and they are beginning to re-form – to congeal, in a sense. I've said many times that you have sprung out of a few souls, that a few souls departed from Creator on this original voyage to gain knowledge and to literally re-create with Creator. And that all these souls broke off into other fragments and still more and more fragments, expanding until they peopled the universe of the Creator.

It has to happen, doesn't it? You originally left the Creator, so how are you going to go back? Do just a few of you go back, leaving the rest in the universe? No. You need to reconvene your board, as it were, and come back together. You need to come to communion with the original soul groups, and eventually all soul groups will combine. Only then will you be ready to move toward creatorship status.

So precognition is now something that is associated with your entire soul group. There are not that many soul groups. I'm not going to put a number on it, although others have done so. I will simply say that there are more than twelve and fewer than sixteen. (How's that for playing a game?)

Changes in Physical Bodies and Perceptions

There are other aspects of the change you are feeling by now, and one of them is the higher-frequency change in your body organs. You will find that your metabolism, your endocrine system, your blood-regulating system, your breathing apparatus – all of these functions of the physical self – are also being exposed to energies from the future. Your energy bodies are literally reaching into the future and bringing back energy from the higher dimension to you.

One of these changes, for example, will be that your digestive system might not work the way it once did. If you wish to use herbs to help your digestion, by all means do so, especially if they grow wild. The growth of organic herbs in conditions of benefit to humankind is valuable and a worthy thing to do. Yet it is always better to gather the herbs on your own or with a guide who knows these things. You might gather an herb yourself in a sacred way that blesses and thanks the plant, where you felt (your feelings, not your thoughts) the plant welcomed you and offered it to you.

You might approach the plant with sincerity and ask it for a leaf or a short stalk. If the plant welcomes you, you will feel it (like *you* might feel going into a room of people you don't know); that's how the plant embraces you. If you feel uncomfortable, the plant would be saying,

"No, I need my whole body," and you would go on to another until you found one where you would continue to feel good as you reached out to tear off a leaf or stalk. Then after you have gathered it, you thank the plant and give it your blessing; then you go on.

One milligram of an herb gathered this way would be 500 times more helpful than taking, say, 500 mg. of that same plant grown in the best possible circumstances by people who care. You see, when a plant grows wild and free in nature, it grows only where it is welcomed by other plants and animals and by the soil, so it has the strength and freedom associated with true love. Just a little bit of that goes a long way.

Using this type of herb to help your digestion (for instance) is worthy. If you cannot gather your own, then take herbs gathered by others in the best possible circumstances or perhaps some that were raised. But do understand that your body might need help now. In the coming days you might find that you will go through a bridging period (like a rainbow bridge) in which your bodies do not function quite right. Don't worry, you are immortal. I am not suggesting that your bodies are immortal and will live forever in their present form, but that your personality—what you know yourself to be—is absolutely immortal.

So have patience with yourself; appreciate and value yourself and others. Know that you will experience a new positioning of the soul groups through your light and sound connections. Some of you are hearing unusual sounds, some of which are associated with that connection. Some of you will even see the face or feel or have a sensation of someone on the other side of the Earth. (If you see someone, that person is usually thousands of miles away—maybe a child, an old person or a young person.) These people will often be in your soul group. What they know and what they think will gradually become known to you when you need it, sometimes as an unexpected but pleasant vision. Know that this is associated with the rejoining, with the true communion.

In *The Explorer Race* I spoke of many things that affect many different aspects of your life, and I spoke largely to your minds to prepare you. In this book I speak to your minds, but also to your hearts and to your ability to visualize and imagine what things might look like. In this book I have also spoken a great deal about the origins of the Creator, something about my own origins, and a little about your origins. In the next book we will go on to other things. We will speak to other parts of you (maybe including parts we have already spoken to) to help you feel the true thread that connects all life.

There will also be special treats in the next book. There will be illustrations of ways to place your hands, and motions. There will be photographs or drawings, things you can actually do step by step with your hands or your arms in order to embrace the light and the age of change for the benefit of all. I will also give you as much homework as I can so you can prepare yourself—and in time, should it please you, help prepare others.

For now I trust that this is enough to give you insight into the origins of Creator, a little bit into my own origins and a little bit into some of your origins. I might talk further about origins of other groups of human beings as well. If I do, I will do it in a way that will integrate with your exercises to move, to make sounds and to prepare for the change.

For now, good night. This is the end of this volume, but the beginning of our true work together.

Trading Planets, Restless Souls

APPENDIX

The Body of Man*

July 25, 1993

The body of man and the initiation of man: we will start out talking about the origin of the masculine body, then about why the human being in the masculine form looks the way he does – what he has, what is missing – and what he needs and where he is going to get it.

As many of you know, I have said for some time that the origin and focus of man is Orion. This does not mean that beings who look human and who have your particular design did not exist in some other place and time before you. But on your souls' journey and in your current quest to discover, to evolve – and to some extent to devolve (meaning to return to slightly before your point of origin) – some time ago you routed yourself through Orion. When you went through Orion the civilization there was somewhat polarized, according to your understanding of polarization. When you went through, it was an energy screen, for your evolutionary cycle began a long time ago. I will not go through all of the history; you can find it in *The Explorer Race.*

Feminine Avoidance of Polarity

When all of the souls arrived in Orion who would be reincarnating at some point during your cycle of incarnation as human beings on Earth, it was just about then that what would be the source of the feminine needed to refocus and move. It was originally intended that Orion

* Originally appeared in the December 1993 issue of the *Sedona Journal of Emergence.*

would also be the source for the feminine. At the time of the passing of your mass soul through Orion, there was some polarity there, but since the feminine needed to have a nonpolarized focus, part of your souls' source split off and went to Sirius. (As you know, Sirius is the point of origin of the feminine human being.)

That was the beginning of the polarization of the sexes. In previous times you had always tended to have an androgynous representation of yourself, a representation of your soul-self in bodies that combined the sexes. I can't say the split-off was an accident, because there are no accidents; it happened because of the timing of your souls' arrival in Orion. When your souls arrived they were detected by the civilization there through a device that could detect what would be perceived as an invasion of foreign matter. (This is not unlike some of the defense systems you have now on Earth that can tell if something is approaching that isn't the norm, such as an unidentified flying object. In a sense, your souls were the UFOs of that time.) They could also detect the split-off to Sirius.

The energy arrived, was detected, and did not go away. Then there was a visitation by a representative of the divine to Orion. This divine emissary went to the ruling bodies of all the Orion planets that you had saturated and announced your presence. She said that it had been decided that Orion would be the point of origin for a new species that would not remain in Orion but would be partially developed there. It would eventually be totally developed on a planet that would come to be present in a nearby solar system.

Now, what she meant by "nearby" is an interesting point. At that time Earth did not exist in your solar system. What you might call the thought form, the spirit of Earth, was there because it had been seeded by the Creator. There was matter present, but it wasn't Earth. At that time the third planet from the Sun was something else entirely.

During the formative stages of what would come to be known as the human male being, a negotiation was going on between another representative of the divine on Sirius. It was at this point that the feminine point of origin – the feminine source souls who would come to be known as the human being – began to negotiate to have a more feminine planet. They were concerned that the masculine source souls were going to be warlike, because at that time in Orion's history the Warrior League prevailed. The Warrior League did not make war so much as constantly prepare for it, and their ethic of getting along and of ruling was rugged. It wasn't their job to rule as much as to keep the peace, from their point of view.

So the feminine source souls on Sirius proposed an idea. You might say they got together with their divine representative on Sirius and said, "What can we do, now that we have become polarized and it appears there are going to be two different types of bodies rather than the androgynous body we are used to? It is quite obvious to us," the feminine source souls said, "that there will be a feminine being and a masculine being. What can we do to make certain that there will be a means of bringing back into our souls' magnetic field the combination and ordination of feminine/masculine traits in *both* beings?"

An Exchange of Planets during Renovation

It was decided that a planet from the Sirius system that was not over-inhabited would be loaned to what is your present solar system. This is an interesting point, because the planet that was here then looked nothing like the one on Sirius. The planet that was then the third planet from the Sun in this solar system was a warm, yellow planet. It had a fairly thin atmosphere, considerably less gravity (and of course less mass) and nitrogen in its atmosphere.

Had that been your planet, your bodies would have lasted a very long time. But it was the Creator's desire that your bodies not last very long so that you would have a sense of urgency to find out who you are. As you know, when you are discovering things about yourself, there is always that sense of urgency. You are true to the moment: When you discover something new, you are happy about it, but it doesn't last because there is that urgency to go on and find out more.

It was decided through negotiations that a planet from the Sirius system would replace that third planet from the Sun in your present solar system. It was almost like receiving collateral on a debt: they would take the third planet from the Sun to Sirius (which is done by traveling through light – they didn't pick it up and carry it) and convert that desert planet into a planet like their own, for they have many water planets.

Water did not exist in your solar system to any great extent before then, and you might have noticed in your travels thus far that water is hard to find on planets. It makes you wonder where this planet's water came from. Your fossil history tells you that water has been on this planet for a very long time, certainly long before the arrival of the human being. That brings up another interesting fact: dinosaurs and all of the so-called prehistoric animals are not native to your galaxy [system] at all, but are from Sirius. Earth existed someplace else and was transferred with the few forms of life that were on it because there was no reason to destroy them. The vegetarian animals, by the way, were very enlightened – not quite fully realized, as you say, but very close.

The male source souls were now in Orion. The plan was discussed by the divine emissary and various ruling bodies present there, especially the Warrior League, which was concerned about what was an invasion, from their point of view. The male form would be developed, but it would occur primarily genetically. The human male being would not be created in prototype form, as one might expect in the creation of anything new. Earth would have the very first one, so this male being would not occur in Orion then or anytime in the future. However, certain beings from Orion were encouraged to join the council on the human being, and their scientific, genetic and philosophical input was sought. The Orion Warrior League had a philosophy similar to that of a warrior league that later sprang up on Earth – the samurai – which was not associated with the Orions but was built into your genetic structure as the Warrior League's contribution.

Masculine Source Souls Arrive on Earth

The beings of Orion made many contributions. The divine emissary stayed there off and on for the next hundred years or so and was welcomed and appreciated, especially by the Warrior League. After that the human-being male source souls were launched, as it were, toward what they expected to find – a warm, yellow planet. Surprise! By the time they arrived here, the source male beings merged in soul form with what was then on Earth in a very primordial form. Thus it was quite some time before the human male body was developed in its substance and form as you know it now.

Meanwhile, there were a lot of major negotiations going on in Sirius. After the hundred years' negotiation, but millions of years before the human male body was developed, the human female body was being developed in Sirius. I don't want to put down the nice Bible story about God reaching into the male body, pulling out a rib and making the female, but in all honesty, that story was concocted for political reasons. The female human body existed a long time before the male body. During that time, millions of years, the male source souls floated about in the general atmosphere over Earth, getting used to it. They had prepared themselves for a desert planet, so it took them a while to adjust once they discovered that this was a water planet. That's why there was a delay in the development of the male body.

During this period when they were separated from their feminine selves (which were in Sirius), they had the opportunity to accustom themselves to the gentle, feminine nature of water. (Water is the primary element of the female, whereas fire is the primary element of the male.) They felt comforted because they were close to a symbolic source of the feminine.

During those millions of years the female body was developed in many variations. When it was tried out, the variations that were too strong were set aside. It was quite well known by the Creator at that time and by the Creator's emissaries that it was the female's job to initiate the male and to reinitiate him to become a more receptive source for the feminine. It would be the responsibility of the feminine being to initiate all human beings onto the planet. [See chapter 6 in *The Explorer Race.*]

Now, why would the Creator want to do that? After all, before the mass of soul beings arrived in Orion and split, some going to Sirius and some staying in Orion, you had a merged being that was both masculine and feminine. What was it that the Creator had in mind?

The Origin of the Male's Restlessness

The Creator knew that there was an inborn restlessness in the masculine side of that being. This restlessness came about because of something that had occurred a long time before.

At the beginning of the entire loop of time you are in, before the development of the human being, even before the development of your souls, a council was held. All those council members were male beings. The divine emissary was a female being – from my point of view she was God's daughter – who visited that council. [See chapter 21 in *The Explorer Race.*] The council was, of course, very respectful and appreciative, but they made a decision to allow polarity to develop on Earth even if it would go to extremes, because they needed to find a place for negative energy to develop its maximum potential for good. They were blind to the idea that it would also develop its maximum potential for bad. The divine emissary pleaded that there be more feminine energy to support the transformation of what is now called negative energy into its divine state. This was denied by the council, whose members thought, No, we will put it in there and allow it to sink or swim on its own! But they didn't say that. Then the divine emissary (otherwise known as God's daughter) left.

About a hundred years later (which, compared to billions of years, is a pretty short time) the council realized they had made a mistake. Since they were all male-oriented, it made them nervous. That is the origin of the restlessness of the male being. The Creator needed to change that before you completed this loop of time that had been created so you could experience what a polarized world could become and discover what to do with negative energy.

Eventually you will return to the time *before* the council makes that decision, and they will make a different decision. It was necessary for the Creator to put that right before you arrived back then, so the council would have the opportunity to decide differently. That council represented source souls. There was a secondary council present whose members were feminine beings, but their votes were not counted on. They would be the means to create peace within the masculine being so that when you arrived back, the council would vote differently.

To summarize, the feminine being was developed first. The planet you now know as the third planet from the Sun – Earth – was transferred from Sirius, and its energy was feminine. It was understood clearly that negative energy, the energy of discomfort, would go to an extreme. That is not normal, by the way. It was originally intended that negative energy would be no more than 2% of your experience on Earth, but because the council voted the way it did, it has been allowed to become extreme, even bizarre in its extremity. Of course, all of your personalities are immortal, just as your souls are immortal. Nevertheless, it does not excuse all of the suffering you have had to experience in body and soul as you have cycled through Earth.

The Arrival on Earth of the Sirian Feminine Beings

The Creator needed to put a little spin on the ball, as it were, a little twist, so that there would be at all times a feminine influence on all beings no matter what they did. Because the feminine human being had already been up and active a long time before the masculine being was put together, it was decided through negotiations that feminine beings from Sirius would go to Earth in a large vehicle. This vehicle had a capacity of about 100,000. The individual at the top would be primarily involved in the creation of the masculine being on Earth and would be a masculine being himself. He was a social geneticist and he came to Earth in this vehicle (you might say that he came from one point and met himself coming from another point, sourcing from two places) to create the genetic engineering for the male being. [See chapter 1 in *The Explorer Race.*]

The female being who was created on Sirius was brought to Earth with about 100,000 people on that ship. Many of them had volunteered to come and be citizens of Earth. These people had a very advanced civilization, and this is why in a distant past you had civilizations that were far more advanced spiritually, technologically, intellectually and so on than the civilization you now have. That does not make you foolish; it simply means that so much suffering and travail has resulted from the extremely polarized negative energy that the highly advanced civilizations had to either go underground or be destroyed.

The feminine being was released on Earth first. The masculine being was created in an entirely different way, from material passing through Orion, thus being influenced by that civilization. Because the masculine arrived on Earth (which was from Sirius) after the planet had been in your solar system for quite a while, the energy of your solar system was present when the masculine being was formed. This energy the feminine being did not have, having been sourced, created and evolved entirely on Sirius. She was not influenced by any energy whatsoever of the solar system you are now in – a crucial point. Thus one might say that the masculine being is more at home in your solar system than the feminine being.

Decisions about Female Gestation, Procreation and Sexual Bonding

As I explained, there was a restlessness already present because of the male council. In addition, because the masculine being that was eventually created had been removed from its feminine self for so long that it felt alienated from the feminine self and became even more restless. It was decided that the only way to bring the masculine being into any form of balance so that the restlessness would not overwhelm him was to birth all people through the female.

The birth cycle takes quite a while so that the soul can accommodate itself. It is not just a biological or a physiological thing. It doesn't have to take that long for a feminine being, but a masculine being needs to spend as much time as possible inside the water in the feminine chamber to help remove the overwhelming restlessness. The nine-month gestation period helps. That is why you do not have cloning here. Even though there have been highly advanced societies living here, cloning was done in only a few. The males who were cloned were very restless.

The male body was designed to be more compact – another contribution of the Warrior League. The original design was to make the male beings small enough to wear body armor, as a warrior might, without being overburdened. Through the evolutionary cycle, the male being turned out taller than the female being, but not necessarily bigger (until later). The first male was not very muscular because the beings then on Orion did not need a lot of muscle; they wore body armor.

The female beings who had been placed on Earth earlier were in total harmony and sympathetic vibration with Earth because it was a Sirian planet and they were from Sirius. They were also in total sympathetic harmony with all animal life forms that existed then.

The masculine being would be born restless, so he had to be created in a way that would allow him to experience the female being. A polarized masculine being could potentially develop a cloning device and not need women at all – for there are planets with only male humanoids, just as there are planets with only females. So it was decided that for man to need woman, he would have to join with her to create offspring. (Remember, it is woman's job to initiate man and help him get past his restlessness so he can arrive back in the soul source form.) It was necessary that the masculine being have a means of reentering woman that he was born through (although, of course, it would not be the same female). This explains the masculine and the feminine sexual organs. The feminine being had to take the masculine being inside of her just as she had had him inside her before birth in order to reorient him. Thus the easiest way to create offspring (to say nothing of sexual pleasure) for the masculine being would be to bond with the feminine being.

You now understand my view of how you got to be on Earth, why masculine beings look the way they do, why they are somewhat restless and why feminine beings had to be disguised to look like Earth beings.

The feminine beings who were discharged from the ships had a highly advanced Sirian culture and were as close to being fully realized and totally feminine as possible so that the planet would have a bias toward the feminine. The masculine beings did not have the same opportunity of millions of years of evolution. They suddenly found themselves in bodies that took a little getting used to and in a situation in which they were forced to interact with beings who could be called extraterrestrials.

The War between the Sexes

The feminine being is the repository of spirit and emotions, which allows her to attain spiritual mastery more easily. The masculine being is the repository of the physical and the intellectual, which allows him to achieve material mastery. The feminine being had the edge, because material mastery is much easier for one who has spiritual mastery.

It was unintentional, but the masculine being felt somewhat resentful of the feminine being because she had the propensity and ability to do certain spiritual and emotional things more easily than he did. It was at this point that the so-called war between the sexes began to evolve. Of course, masculine beings in your time or any other can become spiritual masters and become highly sensitive in their emotional and spiritual bodies. Men as a species are not somewhat lower beings, but it is important to understand how your different points of origin and means of creation have contributed somewhat to your present experience so that you will not feel in this life that somehow you have made a mistake. Although the masculine being felt somewhat resentful, he needed the feminine for all she could provide.

Do you have any questions so far?

What did you mean when you said it was woman's job to initiate men?

Two meanings apply: to initiate is to begin, to start something. The female starts the male. The Creator, of course, provides the spark of life (the soul) and your planet provides the matter, the substance of the body. Even though the male provides a certain element, that sperm could do little if it weren't for the female. The female is not just the incubator; she *starts* the male—and the female too, for that matter.

Then there is a larger sense of initiation. It is necessary to re-create the male being *without* restlessness. When you make your shift from what you call the third dimension to the fourth dimension, there will still be a residual propensity toward restlessness in the male being, although it will be greatly reduced. This propensity is hormonal. If the female being were to take the same amount of testosterone that the male being has, she too would be restless. So it is necessary for the female to add what she can to feminize the male, in a sense, although the male must remain a masculine being in his own right.

The polarized male being will not actually release his restlessness until you evolve to the fifth dimension. He is going to lose a lot of it by the time he gets to the fourth dimension, so there won't be the same action of the testosterone. There will still be the same means of reproduction available, but it might not be the only means.

To what dimension are bodies still polarized?

Bodies really do not have to be polarized even now; cloning is right around the corner. But there is an advantage in terms of your evolution. By the time you get to the ninth dimension you will probably no longer be choosing polarized selves, or even by the sixth dimension. What I am suggesting here is that you don't *have* to have it this way or that way. So let's say anytime from the sixth dimension on.

I thought that souls reincarnate sometimes in a male body and sometimes in a female body.

Yes, that is true. I realize that my story is making it sound as if some souls are always male beings and some souls are always female beings. Thank you for clarifying that point, because it is not my intent to leave that impression. I am talking about the arrival of the original source soul beings. In initial creation, the source soul male beings were male and source soul female beings were female. Immediately after that, the energy could then become either sex in the birth cycle.

Was this the ideal female being for this male being, or could it have been another feminine source soul from another place?

No, she wasn't ideal. If she had gone to some other place at some other time in some other dimension, she would have been quite different — a better female being, one more compatible with the male being, one who would have allowed him to become initiated in a short time. However, this was not available at that time. She had to go to what was available within the same dimension and within a certain proximity — and the distances between Orion and Sirius and your galaxy are not that great. Nor was she allowed to change the time sequence. Time as you experience it was evolving. She was not allowed to step outside of time and go into the future to create. This she would have much preferred to do. Had she been able to go to the future of Sirius to create herself and then devolve back from the future to the time when she would set foot on Earth, her body would have been infinitely better, because she would have known all possible consequences of what would develop in the polarized society on Earth.

In a sense, the female being started off without any genetic impact. Her effect was on the soul level and the unconscious level. Thus the female being had a feeling of not quite being able to measure up. Going to Earth to initiate the male had not been put into her cellular structure.

Over the years, since very early on when you arrived from the stars, certain people, mainly from Zeta Reticuli – scientists and geneticists extraordinaire – have been involved in keeping an eye on the genetic experiment on Earth, of which you, the human being, are the result. They were requested to do so. Much later they found out you were their past lives, but that's another story.

Current Female Genetic Changes

Female beings have had a lot more educational and genetic experiences with extraterrestrials than male beings. I am bringing this up because the Creator knew that the ideal feminine body would be created in the future with full genetic knowledge, at least on a cellular level, of what would happen here, and the female could make more instant adaptations. The source female would have been able to head off on the cellular level certain avenues that the male beings took – that is, the propensity toward being a warrior when it is unnecessary – and allow the genetic experiment on Earth to conclude much more quickly had not linear time been evolving. You had to move *forward.*

The cosmic plan for this place is trying to correct that. What can be done to bring female beings genetically to the place where they will have more innate knowledge? They might not necessarily be able tap it, but if they get in touch with their spiritual power they will know what to do.

It is sort of like trying to correct the genetic soup after it is already off the stove. When female beings are taken aboard ships, genetic material is put into them that is not put into male bodies. Those women attempt very gradually to alter genetically the feminine cellular structure so it would stimulate your instinct, your spirit, your source self, from the future to merge into the past (which is your present). It is part of the reason feminine beings are waking up faster, why feminine beings are more spiritual.

Feminine beings are generally ahead of masculine beings in your evolution toward the fourth dimension because the genetic structure has been changed to allow the feminine being to be pulled and pushed forward at an accelerated rate. Being able to be a feminine being in one life and a masculine being in another life helps create balance in the soul. It is one thing to have restlessness in the body, but we don't want it in the soul, so we create balance.

The A.D. 2050 Boomerang for Females

In the year 2025 an interesting thing will happen. All feminine beings on Earth will experience what I would like to call *the boomerang effect.* This will happen in a very small fold of time whether you are awake or asleep. All feminine beings will suddenly snap into the future and then back into the present. This will allow you to be much more able and have much more influence. It will allow you to have the genetic

magnetic to balance itself. I am not justifying, by the way, the male's having extramarital affairs. I feel it's valuable that you have the structure you have. The reason the male is more pulled to do so is that he was designed to seek that missing portion of himself and would have to get along with the feminine.

In this talk on the body of man we have discussed what the male being has to gain from the female being. Next we will discuss what the female being has to gain from the male.

Descent to Earth

The Body of Woman*

August 15, 1993

All right! Zoosh speaking.

Why does the body of woman look the way it does?

The source of all feminine energy is rooted in the star system of Sirius. In that star system there is a central core not unlike the core one might find in a flower, the point from which all the petals fold out. In this central core live the eternal ones, the elders of Sirius. These elders have lived indefinitely, as far as I can tell. I am not able to see their point of origin, so apparently they have always been. Possibly they are the seeds for the star system itself. These individuals have designed the body of woman to look as you now see it, though there were some trial runs that looked a little different. [See chapter 11 in *The Explorer Race.*] The jaw used to be a little longer, a little wider sometimes. Those genes still run in a latent way in your gene pool. Occasionally, a woman is born looking like that.

The shape of a woman's body suggests certain intriguing possibilities. If one is going to have two versions of a being, as in the male and female human being, it would seem only natural that they would fit together as a set. Of course, sexually they do, but they also don't, because one might expect the male body to be somewhat indented in the chest area to form a complete fit of the set! It was originally intended to be so, and that latent gene is in men. (Occasionally, men are born

* Originally appeared in the January 1994 issue of the *Sedona Journal of Emergence.*

with a bone structure in which the chest bones actually dip inward.) I mention these curiosities because the original ideas have been modified over the years.

Creator Traits in Woman

Woman's body has always been designed to have the focus of the core of creation. So of course, since woman gives birth through her symbolic core of creation, woman's body is designed to be the eternal creator. Now, before women get up on a pedestal, I might remind you that this carries certain responsibilities.

Women over the years have been more allowing than men, allowing things to be – not necessarily approving, but allowing. This has to do with the essence of the Creator. Women also have chosen – not always, but frequently – to be the conscience. The interesting thing about being the conscience of humanity is that women have had to advocate certain points of view. Thus women are really and truly political beings. Although you are only now in your modern society beginning to see women taking official political stances, they have always been active in campaigning for certain conditions, to some extent universally.

Woman's body is intended to entice. Man is constantly seeking the Creator. Woman does not have quite the same drive to find the Creator because woman is, physically speaking, in many ways the Creator's representative. With that idea in mind, woman has certain responsibilities. Man finds woman fair, as they say in the biblical texts, and is attracted to woman. In large part man seeks woman because man seeks the Creator.

I do not want to divvy up the species according to the Creator and that which desires to be anointed by the Creator, but since everyone who comes to Earth to have an Earth life usually has a least two, they will appear on Earth as both man and woman. Consider it a set. You will have at least one each of those lives.

The suggestion here is that there is a desire of the Creator for the shape, the emotions, the drives of man to be integrated, not tamed, with something that is apparently polarized. Woman is apparently polarized from man. In reality, you all know that you have your feminine side and your masculine side whether you are a man or a woman. Nevertheless, since man is drawn to woman, you have man being drawn to the Creator. But there is more.

Since the Creator is attempting to manifest something here in this polarized world that has not been manifested elsewhere, you have to ask yourself, What is the meaning of life? What is the Creator's intention? The intention here in a polarized society in which extremes are encouraged might be extremely artistic, totally nonartistic or anything in between. What is the Creator's intention? The Creator intends to focus on the lesser characteristics of man; that is to say, aggression to the point of self-destruction. Self-destruction means destruction or harm to the self or to anyone else, because everyone is united. Everyone is *one thing*. If you destroy someone else, you are destroying part of yourself.

The End of Self-Destruction

Man has been given the responsibility of resolving self-destructive impulses. He is confronted not only with the need for woman but also with woman as representative of the Creator. Even if man isn't driven to be with a woman, he is born through her. There is no getting away from it, gentlemen: The Creator is intending to bring about an end to the drive toward self-destruction.

One might say, from this point of view, that since self-destruction exists and everything is a part of the Creator, then self-destruction is part of the Creator. But this is the point, if I may be so bold, of the Creator's lesson. The Creator *does not choose* to possess self-destruction, and would like to eliminate it as a factor. The only way to eliminate it is to stage an experience in which self-destruction as a drive is separated out, focused and then reoriented. This is not such a strange phenomenon. After all, when psychologists or psychiatrists work with a patient, they will often try either to separate out the self-destructive impulses and reorient them toward being constructive, or if there are many of them, separate out the positive, constructive forces and concentrate on building them to make a new life for the patient. This behavior imitative of the Creator permeates your society on many levels.

The Creator then chooses to stage the drama of mankind's struggle to rediscover himself and herself—yes, to discover who you are. This knowledge is taken from you so that you will have the drive to discover it—to look everywhere and overlook nothing. The Creator is not taking any chances here, so that self-destruction will become an *uncreation*. That's an interesting word, isn't it? How do you uncreate self-destruction? In order to uncreate it, you have to do more than fold time. You have to go to the core of creation itself, turn it inside out and hold it there for a moment. Creation is itself turned inside out and spun, separating out the less desirable elements. Then the core of creation is flipped back into its natural state of being. Picture this as a hat being flipped inside out. What happens to all that self-destructive stuff? This the Creator is unsure of. (Yes, the Creator has a few blind spots.)

The Creator became aware of Its existence not unlike the way you have become aware of yours. You are suddenly there, you become aware of your personality and then you start exploring. This leads the Creator to consider the possibility that someone created the Creator. Since you became aware of your existence the same way Creator did, you can see how the Creator might have come to this conclusion. The

Creator hopes to separate out all self-destructive impulses in all creation, which gives you the bonus of uncluttering *your* entire life span and all your lives of all the discomfort, self-destruction and misery that has ever been. But once the Creator calls out all of the self-destructive impulse in its many forms, the Creator then hopes that a miracle will happen. Thus the Creator finds Itself in a dilemma. The Creator knows that It wants to be more, but in order to be more, It believes It has to be less. It's a typical paradox. So the Creator is striving to eliminate the last portion of Itself that It sees as holding Itself back.

The body of woman is designed to initiate. After all, life is initiated physically through woman. Yes, man naturally participates, being the other half of the complete being, but man can initiate all he wants and he still isn't going to give birth. Woman is designed to enlighten, to broadcast love. (I would like to use the words *love* and *light* together, but they are not exactly the same.) She nurses the baby—she is designed to give love functionally, physically, out of necessity for the survival of the child. For the most part, nursing a baby is a pleasurable experience not only for baby but for woman.

Man is designed to construct, generally speaking. I am not being hard on men here. I am saying man has been visited with the challenge of moving beyond his own self-destructive impulses. Yes, some women have these impulses, but every woman who has ever known anything self-destructive is accessing her male side. Please do not get upset. Man has been visited with self-destruction in the same way that laboratory machines spin substances to separate one thing from another. This has all been heaped on man, and that is why man can be more self-destructive than woman and why the self-destructive impulse is masculine.

That impulse is designed to be like iron filings, and woman is designed to be the magnet. Man is designed to be drawn to something. The intention is that at some point man moves toward the initiating being—woman—and leaves the self-destructive impulse behind *without himself becoming polarized to the feminine.* It is as if the Creator has taken all the self-destructive impulse of the universe in all existence of all time and poured it onto this planet with the intention of separating it all out.

Woman's responsibility is to initiate man not only into the light, but also into the love. This is why woman has often been placed in the role of lover and mother rather than love and mother creator, as compared to the builder role of man or the destroyer role, its opposite. In order to move woman toward her role she has been given a body that draws man.

What is the consequence of all this?

Obviously, the Creator has entrusted this huge thing to all of you here on Earth to resolve. It has chosen individuals, at this time as well as previously and in the future, whom He intends to be as deep as

possible mentally, spiritually and emotionally. I did not include physically because It has chosen Mother Earth to give a portion of her body, which is the material-mastery self. Many of what are being called the new children are being born now to bring about this change more easily. It is precisely the energy of these children that will do that.

When something is going to be sprung, you have to pull the slingshot all the way back. This is why your societies are at an extreme point. You can look around and say, "Whoops, five or ten more years of this and we are out of here," because Mother Earth cannot tolerate much more in terms of population, to say nothing of pollution and the extremes of emotions.

Animals, Carriers of Knowledge

For example, individuals who have been carrying knowledge of the eternal spirits are beginning to leave the planet. These individuals I refer to as either animals or animal people. They have been leaving for a while; they are beginning to say good-bye. This is because you have released limits as a physical fact; as a result, you are no longer experiencing the protection of limits. You are no longer kept from having your full knowledge. Many animal species have been carrying knowledge for you, but they are no longer required to do so because it is now being visited upon you, albeit a little quickly, suddenly and disturbingly in some ways.

I will define "disturbingly." Has anybody been having nightmares lately, difficult dreams? The struggle with dreams occurs because it is as if your mind is being forced to absorb knowledge while you are asleep. While you are asleep the mind is normally at rest and the spirit travels. But the mind is not allowed to rest now because you are being programmed with knowledge that has been carried for you in the past by the animals. This was so that you wouldn't be burdened by too much knowledge and so you could re-create your reality into something more pleasant. They have done their job now, so they are leaving.

One could say, "But Zoosh, *pollution* is killing the animals." There is something to that; certainly it is true. Yet the timing of it, you know, is a critical factor. There have been times when pollution has been extreme, although caused by Mother Earth – for example, volcanic blasts. A few survivors made it because they had to pass on that knowledge to you some day. Now they are passing it on. It is no longer necessary for them to reincarnate here, so they are leaving.

I regress a little because I want you to see where all of this fits into the plan – the big picture, if you would. The Creator wants you to move beyond your limits because It wants to move beyond *Its* limits. So here you are as representatives or portions of the Creator, and you are given this tremendous task of representing the Creator because It cannot, as

Itself, do what you are doing without destroying Itself. It can be done only as a drama outside the central core of the Creator's being.

Man, Absorber of More Negativity and More Light

So we come back to another fine point. I am approaching this gingerly because I realize it could be upsetting to some men. What you call negative energy and what I refer to as the self-destructive impulse has been visited upon man (men, if you like). This does not suggest that there aren't some women who are self-destructive or negative, but man is dealing with it regularly, so this is his responsibility and he deals readily, if not happily, with the consequences. Man has a tendency to draw to himself negative energy. If we were to shine a light of love and beauty in a room and could quantify the level of light and love absorbed by woman as compared to that absorbed by man, we would discover something most startling: *Man absorbs more than woman!*

Woman is associated with the core of the Creator and does not need as much, but man is seeking the Creator, is driven to find Him. Woman desires to be close to the Creator, to be close to home. Man is desiring to touch the face of the Creator to feel loved, appreciated and forgiven, although there is nothing to forgive. So if we had this light shining in a room, man would absorb huge quantities of it because that which is imbued with the self-destructive impulse would constantly seek the light or the love, absorbing every last drop it can.

Now, some of you women will notice that in relationships you are giving, giving, giving, giving. This is not just because of the social arrangement of your society. It is also because the man in your relationship needs that love and light, and he can accept it from you more easily than he can from the world at large for obvious reasons – trust and so on. But here we constantly face a dichotomy and a challenge. If man is carrying the burden of the self-destructive impulse and woman is carrying the responsibility of creation itself, how can men release the self-destructive impulse so that it might be separated out? The Creator hopes that that which has created the Creator will do something. Creator believes that It is waiting for that miracle, as it were.

Moving through the Dimensions and Screening Out Self-Destruction and Negativity

How is all of this going to take place?

Many times we have discussed the shift between the third and fourth dimensions taking place while you are alive. It is very easy to shift dimensions when you pass out of your physical body; you immediately shift dimensions then. But the Creator, for Its purposes as well as yours, needs to have you, Creator's representatives, be *alive* while shifting dimensions so that you can go through a veil that will function much like a screen. It is desired that as you move through the veil between

dimensions while you are alive, the self-destructive impulse will be filtered out.

I have said on previous occasions that you will have about 2% negative energy in the fourth dimension, but that will be culled out again because you will have another conscious shift from fourth-dimensional to fifth-dimensional Earth. The same souls who have been here all this time are going to go through another filter, a veil between the fourth and fifth dimensions, where you will filter out that 2% negative energy.

This takes a long time. It is like when you filter the coffee grounds from the coffee. You leave something behind, but you take something with you as well. Now, you can say, "But Zoosh, the Creator has the ability to eliminate all negative energy and make this place heaven, the Garden of Eden." On the surface that is so. But you are fulfilling a job for the Creator, a job It needs you or someone to do, and you volunteered. Oh yes, you did! That is why your lives are reasonably short. On other planets you live to 250, 500, 600, 700 years, but can you imagine living the extremes of life that you are living here for 500, 600, 700 years? That's too much. So the Creator gave you a gift – a short life.

The Creator needs you to do what you are doing so that It doesn't have to wind up having to do what It has done many times before – take pity on you and pull out the self-destructive experience you sometimes call evil, condense it someplace, set it up on a shelf, look at it and say, "*Now* what do I do with it?" That's happened a lot of times, especially on Earth. Civilizations have risen and fallen. Very advanced civilizations have existed here before, but they all came to the same end: The Creator said, "I can't let you suffer any longer," pulled out the self-destructive impulse and everybody chose to leave. But this hasn't happened for a while.

You Souls Were Carefully Chosen

The Creator was very careful in choosing souls to come here during the past hundred years or so in preparation for the shift from the third to the fourth dimension. He was very careful to allow people who have achieved multiple levels of spiritual mastery, who had some experience with material mastery and many who had experience as teaching masters, combining spiritual and material mastery into teaching mastery. Many of the children being born now are teaching masters. After all is said and done, even with the self-destruction, at least 30% of those children are going to live to be sixty or seventy years old. Allowing for the extremes in today's world, that isn't bad.

So what you have here, then, is a circumstance in which you have volunteered to do something for the Creator that isn't done yet. It is going to take a while to do it, but until it is completed you are going to stay with it. In order for woman to have completion as woman, she does not have to have children, but she does have to provide, either intimately or generally, at least one man with some form of love and light. (I am using the word "light" because that is your term.) This does not mean the woman has to have sex with a least one man, but that for woman to be complete, she has to function as the allowing creator for a least one man in her life.

Dare I make a prediction? Why not! By the year 1999 every female who is alive at that time, including children, will have done what I am talking about. How many of you have ever put your finger in a baby's hand and felt its fingers close around yours? That's how a baby gives a hug. It seems easy for a woman to provide a little love, you know, a little hug – that's love. We don't have to have anything too elaborate. Even allowing for all the different situations in which women find themselves, sometimes not having the opportunity to be around men at all, it is still going to happen by 1999. Every woman who is alive at that time will have done this.

When that happens you will then have set the stage for the shift into the fourth dimension. At that stage it will take a little more accumulated experience. But once the stage is set, it will begin the shift into the fourth dimension and also the flipping of the universe inside out. What I suggest to women who read this is that if you haven't done this already, find an excuse to give a little love to some man. Give a little hug or something and tell him everything will be all right.

Origins of the Creator

Beginning This Creation

January 9, 1996

How did life work before the friends of the Creator got here? [See the final chapter of The Explorer Race, Book I.*] I'd like to expand that whole end of it, if I may. How does it work? Were they assigned, or did they volunteer? Where were they before that? They had to have been in other creations.*

There are certain core personality types that exist on Earth and everyplace else, although there are many shades of these personalities. Each of those who came here with Creator represented a core personality type that was necessary to invest in all beings created. One of these was the renegade or rebel. If you look at civilizations and personality types – even combinations of personality types, which is more common – you will find the rebel in a great many personality types, but not all. For instance, the rebel is common in the civilization established in the United States, but less common in other civilizations, for instance Peru. You asked where they came from before that. They existed in, as best I can describe it, the place of all things.

Even the Creator?

Yes, that's right.

But He'd been someplace else before that.

It's hard to describe it. Picture a bowl of alphabet soup with the different letters floating about – times the infinite. This would represent a place of all things at once, a vast bowl of soup, if you like. Creation soup. In order to create what Creator had – I don't want to say "in mind," but lacking a better term – it was necessary to go there with certain core resonances.

Gathering the Core Resonances

So Creator simply put out a pulse (that's how it works) that attracted other core resonances of the same general pulse. Not all these core resonances came with Creator, but they were given by Creator the option of whether to involve themselves in the project that He had in mind. Those core resonances who decided to participate in the project directly came with Him, all going through the veil. Other core resonances who decided to participate indirectly remained where they were, but they remained available to Creator for consultation, although not directly available to beings He would generate.

The core resonances who were not interested in the project either had other projects they were interested in, were simply doing something else, or just weren't interested. This was quite a few. This is partly why some things are missing, as it were. For example, a pseudoscience that ought to be an *exact* science is astrology. It can utilize only the known, but it can also utilize the accumulation of what has been known, like an almanac, which accumulates wisdom. Parts are missing, so astrology can only be a pseudoscience, because it can give you no more than its best guess. It cannot tell you exactly, although sometimes it might hit on something exact.

Because some core resonances didn't want to come.

That's right. If the other core resonances had come along, or at least made themselves available to Creator as a consultant, then astrology (for example) would be an exact science. But you see, Creator felt it was all right to continue the project without the participation of all the core resonances, because it was Her intention to allow the beings She generated to become somewhat self-fulfilling – meaning they would seek their own destiny rather than have their destiny provided to them through exact and specific means. Creator wanted you all to evolve and have available to you trial and error (and on some planets such as this one, ignorance). To a lesser degree other planets would have the opportunity to make errors as a society and learn from them.

The lack of some of the core resonances was not perceived as a problem by Creator. Instead, it was understood as a favor, so there were no hard feelings. This necessitated that Creator put a different spin on Its creation. Its original idea was to create something that quickly evolved into a replacement for Itself, but it has taken much more experiential time. Given this longer brewing time, as it were, the knowledge and wisdom attained by those generated by the Creator would be more ingrown – that is, more self-fulfilled, more accomplished by individual and combined entities (once you go past the veil of death, you become combined with your total self). Creator felt that it would be better in the long run, even though it would take more experiential time.

Of all the friends of the Creator, you're the only one who didn't get enmeshed in the

creation. You said you kept your objectivity.

I kept my distance, yes. All of the friends who came through – people, as it were, core resonances, personalities, if you like – were indirectly or directly involved. With the exception of myself and perhaps my higher self (should there be one), there was no one who could observe, interact, be involved and yet have enough objectivity to give guidance to Creator as a participating observer.

Zoosh, what was your core-resonance quality?

Well, I suppose I'd have to say, *the philosopher.*

What about Hieronymus?

The shapist. All materials have shape or form, which to some extent indicates their function. Hieronymus was a shapist who provided the shapes you have come to understand – triangle, circle, square and so on.

What about Ra?

Ra was involved in a situation similar to my own, taking on more than one thing. He was involved in a historical context, meaning time. Loosely speaking, Ra was involved with time and the definitions of precise personality, which falls loosely under history. Ra took all this on so that he could have a stimulating time. (Ra is not one to sit around idly.)

What about Shiva?

Shiva took on *play and celebration.* Earth is a school, but most other planetary creations use play highly creatively. Play is highly involved in education, even learning about family and becoming a good citizen of their culture. Earth is a place where play is the least involved or tolerated. Do you know that play here on Earth is perhaps at the lowest ebb of anyplace where there are civilizations in your known Creator's universe? So while play might not seem to be such a great thing on Earth, it is absolutely pervasive everyplace else.

Counterpoints of Creator's Companions

Lord Michael, then, as a counterpoint to Shiva and play, must be serious?

No, I wouldn't say that; it is hard to describe a counterpoint there. I would say that Michael is more involved with *intellectual analysis and the eternal.* Analysis gives Michael an opportunity to have an ongoing thought for millennia and explore that thought from every conceivable – or inconceivable – direction. Eternity and analysis, though not exactly opposite of play, are pretty close to it.

Because play is spontaneous. Do you have a counterpoint?

I don't have a counterpoint here, but I might have one in another universe who is quiet. I am [chuckles] not known for being quiet. I am, however, in contact occasionally with a part of myself that is always resting in a quiet place, so I think perhaps my counterpoint is quiet.

When you use the word "universe," is that this creation?

Yes.

So did Creator's friends split into point and counterpoint when coming into this creation? Is that how it works?

Understand that the Creator's intention was to have societies that were for the most part benevolent, but He has always been one to put a little spin on the ball. Creator wanted to have teaching planets such as Earth in which there would be polarity, and that polarity would tend to move outward. Within this immediate solar system polarity has some effect on other planets. Although all planets in this solar system are not specifically polarized, using your instruments they would appear so. Once you get out past Pluto you'd find that polarity is not much of a factor.

To answer your question directly, I'd say that those who came in experienced their counterpoints when they came closer to polarity – with the exception of Shiva, who came in with the knowledge that counterpoints were essential for growth. Because Shiva is the wise one (though also pictured as the celebrant), his awareness of counterpoints is a given and a core resonance in itself. Wisdom is usually portrayed in words, in culture and in stories passed on, yet wisdom is not necessarily ascribed to the celebrant, though it can be. It was always Shiva's intent to show that counterpoints are necessarily united, not opposite. That is really what Eastern religions are attempting to achieve at their core, because they do not deny something so much as attempt to achieve its higher level.

So that wisdom has been given through channels by Shiva?

Yes.

Were any of you creators before this?

Well, we had all done some dabbling in it, as is usually the case, so that there would be a universal awareness of the problems and challenges. But I would have to admit that my experience as a creator is puny compared to what your Creator has done here with this known universe. I can admit to being involved in the generation of certain species of lightbeings.

All right. Let's say there are six or seven beings, and one of them says, "I'm going to be a creator." Is he greater, wiser, smarter, or does he have more courage?

No. He just had the idea first. It's not competitive, you understand. Creator just said, "I've got an idea. What do you think of this project?" then showed a rough idea of how it would be. "We can have a lot of fun," says Creator in his own language, "and within this creation I'd like to work out our own issues – things we are interested in, not problems." Shiva, of course, was very much interested (because of his desire to show the necessary unity between counterpoints) in being involved in a place where there would be polarized planets. Shiva was on board immediately.

When Creator had the idea, He had to go before some council and get permission?

No, it doesn't quite work that way. To have that awareness at all, one believes and one applies. You have the idea, so obviously permission has been given or you wouldn't have that awareness. It is a given; it is accepted, and there is absolute faith that the idea has value or you would never have had it.

Is everybody in this creation generated by Creator except for those few who came in with Him? Did any of the others come in later?

Everybody was generated by Him. There have been a few who have sailed in from other universes, but only with permission from Creator. Creator has the authority to deflect beings who are not going to ultimately prove beneficial to the creation here. There is a fairly broad range of inclusion, but if you fall outside of that broad range, you are not included. You are instead encouraged to go elsewhere, and because there are many other elsewheres to go, it's not as if you're lost.

The Redemption of Negativity

Creator's intent here was to create something finite. Part of the reason Creator was allowed to do all that is going on here, some of which is risky business, is that Creator knew the project would be finite and negativity would eventually be redeemed. He knew that negativity would prove its value when used as a spice, and as such could be trusted to be constructive as itself. The reason Creator's project was so highly thought of, encouraged and supported by so many beings over the "years" was that Creator was going to redeem negativity. It had never been redeemed, only controlled – and everybody knew that control is not redemption. Redemption is, as you know, a higher expression of oneself in which by being oneself, one supports life on at least a lateral plane, meaning life around you and potentially interdimensionally as well.

Creator said, "It is my intention to redeem negativity so that it will support life on a lateral plane and hopefully achieve, in higher dimensions of itself, a constructive expression of itself interdimensionally." Creator is not expecting, in this version of Itself, to achieve interdimensional qualities of goodness with negativity, but hopes *you* will do that.

The Creator started creating beings who in turn started creating beings. You said some of these unresolved issues were trillions of years old.

Yes.

How did those issues get into the Creator?

Bringing Along Conundrums

Creator brought core resonances with Him. Those core resonances had certain points of view, but they also brought with them conundrums, things they had been unable to solve from other experiences. I don't want to say Creator bragged, but it was pretty close. Creator said,

"You are free to pursue through this creation the resolution of your conundrums," which included their own and anybody else's. "But," Creator said, "we will do so only in specific places where my most advanced souls will have the capacity to resolve and/or deal with these conundrums." Everybody said, "Okay!"

The friends of the Creator are all down here on Earth?

They're attracted to Earth because this is the place where these things are allowed to happen, where the most advanced (at least the most experienced, but I like to say advanced) souls reside.

All right. So there were a lot of other beings who came in with conundrums?

Yes.

There were preexistent energies beyond what you call friends of the Creator?

Yes. What might get them into this creation . . .

. . . was the promise of being allowed to work out their issues.

Yes, and also permission to bring something that was a conundrum from some part of creation. If they could arrive with a conundrum about which everybody could say, "Hmm, I've never heard about that one," that alone would almost automatically get them in. It's the foundation of creation. But it also stimulates you, because in asking the question you also learn more about yourself. That is the ultimate intention here of the philosopher.

What was the experience of the friends before the alphabet soup?

Before the alphabet soup, before form or even before the experience of individuality . . . let's just say this: Creator comes through the veil, right? Going back through that veil is the alphabet soup, and before that is formlessness—meaning that all is absolutely unified. In order to get to individualistic personality, though, one has to go through the veil to potential creatordom, then beyond to your project. This works the same in a microcosm as it works in Creator's macrocosm.

Coming through the veil?

Inspiration and Original Thoughts

Yes, because when an inspiration comes through for an idea, for example, the inspiration is the formless energy—that's the circuit. Yet the inspiration *takes form within your consciousness*—that is the individuality. The application of the inspiration then becomes your own creatorship in action.

And that can also apply to a human in a life on Earth or a life on any planet?

Yes.

The friends have been out of the alphabet soup and back before—that's where they got their experience, right?

In the different projects and back into the soup.

So there are billions of creators—is that a true statement?

Oh, yes.

Do all creators take core resonances along with them?

No, it's not a requirement; it is usually done only when a creator desires to be involved in a finite project. You might take other creator beings along with you so that their conundrums, or agendas, would be answerable within the context of a finite answer. Levels of the question would be infinite, but there would be a means by which a given finite amount of its outworking could provide an answer.

This is not always the case; as a matter of fact, it's fairly rare to take others with you. But since inclusiveness was one of Creator's core personality traits, Creator was inclined to include others.

So all these billions of creators go in the alphabet soup and back out, back and forth. This is Creator's first creation; do other creators do it many, many times?

Let's just say that the concept of infinity is involved here. There is no moment during which a creator is not creating.

While it's in its creation, but what about when it goes back into the alphabet soup?

Even then there is creation, because that which is considered nonphysical here is really physical—thought, for example. Say that you have a thought and then you let it go; or you have an imagination or a fantasy. It has a form of its own; it exists for a while and can be picked up elsewhere, can even be tapped; it can even be taken, as has been done. Occasionally people have original thoughts (which is the intention), and if one of those original thoughts is considered worthy enough, someone, perhaps a creator, might pick it up and say, "Oh, this is great; I love it!" and create an entire creation from that thought.

While it's in the middle of its creation?

Anywhere. Let's just suppose that people have thought balloons above their heads as in cartoons and they had a fantasy going on and then said, "Okay, back to work." If there was a creator somewhere in the alphabet soup or even here, your Creator might say, "Oh, I like that. I hadn't considered it that way," reach down, pick up that balloon and expand on it in some creative way.

You're saying that everybody in the alphabet soup can tune in to our thoughts here on this planet, in this creation or in any creation?

Oh, yes.

What a voyeuristic place, eh?

Well, you have to understand that you cannot be at creator level unless someone feeds you. Food for a creator would be an original thought, because a creator could take that thought and expand upon it (physically, nonphysically, whatever) and create something from it. And since all life is intended to interact with all other life, it would go against the core of creation to have those in the alphabet soup excluded from the throwaway creation that goes on here, where people might have a fantasy and come up with some extraordinary combination of events. It would be like denying these creators food to not let them have it.

Is this alphabet soup the ultimate, or is that just a teeny, tiny cereal bowl and off here there's another one?

Individual Personality

As the philosopher I can only answer that question by saying that one can only guess at what is the ultimate. I can speak of what I know, but because I am the philosopher I can extrapolate that there is infinitely more I do *not* know. Part of the reason I can be certain of this is that I am still intact as an individual personality. This must mean that there is a great deal more to know philosophically than I know now. I don't often reveal anything to you about my personality, but why not? It fits in.

Well, that's very cheerful, because it's not ending, then. Let's go back into the alphabet soup. You remained who you are, you stayed unique, right?

Yes, but I am blended with many things at the same time, although I am a core resonance. It's not the same kind of individuality.

But you still feel "I am."

Yes, I have my personality.

Even when you become a creator, you are still that individuality.

Yes, I will have my personality.

So it's got to go back beyond where the Creator has been or where the alphabet soup is, where you lose that personality. Is that true?

You don't actually lose it; you just expand your definition of it. When a mother gives birth to a child, the child's personality blends as it develops. For the first couple of years that child's personality and the mother's are almost the same. This is a microcosmic example of the idea of what makes up your personality. Now, the same could be said for a pet. People are often so attached to their dogs that the dog's personality (the spontaneity, unconditional love and so on) becomes a safe haven, as it were, for those people to express those traits to their dog, although not necessarily to the rest of the world. In that sense the personality traits of the dog are shared with the human.

After a while, as a creator you expand on the definition of your own personality, so that the "individual" (as you might see it now) personalities of other people feel like your own personality. The experience of individuality becomes blurred, not unlike the way watercolors run into each other. There is less a sense of boundaries and more a sense of "where I end you begin."

Okay, but when you wind up the creation and the Creator goes off with his friends to the local soda bar and has a soda and they talk it over, the Creator is more than He was when He started, but He's still who He was plus some, right?

Yes, that's the intent. Though the intention of any individual creator might not be that it becomes more, within life one's experience *does* make one become more, simply from the cumulative effect of that experience. So yes, they are more.

So Creator has literally broken Himself into a billion pieces, and that spark of light inhabits all of us in this creation. Can you talk to Him as a being separate from the totality of His creation? Does He retain some spark of individuality now, while He's a creator?

Remember that I said that one of the Creator's core aspects is *inclusiveness.* I can find the Creator's original personality by focusing in on that inclusiveness and other core resonant traits and speak to the Creator's individual personality before the Creator created anything. Even now as I am speaking with you, I am speaking with the Creator, because you are a portion of Her. Creator has generated you, so I cannot speak to you without also speaking to Her. But if I wish to speak *only* to Creator, I would key in on inclusiveness and speak to *that* as an energy. But I do not do that; generally, if I have something to say to the Creator, I will say it directly through somebody else, because it gets to Her.

What I'm leading up to is that if we want to become creators, it takes a lot of courage, because you literally separate yourself so that you don't exist as a conscious being.

I'd rather say that you expand your definition of what is a conscious being. All of the individual molecules that make up your finger are yours, but if you had the capacity to ask one of these molecules, "Do you know who I am?" it would say, "No, what is this? We have no idea." Yet these molecules are a portion of your physical body. So the molecules themselves, on an individual basis, don't need to know who they are a part of. The biggest factor is that to become a creator you cannot be attached to your creations knowing who you are.

You cannot be attached to their knowing who you are at that point in time.

No, you cannot, because you will be necessarily expanding. The minute you start generating life in any form at all, whether it be molecules or something else, you become more. Since you were the focal point for that generated creation, you necessarily expand who you are. You might not in every moment be in absolute mental communication with all of your creations, but you are *always* in energetic connection with them no matter what form they take.

So when the Creator is through here and goes out for a coffee break, is He going to remember all of these . . .

Absolutely.

He's going to remember if He chooses to focus on every being who inhabited . . .

At any time, and every molecule of every being.

Far out, what a gift!

Well, you know, that's the purpose of the project, so that if you missed something as a creator, one of those individual molecules or even atoms might have attained it. So as you examine one thing after another (not linearly of course), you might find something you overlooked.

Hmm. So the Creator goes off and does quite a bit of introspection for a while when He gets this all together.

Cartoonly speaking, yes, but life goes on nevertheless. Creator is exploring.

Are you thinking of doing it next time?

Well, I like my role as the philosopher; it allows me to put my feet up on the pickle barrel of life and tell stories.

[Laughter.] Come on; no guts, no glory!

Well, that's what you've always said.

All right. We will continue later, and someday at least a portion of these conversations will end up somewhere. The intention here in the long run is to help people to see the big picture so they can be cognizant of how similar it is to the little picture. Good night.

Creating with Core Resonances

January 30, 1996

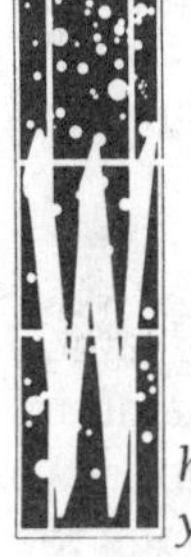

What are we talking about as far as this Creator is concerned – our Creator, your friend?

We are talking about all that exists.

But there are a billion creators!

I understand, but it's like this: Within the scope of creation – all that you can imagine, all that you know about or can even fantasize imagining about, even your ability to imagine the unknown, all of this – Creator came from someplace to this place. I'm using your language as best I can, yet I must tell you that this place you are in is *everywhere.* It's almost impossible to say that it does not include everything.

The data from the Hubble space telescope show that there are probably fifty billion times fifty billion galaxies.

More than that. They are just trying to help people stretch their imaginations.

Bentov talked about forty-nine universes and three chakras in a cosmos – that was one creator.

Bentov's and everyone else's ideas are an attempt to make sense out of something that is beyond sensibility. Those are all interesting models – making the universe smaller so that it becomes manageable and understandable. Science in general is an attempt to make things smaller so that they make sense and can be understood more easily.

Okay, but you're using the word "universe" – one, one verse. Does "universe" encompass what this Creator has created?

We're struggling with words here, but we know that life exists dimensionally right here. Here you are in 3.47, yet second dimension or eighth dimension is here also. The best way I can describe it is to say that everything you are – in all of the dimensions in this place and everything this encompasses, in which all those portions of yourself can even begin to imagine – has been created by the Creator that you know.

Before this creation we were generated out of the last creation we were in. This Creator came forth to do this creation and there are thirty billion creations . . .

No, no. This question is phrased repeatedly with the idea of starting and stopping, but we cannot do that. This is the problem: The language and philosophy on this planet are based on the finite. If you examine that question, you can see that; but you must base your question on the infinite. If we blow up a bubble and all around it is unutilized matter, that bubble goes only so far. Then we blow another bubble. That's the best way to describe universes that I can give, because it includes the apparently finite, yet there's a clarity between the expanded bubble – rings, if you like.

The Finite within the Infinite

If you look at your universe, the idea of rings – things getting larger, but in concentric rings – is repeated over and over again in literature, art, especially in symbology. This suggests, of course, that the constant repetition not only rings true through all the repetitions, but through the infinite. That which you are is that which you are infinitely. It's very difficult here to define infinity within a finite language and philosophy.

I'm going to have to answer your question like that. We can't say this creator and that creator, because if we did, we would necessarily be claiming to have "the truth," thus claiming that our truth is infinite and the truth of others is finite. If we do that, we are simply being demagogic.

Do you know the vanishing point on the highway, where the highway looks like it comes to a point in the distance? Imagine yourself inside a coil and looking downward. That coil apparently comes to a point, yet when you move to that point the coil is still there. That is infinity. The vanishing point is an illusion that is quickly revealed once we get there. Using this analogy, we can measure in a finite way this segment of infinity, but it's very important that the idea of illusion is made clear.

In the philosophy of the illusion, the possibility of expansion is a given, so one is not trapped in one's reality. That reality becomes pliable when you apply the philosophy that reality is an illusion, meaning that you are here; yet when you move further on in your life you are no longer here, but *there*. The future is totally veiled to you, but when you get up tomorrow and whatever happens takes place, that veil is moved forward in the same way that the vanishing point moves forward when you go down the highway.

Core Resonances – The Key of G

I am curious about where the core resonances were before they decided to come here.

Really, the best analogy for core resonances is musical tones, without suggesting that musical tones themselves are core resonances. You can strike a tuning fork and measure or quantify the vibration. You cannot say that the tone does not exist when you no longer hear the fork ringing. You know that the tone is in the fork because whenever you strike it, it sounds again. This tells you that the moment of striking is the moment of creation. At that moment the tuning fork re-creates that tone, and even though you no longer hear it when you haven't struck it for a while, the creation is still there. We're dancing around with analogies here, but that's the closest I can get to help you understand the idea of core resonance.

Let me try to get in words the little bit I understand of what you have told me. Creator had the idea to resolve the unresolved differences of past creations. He evidently had had great experience and understanding and wisdom, but He didn't have all the little pieces He needed for His creation. You said other beings came in with Him to provide those core resonances. How many? Six, seven, eight?

I will use the tonal scale again, because all physical life and, as far as I can tell, all spirit life can be created through various resonant tones. I won't tell you exactly how because I don't want anybody re-creating the universe at this time. I will say that Creator brought with Him the key of G.

He was the key of G?

No, he brought it in; He invited the key of G to come with Him. Basically He didn't have those tones. When I put it to you like that, it is something that could be illustrated, even somewhat humorously.

I was thinking of qualities.

I understand, but the reason I want to create it as musical tones is that it steers people more toward music, which is infinitely more important than you know, and is why I went through the whole tuning-fork analogy. We know that you can play one clear note, and although it is beautiful, you long to hear other notes combined with it. The combination of notes has to do with creation of music and melody, but it also is a clear-cut analogy for how people of different vibrations can create better when they come together. As long as they are in harmony, they can create something similar to the way Creator does. They can strike clear and perfect resonant tones in themselves by being totally themselves meditationally and having the clear thought, or by somebody standing outside them who directs the orchestra, stating, "This is what we wish to happen."

With a meditation totally in oneself (eastern or new-age style) the vibrational element should be compatibile. If there is not compatibility in a roomful of people, then you try different combinations until you have total compatibility. When that compatibility is there and everybody is in total focus within their own resonance, then the director has you maintain that position and speaks what you all would like to have happen. Or you might have the thought, but it's difficult to maintain your core resonant self (focused totally in your core self – that which is completely you) as a meditative state and simultaneously maintain a thought. It's easier for the conductor to speak the thought *for* you.

You're saying that's how you create a universe?

Core Resonant Tone As the Feminine Principle

No, I'm not talking about Creator; I'm talking about individual human beings, what they can do right now. I realize you're trying to get a creation sequence here, how it's all done. Let's just say that before there was light in and even beyond the visual spectrum, before there was anything physical, there was space. Yet within that space – even if we don't have Creator there – is a core resonant vibration, otherwise known as a tone. That tone is *there*, all right? Arthur Clarke tried to bring this out, you know, and so did the director of the movie *2001: A Space Odyssey*. It wasn't very clear in the movie, but the intention was to show that even something that appears to be nothing has a tone. When the tone is there, it acts as the feminine principle, which means that the space is ready to receive anything connected with that tone.

Before Creator stepped through the veil of where He was before, He was invited to the space He arrived in because the core resonant tone of that space, the feminine, invited the mass to create within its space. Let's say that there are other spaces, regardless of the dimension. If there are other tones, the creator who is in harmony with the tones in a particular space will go to that space. That's as close as I can come to describing it in the language and philosophy of this society.

Okay. So this Creator and Its friends . . .

He/She/It.

. . . have an intention to create.

Simultaneous Conscious and Unconscious Creation

Yes, but the interesting thing is that Creator is not unlike you, in that even without an intention to create something specific, It is nevertheless creating in every moment. You are this way also because you are portions of the Creator. Creator can mold you up to a point, but It's still molding you with the substance of Itself.

Even when Creator is not specifically creating something, there is creation going on. When you're asleep you have your dream life; your body has its physiological functions, things go on, but you are asleep.

Creator does not sleep, but while Creator is creating all these wondrous things, other creation is still going on. It is in the nature of Creator, combined with the feminine space that receives creation, that this creation takes place *on its own*.

The conundrum for Creator is that when one is creating, one desires to know everything one is creating. You have asked, "Where is Creator going to go?" Creator is going to go to the next place that It would go to, but the reason It knows there *is* a next place is that Creator can feel Itself creating even though it is not specifically working on it. Creator is working on many, many projects at the same time, to say nothing of individuals within the mass of planets and so on. Creator is running on automatic, so something else is also being created, which is basically the tether that goes to the next creation, where Creator is naturally creating even now as we speak because of Its impetus to create.

This you are also doing; this is how you as an individual can function in more than one dimension at once. Even though you are here and are focused in this life, you have counterparts, other lives living in other dimensions and in some cases in other times. What you do in this life necessarily affects them, just as what they do in their lives necessarily affects you – in the long run usually for your benefit.

How do you connect to these other lives? You, being portions of the Creator, are tethered to all creation; and all creation, in your chain of lives, is a microcosm of what goes on for Creator in the universal macrocosm. This is something that Creator did not build into you; it is something that happens because you are made up of the stuff of Creator. To use another analogy, you can do things with your body while your autonomic nervous system and certain portions of your body run themselves. You're not *thinking* that you want your diaphragm to go in and out, but it simply does it when you breathe.

I'm trying to explain the mechanics of creation. This suggests strongly that since there is something going on that is a portion of Creator that Creator is not directly involved in, Creator obviously has a higher self. Creator must ask Itself, "If I am creating beyond that which I am intending to create, then who and what gives me the authority and the ability to do this?" This has been Creator's question for a long time, and the Explorer Race will make it possible for Creator to find out at least the next step. That's why what's going on here is so important. It's the pivot upon which all creation in this Creator's universe hinges, because it will allow you to replace Creator and also allow Creator to move on to the next level, to which it is literally tethered.

If there are fifty times fifty billion planets in this universe and they all have sentient life, and if all sentient life is part of the Creator, why are these few souls on the planet Earth so important?

Creator's Conundrums

Before this time no group of souls successfully completed—or at least came as far as you have toward successfully completing—the study of all applications, consequences and creations. Almost everybody else is living a relatively uncomplicated life compared to yours. Life here is necessarily complicated so that you will have the opportunity, at least within the microcosm of your individual daily lives, to experience responsibility, consequences, creation—all the stuff I have talked about. That's why it's so important.

Creator has in mind that individuals can function—have daily lives with meaning and purpose and lovers and business and children and so on—while they are training in every moment to be a creator. In other words, is it possible to be everything within a moment and also to be finite? Another conundrum that Creator had was, how could you have a finite thing within the infinite? Yet you know you can have it, because a star will burn for just so long, then it changes form. You could say, "The mass is still there, but it's in another form." That might be so, yet it is no longer burning. The burn time is finite, yet all life is infinite.

It was a conundrum the Creator was not comfortable living with, so He had to create something that would be overwhelmingly finite within the infinite. All mass changes form; yet the form it takes is necessarily finite and measurable for its duration. Creator said that it would be necessary to have a planet where as much of this can be explored as possible without overwhelming people. Creator asked, "How can I do this?" and answered, "For starters, I've got to limit their intellectual power."

Now, your normal intellectual power in your lightbodies is not measurable by your standard IQ tests. To give you an idea, in your normal lightbodies you have the capacity to tap into whatever you need to know when you need to know it, so your IQ would be infinite, way off the scale. Creator said, "We can't let you have that in your day-to-day life, or you'll never re-create anything. You'll simply continue to create what you know to be true up to that point of time in creation. We need to have you forget who you are so you can re-create from a position of ignorance. Therefore, we will give you a conscious mind that is necessarily limited or that can be measured through a finite test.

It's not easy to measure a quantity of thought. How many thoughts does it take to make a pound? So Creator necessarily has to limit your capacity, limit your body functions so that you do not live too long in this school. Creator has certain problems that She's been fascinated with for years, and as a result, She needs to know how much She can do Herself and how much Her offspring (you, Explorer Race) can accomplish in Her stead—meaning when you replace Her. And when you replace Creator and He goes on, that which is still unresolved in Creator's knowingness (both yours and His) . . .

Has this ever been done in a previous creation, where a creator is replaced within the creation and moves up?

It has been done on a smaller scale, but not on such a massive scale.

How many cosmic days, which Vywamus says are 4.6 billion years, has this creation been . . .

The Circles of Creations

Again, we're talking about creation in a circle. How many can you imagine? Let's say that creation is a circle, but in the form of a tube. As you keep going around, the tube gets thicker and thicker. At every complete loop it gets thicker because of the accumulated experience, yet we cover the same ground. That's my best analogy to date.

Let me try. There's a creation, like the rings inside a tree trunk. Another creation uses that as part of it, but expands it. Then the next one uses everything that is there and expands upon that?

Yes. This cannot be measured finitely. The best way to think of it is that instead of cosmic days, it's just around and around, continually accumulating; the tube gets thicker and thicker—or, if you prefer, the tree trunk gets thicker and thicker.

So this creation we're part of is just one ring in the tree trunk?

Not necessarily. Because you use all that has been used before, it's hard to say how many rings around the tube that has been—and maybe you go more than one time around the tube. Understand that I cannot fully agree that *x* number of cosmic days are involved in creation, because who's to say what a day is, anyway?

All right, let's back up a little. The Creator had an idea, an experience that led Him to believe He could be a creator. He had to go to a council to get permission.

Yes.

So this council is in the center of creation? Beyond? Where is it?

This council is not really a physical place.

Of course not, but . . .

Creator As Solver of Conundrums

It is all that has ever existed and more—which encompasses all that might exist, all that could exist, probabilities ad infinitum. In that sense, then, Creator was given permission to do this only because Creator said, "I will attempt to solve conundrums."

How many applicants does this council get? Is there a long line? Is that what they do during their . . .

No, it's a means by which creation is overseen so that someone does not start something capriciously. Understand that although one individual might consider something very important, somebody else might consider it entirely capricious. You need to have a body of many to decide whether something is capricious or not, because someone might decide, for example, that in order to resolve all pain everywhere you need to experience every aspect of pain there is, so that the full tonality of pain can be understood and then chimed out of existence, as it were. I'm not saying I want that to happen; I'm saying that someone *could* come up with that idea and say, "This is really worth doing."

Yet a council will say, "Well, we don't think so. Look at all the suffering that will happen." Do you understand?

Vywamus said once that there is a dimension someplace that completes cosmic days that didn't work out.

Yes, it picks up the pieces.

And that everyone in creation gets a little boost once something is resolved.

You know, the Bible, western religion, has tried to resolve this by referring to purgatory or the afterlife experience (depending on your philosophy). But basically everything that is created (even in your fantasies, which do not usually have an ending) must be resolved in some way. There is a means to resolve it, so I agree with that being to that extent. There is a tonality that is rippled in its tones, that vibrates everything into some kind of balanced (for lack of a better word) order, sort of like taking a handful of coins and dropping them into a coin shaker, where they wind up in the proper stacks.

Creator As Conductor in Scale of G

You said that some of those who were invited didn't come, so Creator didn't get all the notes He wanted. Are they in the key of G, too?

No. Creator was coming into a space that was welcoming Her, and She wanted to bring all that She could into this nurturing, feminine space. But Creator would have liked to have brought more balanced energy—balance between the masculine, the feminine, the third sex and what is best described as the quadrangle sex (which is probably not something you will ever see on Earth or even run into as the Explorer Race, so I'm not going to go into it). Creator would like to have brought more of that third-sex energy, but those beings were not interested in participating. If Creator had been able to bring more of that combined sex (which is not the androgynous being, but the being that has both), chances are you would have been able to succeed in the genetic experiment on Earth in about two-thirds of the experiential time you've used to do it.

If the quadrangle sex could have come in, you probably would have been able to accomplish it in one-tenth the experiential time. The quadrangle-sex energy did *not* come in because it believed, knowing your Creator as a friend, that Creator would enjoy doing it longer, so it just butted out.

All right. What did they find more interesting?

They are involved in another creation, plus they have part-time duties on the Council of Allowance, which has to do with where Creator went to get permission.

An Example of a Renegade Core Resonance

One of the core resonances is the renegade energy, which is essential in creation. Think about it; you know, for example, that Robert Fulton was considered a renegade in his time because he flew in the face of what people told him could be done. Many great inventors have been renegades or were considered renegades in their time. To choose a more current example, Dr. Kevorkian is a renegade. He is helping suffering people to die. I'm going to tell you something that you might not want to hear, but it's true: Within twenty or twenty-five years that will be a major growth business.

The human will have a choice to . . .

If people are suffering and don't want to go on suffering for any reason, they will have the choice. They will simply go to a clinic; by that time Dr. Kevorkian will be considered a saint, a great man in his time, a pioneer—but not by the church. There have been many others who have done this, but Kevorkian is doing this publicly and is therefore a renegade. He's not doing it quietly like other physicians have done. He's trying to find ways in which people can die comfortably without suffering because he wants to help people end their suffering. His purpose is not to help people die—that's only the way he's portrayed. His purpose is to help people to *end their suffering*—that is his entire purpose. This, as I say, will be a huge industry in twenty to twenty-five years.

The injunction against suicide, then, was, like some of these other commandments, just to control people? If everybody thought they could get out, they couldn't be controlled, right?

Well, that really has to do with order in a practical society. Obviously, some people, if they weren't conditioned gradually to be allowed to let go of their lives physically, they . . . people do crazy things, you know. Maybe the pilot of a plane says, "I want to die now," and dives straight into a mountain. That would be an exception, but it could happen. Or a person might drive down a highway and suddenly turn his car into a bridge abutment with the idea of killing himself, but the car careens off and takes out five or six other people on the way. Understand that these rules are good. People need to be conditioned gradually into understanding that if they want to end their lives, they can go to a clinic where there is music and love and nurturing. They are allowed to end their lives in the most benevolent way, and the ceremony afterward takes into account their family. Everything is done in a loving way. That's coming. You might live to see it—no pun intended.

All right. I still have this consuming curiosity about these beings who came in with the

Creator. I was trying to get some words for the flavors they added to the Creator.

Really, the best way to describe it is by tones. Readers can listen to the G scale. The purpose of these books is not so much to inform as to stimulate, so we're giving readers something they can use. That's why the book is a manual, not a proclamation that says, "This is how it is; you're stuck with it and that's the end." We don't want to do that.

All right. You have had creations in the keys of C, B, E, A – all the others?

Using our analogy, yes. This tells you something, that . . .

Each one is focused on a particular note, then. Give me some of the others. I'm trying to get words to understand each one.

Hmm. There is the peaceful one; there is excitement, happiness – you can use words like that. Let your imagination be your guide. The foundational element of creation is *feelings*.

In people's near-death experiences, when they are revived and come back to life, their overwhelming point of agreement isn't just the light – it's the *feeling*. They all experience this wonderful feeling of unconditional love. So creation has to do with feelings first and the physical stuff or whatever afterward. Everything must be within a certain precise range of feelings for creation to take place. This includes all possible creations within that specific creation, with complete accountability as well as applications, responsibilities and so on. But the foundation for the creation is the feeling.

So could you say this creation is more oriented to feeling or more to love? There's more love in this particular creation than there might have been in some of the others – perhaps some of the others were more mental or more focused in other . . .

In my experience all creations are founded in love energy. This creation has had more of the finite involved in it. There is that conundrum again: living within the finite as a portion of the infinite. This had to be included. If you're going to have a foundation of love, you've got to have love in all of its faces – tough love, for example – because there's the infinite and the finite. One might experience one aspect of love, yet all aspects of love are available. This has to do with creator training. If you have, say, another planet and a more benevolent society, love is characterized by love, beauty, nurturing – you know, hugs and kisses; there won't be any tough love. There won't be anybody talking at you on the street who seems like a totally crazy person who claims to love you, and in your mind you know they're nuts.

So could you say that one fundamental attribute for a creator is the ability to love?

That is bedrock. You must be able to have absolute, total, unconditional love.

But different creations, then, structure purposes and intentions on top of that.

Yes, according to the agenda there.

Creator's Remembered History

How many creations does the Creator remember?

A moment. A hundred and twenty-three. He's older than I am. I kid him about that. I think that there are some "people" (creator people, you understand) that Creator brought along who remember up to twenty-three. I think twenty-three is the top number.

Can I ask which one that is? Shiva?

Shiva remembers nineteen creations.

Nineteen? Is the first creation you remember the one where you were generated, where you woke up?

It's more like you remember being less complicated, a simple tone, as it were. You remember being one thing, and then you became more. The best you can do is trace yourself back to the one thing you remember yourself being. The reason you can't trace yourself back before that is that whoever created you suddenly said, "I'm going to start something new," and they did something—boing! It starts with *one thing*, one main focus, and after that you accumulate more. In other words, you start out as less (if you don't mind my use of that term) and then become more.

So why didn't the Creator bring in people He remembered 100 or 110 creations ago? Why such new, young helpers here?

Understand that we're using the word "young" here euphemistically; it's a joke! I'm not going to treat your joke seriously. [Laughter.]

All right. I just wondered why He didn't have helpers who had . . .

The Creator had the help of the Council of Allowance, and Creator Itself is tethered to something more. So Creator knows that It's having help, but It doesn't know exactly what is going on. You might say that Creator is tethered to Its unconscious higher self, which It can't wait to be with—and for all Creator knows, there's a tether beyond that. The main thing is that Creator knows where It's going, because It can follow the tether. It is not unlike the tether all of you have to Creator.

Music's Impact

You know, music is astonishingly more important than people realize. Unfortunately, it really does not have its maximum impact unless you're experiencing the instruments themselves. You get a minor impact from a speaker moving the air, but it is nothing like having the clarinet in your room. Go to a symphony sometime—but only to hear music you like. If somebody else's expression, regardless of its motivation, is music you really like, and if the people are quiet while they are listening, that's good. Ideally, the musician or the orchestra is playing acoustical instruments. That's not difficult to find, especially for classical threesomes. But with new-age music there's always the hazard of synthesizers; although it is beautiful, it does not move the . . .

There's something in the bodies of the human, whether spiritual, etheric, astral, physical, whatever, that vibrates to that music – is that what you're saying?

Yes. Music is a language, the universal language; musicians will tell you that. You can play the same music on the clarinet in Portugal or Iceland or in downtown Moscow, and everybody will love it. It doesn't make any difference that you can't speak the language.

Does the piano do the same thing?

Yes, as long as it's acoustical. The instrument has to be acoustical to have its maximum effect; this does not mean that electronic instruments do not have an effect, but it's perhaps 10% of that of an acoustical instrument.

If the music, the words, the expression and the emotions being expressed are benevolent (which rock music is not known for), you would get something. If it's antilife, which some rock music is . . . interestingly enough, if antilife music is played on acoustical instruments and everybody is quiet (not yelling and screaming), it would probably have significantly more negative impact. In that sense we can be happy that rock music is electronic.

Whenever creation is discussed in a finite world (we're discussing infinity), there is necessarily a certain struggle to try and grasp it within the context of your illusory, finite mental process. So the idea of having a back and forth is really okay there, because a lot of people would perhaps have even more difficulty in framing questions. Some people might be able to ask the ultimate philosophical question.

What is the ultimate philosophical question?

The ultimate question is not really a question at all; it's a statement.

If you know the ultimate philosophical question, you're also a philosopher.

The interesting thing is that the ultimate philosophical question is a statement that can be perceived as a question.

The Forms of Creator's Companions

All right. So the Creator is dreaming, yet you can talk to Him and He knows what's going on in all hearts and souls. What forms did His companions have at this point? Were they just light, or did they actually have forms before the creation?

They were personalities, a persona, a portion of a larger personality.

Did they have shape and form? Or were they just light?

Basically they were whatever shape of light, sound and feeling they cared to express themselves as.

So if we looked at them, we would see flowing colors and hear musical notes and they would be a flowing shape?

Perhaps. It would depend on what their mood was. Different emotions, different feelings might be expressed as different forms or colors or sounds or even tactile sensations.

If there's an inner circle and then another circle around that and another and another,

are those other creations ongoing, and is that creator still having to work at that creation? Or does he take a coffee break?

Let's just say it's a labor of love. Would you want to leave a labor of love?

But at some point don't they close up shop? And once everything comes back into the creator, isn't the creation basically over?

Yes, unless you had not resolved the conundrums you started with, in which case you would have the option to re-create it in a different way. In the case of your Creator (who has decided to "leave it to Beaver," as it were), It has decided to allow you to take over. He'll just go on up the tether to the next level, and you'll put your spin on the ball with finite and infinite, should Creator not come to what She feels is a happy conclusion here.

The Number of Members in the Explorer Race

Will there be five billion souls doing this, or will we all somehow coalesce into one being?

There might very well be more than five billion, because we have to include *all* the souls who ever incarnated on your path, which is significantly more than five billion. When you were birthed out of Creator, someone might have had a life right then for whatever time, then everybody thundered on down the road of the Explorer Race. Maybe they didn't have any other lives, yet they had that life. So we're talking about a number significantly more than five billion. Yes, there is a process of coalescing toward the end; you originally went out and became the many, but you will come back and become the few.

So there would be a council or a committee running this creation, is that what you're saying? There'd be a hierarchy where somebody is in charge?

Well, it's hard to say. If you had to put a personality on it, you wouldn't be able to individualize the personality and come back and find that same personality each time. In one moment it might seem highly complex, involved and intellectually stimulating, but in the next moment you might think you were at the Mad Hatter's tea party.

Might you call it a very large symphony orchestra, where the members of the orchestra are playing solo?

Yes. Very good.

Has this been done before, where the Creator has left and turned it over to Its creation?

Not that I'm aware of. Although I am an infinite being, my perception is limited to what I have experienced.

So this might be an original idea?

As far as I know. Even tapping into the beyond, I'm not aware of this ever having occurred before. That's another reason the Council of Allowance let Creator do it. Here was something genuinely new, an attempt to create something of value. That's the main thing—is it of value? Creating something new is not good enough. It also has to be of value.

So He didn't have this idea before the creation. It was only in the middle of trying to resolve everything He said He would resolve that He got the idea.

Yes, but it's not actually an idea. It's an evolutionary process. If you as an individual start out as a child and learn more and more and become more and more throughout your life, it is the same thing for the Creator. As you become more and more, gradually your perception changes.

Hopefully.

Yes, hopefully.

We don't get to get into the personalities of the Creator's helpers, then? You're not too keen to do that.

I'm not too keen to do that at this time, but I'm not ruling it out.

Later.

Then I'll say good night.

Jesus, the Master Teacher

February 6, 1997

What I'd like to talk about is whether the being known as Jesus or Sananda was generated, or a friend of the Creator. I mean, he was either preexistent to the creation or part of the creation.

Compared to how he is portrayed, his actual personality was not preexistent. The big difference between the average person on the pathway, if you would, and Jesus when he was here, is that *Jesus remembered who he was*—and that was that. Jesus was the exception to the rule. He was seen as a teacher, a guide, someone to stimulate something and also to accelerate negativity, you might say. I know some people would throw up their arms and say, "Jesus was here to accelerate negativity?" Not intentionally. Jesus was here to offer the potential for accelerating the planet and everybody on it to a higher level 3000 years earlier than expected.

Plan B was that if it didn't work, it would heighten certain negative aspects of control, fear, domination and manipulation that had already greatly influenced history on Earth and, to a lesser degree, other planets. An accelerated curve could take place so that instead of having another million years or so of gradually increasing misery, you could have a couple of thousand years or so of heightened and extreme negativity. I know this sounds like Creator must be a bad guy, but Creator isn't a bad guy (or gal, for that matter). It's like, "Let's get it over with."

So plan B was, "We're going to get it over with; we're going to go through it fast. We're going to have only the most advanced souls come to Earth during this time so that they can deal with it on the soul level, and the souls themselves will not be bruised as a result of their experience" – even, for an example, a minor experience of being in a traffic jam when cars honk at you, you honk at other cars, that kind of aggravation.

Take my minor example: If Jesus hadn't come when he did, that kind of an aggravating traffic-jam experience probably wouldn't have developed for another hundred thousand years, because you are on a very slow curve. But since plan A didn't work and since a lot of what Jesus came here to do was turned into the antithesis of why he came here (he didn't want to become a god; he wanted to be a teacher), you were then put on track B. Track B is just about over now. The universe was reasonably happy about it because instead of Earth and its denizens going through a million years or so of gradually increasing negativity, everybody can get it over with in a couple thousand years. Citizens on other planets were pretty happy about it and were very supportive of you – and still are.

Okay. I'm interested in the being who called himself Jesus, who, according to* The Bird Tribe *by Ken Carey, has been here many, many times. He started the Iroquois Federation in the Americas. I'd like to trace his evolution. When did he start with the human race? How did he get involved with where we are now – tracing it through. Is there another name we can call him? Does he have several names? He's been called many, many names.

Let's not add another name to the hash; we might add it later.

He seems to be an extraordinary, important part of the Explorer Race, but I don't know how far back he goes.

He goes back a long way in terms of his spirit, yes. But in terms of his manifestation as a physical being, I'd say he probably doesn't go back much further than about 2500 years. There were other people – it wasn't as if he had to carry the ball himself and was all alone. Before his birth in his most famous life, he hung around for a while in other physical lives. He had a life here and there on Earth. (I'm going to confine this to his lives on Earth; I won't go into lives he had three million years ago on Arcturus, because they don't affect you.)

Is there some way we can trace back the thread of his incredible compassion for humanity? Where did it start?

From the Fount of Compassion and Love

If you want to trace it right back to the source, within Creator there is a fount of compassion and love, just like there is a fount of intellectual study. And Jesus' personality, the man you know as Jesus, was literally pulled right out of that fount and launched as that personality. He did not live only as a man; for example, he was a very well-known and respected medicine woman in both southern Africa and the northern United States.

Was he White Buffalo Calf Woman?

I don't want to give names, because the minute I give names people say, "Oh, that's *our* tribe!" So I won't say what tribe in either place, because I don't want people to put a claim on it. I will just say that he had different lives in which cultures were at a crossroads. Sometimes a culture reaches a crossroads at its beginning, at other times later in the course of their society where it could go this way or that way. Most often he/she would show up at that crossroads to guide people toward the more benevolent choice, so that society could mature more benevolently.

True wealth has to do with a society's benevolent directions. Beyond physical Earth we do not measure true growth by technological standards or numbers. It's by how much you become more responsible and benevolent in your applications.

Learning the Future Retroactively

He hasn't been around that long in that form, but he's a physical girl child right now on Earth. "He" is hidden well, so don't even try to find him; it can't be done.

Does he have a particular role or focus?

He's learning the future in retro fashion. You have about a hundred possible futures at the moment for the evolution of your society and culture, so she is retro-learning – from the future back to the present – so that she'll be a master of all possible futures of your society. By the time she comes of age and goes out amongst people to teach, she will know all possible futures of your society and, through inspiration, all possible futures of every person. That's why I say, don't bother trying to find her.

But even though that happens, Sananda is still doing what Sananda does? Or does that keep him from functioning on the higher levels?

No. Remember that Jesus' personality is fully integrated – he always remembers who he or she is, you understand? There is not that gift of ignorance, because there is no need to re-create one's existence. The lessons that Jesus has are not the same as the lessons that you might have as an individual soul. You might be working on spiritual or material mastery, but Jesus' lessons all have to do with teaching mastery.

Jesus is not on the same learning curve, so he does not need to forget who he is. He needs to remember not only everything he is, but also all other possibilities. He must be able to acquire all information, so the gift of ignorance would not be a gift for him.

Do you see that the lives given in Ken Carey's book are accurate?

Let's just say that it's one probable lineage. I can't say it isn't accurate and I can't say it is. If I said it's accurate, I necessarily exclude other explanations of Jesus' life. I won't do that. I can only say that it is one probable lineage, period. If you're looking for (and I know you are) absolute answers, that's the wrong place to look, because the whole existence of that entire being has to do with inclusivity, not exclusivity.

Remember, in teaching mastery you must include as much as possible so you can reach as many people as possible in as many ways as possible. So you might live a life as a dolphin to see that you can reach dolphins, or you might experience life temporarily as a mountain so you can understand the life of a planet.

It just seems that he has been such an integral part of the evolution of the Earth that I wondered how far back his connection with humanity went.

If we go back 2500 years, before that he was a portion of the fount of compassion and love of Creator, but integrated with Creator. I can't separate Creator from other life, but for the sake of our example, Jesus was not separated from Creator by anything before that point.

But he had an existence before that where he had been individual and had been learning and affecting – you know, out there operating?

How much does Creator have to learn? Creator creates reality, yes? A given reality? Yes. What does Creator have to learn about? It's not to learn so much, but to do, then see how it works. So you could say that Jesus was involved in it because he was literally involved in all that Creator is, which we cannot define strictly within these dimensional and linguistic terms.

Understand that as a portion of Creator, a person or an individual fragment of Creator's personality can be everywhere at once. If you were able to stand in front of Creator right now and say, "Creator, where are you?" Creator's going to say "I Am" – meaning everywhere. Creator tends to speak in short terminology, because what Creator has to say is necessarily unlimited, and when your language and context is unlimited, your language tends to be brief. It's only when we're creating limitations that we need more and more words. Interesting, isn't it? That's why your world has paradoxes. (By the way, that has to do with the law of paradoxes. You can write that down; that's one of those fundamentals. Zoosh never forgets that, but you can if you want.)

You see, right now we have Jesus in a physical life retro-studying, but he/she did not always do that. Jesus had lives, particularly one life as a medicine woman, in which she studied all of the probable pasts. When you are an unlimited being and you study a probable past, you don't look in a book and read about it. Why do that if you can go there? It's like having a time machine available to you at any moment.

How did he get to be an unlimited being when the rest of us are such limited beings?

Physical Life Demands Focus

Because it is Jesus' *job* to be a teacher. It is Jesus' evolutionary path to be the master teacher, so he must have all avenues of exploration available so he/she can teach all possible beings in any way they can learn. He has to be unlimited, but he must be able to be a physical being, not just a radiant being of light. He must be able to be focused.

The first law of physical life for a soul is, "Are you ready to focus?"

It is the nature of a soul in its lightbody to be completely dispersed. Remember, a soul comes out of Creator, and Creator is dispersed throughout all life and existence. Therefore a soul's nature is to be dispersed. So you must say to the soul, "Are you ready to focus?" The soul says, "What's that?" Then you say to the soul, "This means you must be in one place at one time, or at least within the context of one being." And the soul says, "You mean I can't be consciously everywhere at once?" And you say, "No, you have to be focused in one place. Are you ready to do that?" Most souls say "No way!" That's why most souls do not create themselves as physical beings!

The percentage of souls who have ever been a physical being in any form is so small as to be less than 1%. Most souls are not ready to give up *even momentarily* what is necessary. (A life of 100 or even 300 years might be momentary.) Within the thunder of time, most souls, as a portion of the Creator, are not willing to give up being everywhere at once to be in just one place even for a short time. Would *you* give that up? It's magnificent—would you think of giving it up for even a moment, much less a lifetime? It is very tough to give that up.

It is only for the responsible souls who (a) want to be more, therefore they come from a portion of the Creator that has to do with being more; or who (b) wish to *do* more for others, therefore they come from *that* portion of Creator. Maybe they're a portion of Creator's creative imagination; can you imagine telling your imagination, "You can only imagine about this, not about that"? It's not possible. In the portion of Creator's consciousness that is the fount of creative imagination there is not a single soul, a single light, a single essence of personality there that has ever incarnated as a physical being anywhere, anytime! They are not willing to give up that freedom and beauty and wonder, nor are they expected to, because that is the function of their light within Creator's light. Focusing as a physical being is, in the larger sense, rare!

Is Jesus part of a group studying to be master teachers, or is there just one?

The Jesus personality intends to be the master teacher for the Explorer Race. That's why he has had so many lives here on your planet. You might say, "Well, if Jesus was not focused as an individual before 2500 years ago, where was he, what was he doing?" We've already gone into that a little bit—he was exploring the past.

The Explorer Race was just beginning to come together as a reasonable mass of motivated souls about 2000 years ago, so there had been no need for Jesus to be doing anything else or for many master teachers to be available. When Jesus learns all he must learn to be a master teacher for all possible focused beings, then he/she will begin to train others. Again we have the idea of the students of Jesus—masculine, feminine, possibly the third sex—who will probably come in this lifetime with Jesus in the feminine body.

Is there a specific reason why Jesus is feminine now? Is it a return to the feminine energy? It's been masculine for such a long time.

Jesus and Her New Students/Master Teachers

It's really simple. The reason Jesus is a feminine being right now is because the feminine energy is rising on Earth and it's much easier to be a feminine being. She will begin to take on apostles, both masculine and feminine, and teach them how to become master teachers. These will be individuals who are human beings like yourselves, and Jesus will need to literally make a connection for them through him/herself. These students cannot possibly learn all within a given lifetime, but can touch into the teaching mastery that Jesus has through this feminine Jesus. So there'd be a link to Jesus. That is the primary means to get the master teaching apostles out there.

These will be new people, but they will necessarily have had lives initially as part of your civilization. They won't arrive here on golden chariots, as it were, to tell you how to live from an attitudinal point of view, because that doesn't work. They've got to come from your ranks so they at least know what you're up against. They can't talk down to you; they've got to talk to you eye to eye, as equals. If they can't do that, what good could they do? We don't want people to give away their power to the idea of a god.

A master teacher, by the way, does not arrive with the golden auric field clearly visible; he/she is basically quite innocuous-looking. A master teacher does not attempt to look grand in any way. It is the master teacher's desire to look plain and simple, because she/he wants to avoid *at all costs* the idea of being elevated to the status of a god!

The master teacher will tend to take on students who have no need to be perceived as anything more than an ordinary person. The students or new apostles will look just like ordinary folks. You'll be able to tell once they start teaching, because they will be magnificent teachers, but you won't be able to tell by looking at them.

Okay. So you're saying that it will probably be thirty or forty years before they're out teaching, by the time she grows up and begins teaching?

No, no. Remember, their initial teaching, familiarization, techniques . . . some things can be taught through inspiration, meaning that the apostle speaks to a group without necessarily knowing what is going to be said—inspiration, yes? Some things can be done that way and other things must be experiential. They'll probably have only two or three years of experiential training before they go out, because the link with Jesus will have been solidly established and they will be able to say and do whatever is needed. In about twenty-five years from now the apostles will be out amongst you.

Only this personality, then; this isn't something that Buddha or other teachers . . .

No, no. It isn't something that Buddha is doing. It's something that Buddha *could* do, but this is not Buddha's field of interest.

What's his interest?

Well, his field of interest remains in the establishment of a thought that links all thought everywhere, so that one could literally have a generated thought (not words) akin to a feeling that links you with all consciousness that can have similar feeling, so you're not linked to something negative or unpleasant. This means that you would be linked with mostly higher intelligent beings or more advanced spiritual beings on Earth. Ninety-nine percent of what you're linked with is off the Earth —extraterrestrial or higher-dimensional.

It is something that Buddha continues to be interested in and continues to support. It has significant value, especially in societies that are focused in the intellectual. Buddha is very active now in Andromeda and Arcturus and other places where thought is very important.

Will he be active on the Earth again?

I think in about a hundred years he'll choose to reincarnate, because by that time you'll be at another crossroads. You'll be deciding what to do with thought.

Will we still have the Andromeda brain?

No, you won't; you'll primarily have the instinctual brain, but you'll be at the point where you will be uncertain as to what to do with collective wisdom —books and so on —whether you should encrypt the information and put it on a disk somewhere, or whether you should pass it on by teaching people. At that point your society will be literally questioning, "Is there any point in teaching people anything if that flow of instantaneous wisdom is there?"

Buddha will come to help people to focus on saving the best to create a library of man, if you like, and also to know what to discard.

Why was the evolution of humanity going so slow? We have had many secret government – the Order, the Brotherhood of Light, the hierarchy of local masters in the astral and whatever. Why were the overlords forcing us to move so slowly?

Well, initially it was believed that no focused soul (meaning incarnated soul being) could deal with extremes —polarities, negative energy and so on —except in a very gradual way. A few billion years ago we all put our feet up around the pickle barrel, as it were, and believe me, no one ever considered that a soul could be born into a life of extreme negativity and survive more than a few weeks. But now people are doing it all the time —and that's part of the reason we need to change it. However, the original belief by everyone was that it wouldn't work.

Was this discussion before everybody started?

It was when we discussed the scenario of how it would work. No one believed it would be possible to gradually build up a tolerance at the soul level for extreme negativity and still have the will to live – nor did anybody wish to create that in the soul. So the original curve was gradual. Then with plan A and plan B working out the way they did in Jesus' lifetime in which he was so well-known, we all discovered that it was suddenly possible for people to do it. They could learn all they could about what value there is in negativity and also what is not good about it. (Negativity does have the ability to encourage growth by creating annoyance and so on.)

What Preceded the Present Crash Course

Now everyone has learned what can be done. It's almost as if the fount of impatience in the Creator said, "Let's hurry this thing up!" and plan B literally did that. It was a crash course whereby a tremendous amount of negative ramifications were explored in a very short amount of time. A couple thousand years is very fast. That's why only the most advanced souls have been allowed to come here in the past 2000 years. You could say, "But Zoosh! We've had some pretty rough characters here during that time!" Even those characters were advanced souls.

We've never talked about why the earlier guiding groups – the White Brotherhood and the Creators of the New Dawn and so on – were so slow. They were almost holding it back.

You have to understand that as the original creation was perceived, no one was in a position to say to the Creator, "It's too slow; what's the matter with you?" You're not going to do that to the being who has essentially sculpted you. Creator always wants you to be more than It is, but in those days that lesson was not widely recognized even by the most advanced beings. It was believed by the most advanced beings who were generated from Creator that it was their job to perpetuate and lift up and encourage all life *to be like Creator*. It was not understood at that time that it was their job to grow and become *more* than Creator. This is being understood now, but it wasn't understood then. So all curves of growth would have necessarily been slow, because even imagining that one might be more than one's Creator was not possible. Even Creator didn't have that concept, so how can you expect these beings to have had that concept? Where could they have gotten it? They came out of the Creator, and Creator didn't have it, so they didn't, either.

Where did it come from?

To share some of Creator's lessons with you, Creator has learned that the accumulation of experience It has had here creates an impetus for growth. We know how Creator's going to grow and how you, as a portion or as children of Creator, will replace It. But Creator did not know this in the beginning. Creator thought He'd come here and make this creation and see what it turned into. He'd take some unsolvable problems from others and work them out and see what would come of it, and when He was done (if He ever *would* be done – that was his point of view in the beginning) or when He was stopped, then He'd see what He had and decide what to do after that.

Creator didn't really envision anything past that – what He would do afterward. As a result, He learned through experience that the impetus of change (the Explorer Race is perhaps the zenith of change because you people have to change twenty times in a day, if not more), the overwhelming energy of change, necessarily affects the individual who generated the change. For example, a mother gives birth to a child, and regardless of what the mother has learned in her lifetime, it is very possible that her child will learn more and that the mother will be affected by that "more" the child learns.

Here we have the Creator giving birth to you, and the infinite number of these children that are focused portions of the Creator's original self all create a forward energy. At some point Creator realized, "I'm growing!" When you grow, the necessary element of growth is change, plus you want to know what's coming. Even Creator wanted to know what was coming. It wasn't a long time before Creator realized that something was coming that He didn't know about. So you're going to replace Creator, and Creator's going to go find out what's next.

This is a lesson for Creator. This is something Creator actually learned: that when you generate something, you focus it, give birth to it, then it necessarily develops its own energy and agenda, and the backwash from that affects *everyone and everything else* – including the one who created it! This is to say that the whole, or mass, becomes greater than the sum of its parts.

Going to the Next Level

Okay. How are those other 99% of great lights going to feel when this little 1% of focused beings takes over for the Creator? Where's that going to leave them?

They're going to be happy because they're with Creator and they're going up to the next level with Creator. They don't know what's coming, but they're looking forward to it: "Wow! What's coming?" They're not sorry. This is the 99% percent of Creator that is unfocused and still remains in a portion of Creator. If we take a portion of Creator, call it a soul, have it focus into being an individual personality that we can say has "stepped out" of Creator (even though it's totally linked and the rest of it is still with Creator), you might say they are souls, but you might also say it's the larger context of Creator. He's heading up there with them.

So 99% of the Creator is moving up and 1% is going to stay here and continue to – what, create?

Create! To do more than Creator did and to develop your own lessons. You will become aware of your lessons as you're going along.

Creator did not realize He had any lessons; He came here to see what He could do. It didn't occur to Creator for a moment that He actually had any lessons, but all beings have lessons. It is like the juices, the flavors of life. If you don't have spices in your food, it is bland indeed.

This brings up some interesting questions, because if 99% of the Creator is going to move up, those unfocused beings – aren't they galaxies and . . .

They are not unfocused beings; they are focused in Creator and part of the whole. They have simply not decided to focus individually, to focus as a personality on a string, as it were, from Creator. But they are focused as what they are.

But if they're focused as rays, as lights, as zodiacs, as galaxies – if the life force goes out of the creation, then what's going to happen to creation?

Remember, it is synchronous. You step up at the same moment Creator steps up. Creator doesn't cut out and then there's a two-week gap before you step in. It doesn't work that way.

This 1% goes out as souls and becomes all the rays . . .

Yes, the 1% that has been focused, that has experienced the responsibility and consequences, moves up to replace Creator. Creator moves up because He's been replaced and can move on and retire to a life of who-knows-what at the next stop.

Right. Will these focused souls who have been physical coalesce into one being, or are we going to be run by committee? How will decisions be made?

Any way you want.

20 Recent Events in Explorer Race History

February 13, 1996

All right. Greetings. The energy is a good connection tonight.

There are eight tones in a scale, so are there eight beings who came in with the Creator?

When we are working with infinity we can "bake it down" into subheadings. But if we can understand the significance of the angelic choir that people hear when they pass through the veils at the end of their cycle, then we can understand that the tone or vibration is truly not only of the infinite, but of the infinite beings who came in with Creator. This angelic choir that one hears and feels has been reported by those who have had near-death experiences, so you know about it.

The Thinning Veils and Melding Dimensions

I mention this now because you are at a time when the veils are very thin between the dimensions. There will be times when some of you will hear something very similar to the angelic choir – hear/feel it, because you will have feelings in your body. I don't want you to be frightened when you hear it, because some of you might identify it with the end of your cycle, and it isn't. It is instead a significant reminder of the infinite in all of you.

It's important to mention this tonight because these connections, these actual meldings between the dimensions, are becoming stronger and stronger. To some extent this is the result of the interaction of the photons and planet Earth and you as her outgrowth. It also has to do with the harmonizing (if I might use a pun) between your physical self, your spiritual self and all of your dimensional selves. These days your experience of spirituality, for those of you who meditate or whatever, is becoming more profound. Even those of you with religious fervor who are practicing a religion of some sort will feel the electricity, the momentum of the experience, so strongly that it will literally sweep you up into some full and total experiential moment of ecstasy.

I mention this again because so many unusual phenomena are going on that you will have the literal experience like those in the sixties had who were on mind-expanding drugs, or like anyone in the deepest levels of meditation. You will see color and feel it at the same time. You will be able to smell color and tones when you hear the angelic choir; there will be a fragrance with it that will be undeniable for some of you. It will smell like the ultimate greenhouse with the most wonderful flowers. These are all experiences that have been reported by people who have near-death experiences, yet what is a near-death experience but a motion beyond the veils? These experiential happenings are going on even now as we speak here in this room, but by the time you read this so many more people will be having them that I feel it is essential to mention it. That's how I'm going to answer that question.

Sometimes we can examine something so closely that it keeps repeating. Scientists used to think that atoms were the smallest division of matter. Then they discovered protons and other particles, then quarks. They would say each time, "This is the smallest there is." Now they've discovered something smaller than quarks. The greater and more powerful the instrument, the more they can revise what they have already learned. This is what's going on. We're at that level with you now where the more we scrutinize something, the more repetitive things tend to be, because they're microcosms of macrocosms.

But I want to look at the macrocosm. People need to understand creation, so let's focus on the 1% that is the Explorer Race. Can you outline how one probability could work out? It's not going to go to the end of the creation; we're going to become fourth-dimensional, fifth, sixth, seventh – at what point do we become an apprentice creator?

Expansion of the Explorer Race Scenario

You know, the whole idea of the Explorer Race has been significantly expanded just in the past few experiential years. It's hard to put it in terms of time when you're going to become that, but at the point when you no longer have any attachment to anything and you are also consistently in unconditional love with responsibility – that's when you will be the Creator. I need to put those criteria on it because evolving is more accessible or reliable than years.

On one hand, some souls will go out as human beings, becoming the Explorer Race, as indicated in *The Explorer Race*, but the whole Explorer

Race scenario has expanded since then. On the other hand, other souls will not be physically exploring the universe in spaceships and doing the Star Trek thing.

Because of significant spiritual advancement (believe it or not) in the people on your planet, you're going to have dimensional experience. Some people, some souls, will go directly to higher dimensions all over the galaxy and teach in various forms of lightbodies—what you would perceive at this point as a lightbody (in their dimension they feel substantive, though their body is light). This will go all the way to the ninth dimension. We'll have some beings who are . . . I don't want to say *ascending*, because the understanding of ascension now is something different. But some souls will focus themselves through Earth and, keeping that focus (not unlike shining a light through a lens), will go in their lightbodies directly to other cultures—Zeta Reticuli, Andromeda, Sagittarius, Virgo, wherever. They'll go to all these constellations and begin to teach the lessons learned by their lives here in a way in which those people can accept them.

What was originally a very narrow focus, in terms of an Explorer Race of people, has now expanded significantly. This is very clearly a demonstration of the awakening spiritual dynamic because of Earth experience and the souls here, but (as you've been told for many years now by many different beings) *as Earth goes, so goes the universe.* As things are improving on Earth, things are also improving in the universe, which means that as you come into your power here, so will individuals who have been living benign, perhaps benevolent lives elsewhere. The more universal aspect of coming into their power has to do with responsibility—to one's self, one's culture, one's fellow citizens—and that responsibility is to learn, so that growth can become an actual experience rather than something merely discussed.

You know, in many benevolent civilizations growth is simply a philosophy to be discussed, whereas on Earth now, growth is a day-to-day experience. It's not always fun, but it's right there. Civilizations where there is benevolence will be able to experience growth on a spiritual level integrated into their physical culture, not tainting the culture with external negativity but simply experiencing adventure—or even annoyance, should that be necessary to create an impetus for growth. The main thing is that the universe evolves toward greater responsibility, and the Explorer Race's job, regardless of the dimension or expression, is to encourage responsibility.

Can you briefly state why the scope of the Explorer Race has expanded so much? Originally it was just going out to bring vigor and enthusiasm and some lust for life to the rest of the cosmos. Was it always to replace the Creator, or is that a recent change?

No. It wasn't always to do that.

So how did all this change in the last few years?

What the Recent Acceleration of Particles Means

Here we go into particle physics. I'm going to try and dodge around that as lightly as possible so we don't get too erudite, but it came about largely as a result of the acceleration of the velocity of particles of existence, which I will simply refer to as the cosmic energy of life. If you were to put a more understandable word to it, I would call it *love,* because it is the energy of love that holds all life together. No two molecules come together to form any version of mass without love. They don't exactly hug and kiss each other, but there needs to be that cosmic energy.

This energy has picked up velocity. To some extent it has to do with the Photon Doctor [see chapter 26 of *Shining the Light II],* which is mainly performing the function of an inoculation. Let's create a model of the universe and put the Earth at the center. As the people who are incarnating on Earth become more spiritual *in spite* of their physical condition (no matter what's going on, if you're in pain, if you're confused, you're still becoming more spiritual), you necessarily move through the moments of transformation.

You've all had moments in which you've had some startling insight that has changed your entire perception and in the process changed your physical reality in some way. These inspirations have a physically transformative effect as well as a mental, emotional and spiritual one. See the universe spinning around the Earth in our model, the light growing brighter in the center where Earth is. Even though there is dark and confused energy there – the energy of that which has not been resolved – the light grows. And as it expands it seems to push the dark energy away.

If you could follow the energy, you would discover that it doesn't become less simply because there are quantifiably fewer measurable particles per square centimeter. As it disperses it becomes transformed. If we were creating a ring, and the ring of this dark energy forced outward by the light is becoming larger in its circumference, it is also transforming – because from the outside of the universe, from the world from which Creator has come (if we can imagine beings surrounding your universe), there is great light. The light contains not only all of the colors of your known spectrum, but colors you don't now see but can almost feel – for example, certain vibrations you can feel but cannot hear, tones above your actual hearing range, though you can have a slight feeling of them. These are colors beyond the visible light spectrum.

So as the energy of confusion is blinded by the light, a couple of things happen. Picture this dark circle radiating out; the light on the inside radiates out and affects all in its path in a good way. Some of you have driven through the woods at night and have come across a deer; when it sees your headlights, it freezes. This unresolved energy of

it (which is the first time that's *ever* happened!) and provided it with a goal in life, which is to be part of the growth curve. Before, negativity had no purpose, no reason for being; it was basically an outcast, hated and despised by everyone, or at least rejected, and it was always angry and upset, fulfilling everyone's prophecy. Now it has changed its image of itself, because it has found that it can be beneficial, not by learning how to be good (which is not what it is) but *by being the core of its existence.* And the core of its existence is in the growth curve. Thus 2% negativity is essential to growth.

It has discovered something very important (and I'm speaking of negativity here as a being, an existence), which is that the reason it got so extreme was that you, as beings working with it, *did not listen to the subtle message*. After all, once negativity has built up to the point of annoyance, it's time for you to act on it, *do* something, *change* the situation, *make it better*. If you don't, it just gets worse and worse and worse – and that is not negativity's fault.

I'm not berating you as a culture, as human beings; I'm just reminding you that things don't have to become extreme if they're taken care of when they are minor, when they are subtle. It is negativity's nature to be subtle, and negativity itself does not feel good about things when it is at its extreme. It wants to be subtle and wants you to respond to its subtlety so it can be part of a benevolent organism called growth.

You on Earth have supported, sustained and, most important, accelerated the learning curve through reintegrating benevolent negativity into the learning curve, thus allowing yourselves to grow very quickly. Because of what you are doing, you are also allowing the universe to grow very quickly. This is why the Explorer Race is able to accomplish much more at many different dimensions rather than the single-minded purpose once envisioned for it.

The Acceleration of Time from 1991

All of this has basically accelerated since you left with that President in 1991 [see chapter 2 in Shining the Light I*], because in 1991 the outcome was foreseen as the expansion out into the universe.*

That's right.

When you first talked about the Creator, we didn't even know the Creator had friends. I think it was December 1994 [chapter 35 in The Explorer Race*] when you first talked about the Explorer Race taking the place of the Creator. So in those three-plus years something triggered this expanded life and love that caused the learning curve. What were some of those triggers?*

Between 1991 and 1994 are how many years?

Three.

No. It *seems* like there is; there *ought* to be – that's mathematics. But remember, time has been speeding up. If we put time back the way it was for the sake of the explanation here, the actual amount of time between 1991 and 1994 was about fifteen years.

Really?

So you accomplished spiritually about as much in those three years as you would have accomplished in fifteen years.

But the fifteen years wouldn't have made any difference. Some other trigger point came in here; something else happened in that time that caused this accelerated expansion.

It's true enough. We're getting into a marginal territory that I can't talk about right now. This will have to wait for another chapter. The main thing that happened is that people in general, on the unconscious level and to some extent on the subconscious level, began to realize that true power has to do with who you are, how you acknowledge what you are, what you do with what you are and how you approve of what you do. This came about to some extent because of the reaction of the "me generation" to major cultural phenomena that provoked the question: how far can an individual go in expressing what he wants to express and not become wildly detrimental to the rest of the civilization?

It was really around 1989 when the momentum began to build to change from the individual to the collective. Even now some of the more significant movements, at least in your western culture, are concerned with collective things—what "we" are doing compared to what "I" am doing. A good example of the "we" movement is the fundamentalist Christian movement. Once upon a time individuals of that belief were few in number. Now they are many, and they are choosing to pursue their beliefs as a group rather than being so fragmented into different interpretations of the Bible that they can't acknowledge the value of each other's existence. Now many different groups have come together to form a "we."

That is one example, but it's an important one, especially in the timing of this book. By the time you read this, the fundamentalists will be moving past an egocentric identification of the "we," meaning a definition of Christian fundamentalism associated with words. It will become associated with feelings, which will be a bridge to bring it into direct association with love and unconditional love. That is coming.

I mention this because these are all steps, and while it seems to be moving like a snail, when you look back you can see it was really stepping right along.

Okay, but isn't there another factor here? The guys upstairs were really upset when the secret government took the President and put him on another time track. Wasn't there some exterior force applied here that caused a really explosive effect or a rapid acceleration?

No. If we had done anything, we would have necessarily changed the outcome. You could say, "Well, the sinister secret government changed the outcome!" As it always happens with them (which is why they're so cautious about doing anything anymore), every time they interfere with something, the backwash is always something more benevolent: The inadvertent backwash allowed the experiment to continue expanding.

In the long run their changing this presidential time line is really at the root of why the inner circle is breaking down now. The energy for the transformation simply built up. It's as if the runners are at the starting line and the starter delays firing his pistol. The tension keeps building up. So the tension built up after that change in time lines, and it built more light, more spiritual momentum. It was largely responsible for the breakdown and fragmentation of the inner circle of the secret government. They know this now.

We could see that this would happen. It was almost laughable, in the sense that the sinister secret government did it, and those of us who were looking at this from a distance were saying, "They really don't know what they're doing! They think they're making this big coup." If they hadn't done that, they would have been able to keep their inner circle intact for another ten years, but that was the end of it.

Keeping the thumb tighter on the energy, keeping them more tightly controlled.

That's right, because when you change things, especially when you're dealing with time, *everything* changes.

The Photon Doctor's Realignment of Earth's Particles

What about the activity of the Photon Belt? We're calling it the Photon Doctor, but that is our higher selves, isn't it, the other-dimensional parts of ourselves?

It is that, but as I've said before, the Photon Belt is basically here to assist Earth and other planets. But it necessarily affects you because you are made up physically of Earth. I won't define the Photon Belt as being made up of your higher selves; if I do that, I necessarily fall into the trap of saying that your higher selves exist apart from your soul that is impregnated within you. I don't want to fall into that trap.

From my point of view, the Photon Doctor is here for Mother Earth, and consequently you are also affected, being physical portions of Mother Earth. The effect of the Photon Doctor on Mother Earth—the actual realignment of the particles of Mother Earth's physical body and her physical extensions (which is you and all other life that grows out of her)—necessarily allows a greater expression of your physical particles. If we can say that you as a soul have an oversoul, that you have a light side and a dark side and all that, we could also say that all those particles are individual portions of some larger thing. Particles are Mother Earth's physical being. Because Mother Earth is a master teacher, she is impelled by her own needs to foster spiritual evolution in you. She will seek to have any improvement in her passed on to you through osmosis, you being a portion of herself.

As she expands, she moves beyond being a master teacher and into being a master of the creation of all forms, which includes the spiritual, material and the teaching levels. This helps you move into the creation of the cosmic love energy itself, since when she expands one level, you also expand one level in your physical particles. (This is a long way to go to answer —I could have just said yes.)

It's interesting that you're using the same term – master teacher – for Mother Earth as you do for Sananda, the being who was Jesus. Is there a relationship?

Not directly. But there is more than one master teacher. You have the spiritual master, then it expands out; you have the material master, it expands out; you have the teaching master, who takes all of this in. The next level for Mother Earth is the mastery of all forms, which allows *you* to become involved in the creation of the particles of cosmic energy, which we're defining as love for the purpose of our model here. So Mother Earth then moves beyond utilizing love to creating love as an energy. If she can do that, it gives you that potential too, whether or not you achieve it. This is another step toward achieving creator status.

You mentioned the world from which the Creator came. Did any of those at this late juncture step in and bring any wisdom, energy, help? Has there been any interceding?

No, of course not. You wouldn't come in at the final moment and say, "Oh well, it looks pretty good; let's join in." You wouldn't do that at that late date because you'd want to wait until your friend Creator says, "Okay! This is it!" But He's not going to say that, because He's already given birth to you and you're going to replace Him. So you're the one who will say to those other beings, "Okay, this is it!" —or more likely, you'll birth someone and head on up, too.

No, it doesn't work like that. Everybody's going to stand back at a distance and enjoy how it's coming along.

The New Potential: Explorer Race Birthing Its Replacement

We've brought in a whole new factor here: the Explorer Race has the potential to become the Creator and then birth another level of itself and move on up?

Yes! You're tethered to the Creator no matter what, so where the Creator goes, you will necessarily follow.

Ah! Maybe not at the end of this creation, maybe before then.

Absolutely, because the cycles of creation are timed not on the basis of some formula worked out in advance, but on the basis of an achieved *feeling,* meaning that at some point Creator feels, "Ah! This is it!" and moves up. The Creator can say, "This is it! I've got the perfect feeling now," and you move up, because *you've* got the feeling that Creator had when He was creator. Then at some point you're going to give birth because you'll want to head on up, too. We just keep on going, see, because as long as it works we're going to keep on going with it.

So it's not like someone from the world from which the Creator came is standing there saying, "I'm next, I'm a creator, I've got my plan."

Oh no, they can do whatever they want someplace else. But if they don't wish to do that, they can just stand back and observe – which is what they are doing because they're interested. Since they are not experiencing finite time, they can watch as long as they'd like. They weren't so interested in participating in the beginning, remember? They weren't interested at all. Now they're interested.

These are the ones who made the music thin because they wouldn't join in?

They weren't interested in participating. It didn't hold any fascination for them, but you have to understand that in the beginning Creator didn't have a definite idea of what She wanted to create. Now She not only has the idea, but the experience of it. Within the experience, then, the other beings can look back and say, "Oho! This is something *else!*"

The Creator had a really good idea: "I've got a way to reclaim all unresolved things and redeem negativity." He didn't have a really complete plan.

When Creator Began to Create

No, because if there had been a plan, Creator would necessarily have limited the experience. Creator had enough awareness of the fundamental art of creation by the time of the beginning of your universe to know that having a plan would inevitably limit what you would do. And since Creator was attempting to resolve conundrums – things that had been unresolved before in other universes . . .

At one point you said some of those unresolved issues went back a trillion trillion years, so we're talking about really ancient ones.

Ancient things, yes – negativity being one of the main ones. Creator did not limit the experience; He just allowed things to unfold, and since Creator is divine love itself, initially things unfolded along that line. It was right around the time when Creator birthed you to eventually replace Him that He had to summon portions of Himself that might be described as negativity, for there is no existence within the creation of this Creator that is not a portion of the Creator Itself.

But you said that there were core resonances that the Creator didn't have.

That's true. But the Creator was essentially creating beautiful, divine planets and beautiful, divine, wonderful beings, and everything was nicey-nicey, as I like to say.

And boring and stagnant.

And not growing at all on its own. It had no impetus on its own before the Creator realized, "Wait a minute, I have to do something here." That's when Creator considered the idea of allowing negativity to be itself. Creator had sprinkled little bits and pieces of it, but was shy about using it. Eventually, after Creator had birthed you all, He needed to take the limits off of negativity in order to go beyond what had been done with it. Creator didn't create Satan or anything like that, but He allowed negativity to achieve its ultimate purpose through the actions of individual beings. Of course, on most planets negativity has not been extreme – it's very rare. But you, all being reasonably experienced souls, were allowed to take it out a little bit further than had been allowed before.

In some sort of measurable time span, which I know you don't like, if you go from A to Z, what percentage of creation had elapsed before Creator birthed what would become the Explorer Race?

Oh, that's a good question; I can answer that. Not to sound too much like a body temperature, it was about 98.6235%.

And how would you describe that? What happened?

The creator beings who originally did not want to be involved in the Creator's creation communicated to Her and said, "Well, this is all very nice, but you know, you still haven't resolved anything. I mean, it's beautiful, but you haven't done anything new."

"You haven't done what you told us you would do."

That's right, "You haven't resolved the unresolvable; you haven't achieved your purpose as you stated it to us before you created all this." So Creator then realized that not only were they correct, but that He/She was unfulfilled as a result of the creation. It appreciated and loved everything and so on, but It did not feel like more than It had been when It started the whole thing – and the whole purpose on a personal level for Creator was to become more. So the mass of creation alone does not make Creator more; Creator *must expand as a being* to be more. It was at this point that Creator decided to let negativity loose to achieve its own level, and since it would not be attracted to places of unconditional love and bright light and all that, it filtered its way into races of beings on different planets around the universe at that time that had a greater sense of adventure, excitement – those kinds of emotions having to do largely with a good time.

When Negativity Was Loosed

Is there anything that can help us understand where, when, how? Are there any names, places, anything we've ever heard of? At what point did it come into something that we would recognize?

I suppose it came into something you would recognize as the planet that used to exist in the Pleiades star system, and that was way back.

At least it's in this little part of the . . .

Yes, your little corner of the universe, yes. It never really had a chance to run wild before then. When that planet got blown up, the lightbeings went in and rebuilt all the souls, but could only put them back the way they had been before.

Are there any star systems or constellations or anything we can see out there that came before the Pleiades?

That had an extreme experience of negativity? No.

The Pleiades came even before Orion or anything?

This particular planet in the Pleaides star system was before Orion, yes. Orion managed to utilize negativity in such a way that it was not destructive to the planet, because not all of the citizens on the planet were involved in it. There were just certain groups, warrior clans and so on, that utilized it, integrated it and actually applied it in a way (through the credo of the warrior clan) that allowed them to use it constructively, from their point of view. Most of the citizens were not involved in that at all.

But isn't that where the evil empire story played out?

It is where it played out, but you have to remember that even then, most of the citizens were not involved in any major drama. The citizens of the empire lived relatively benign lives and were not negative beings. As is so often the case with power struggles, there is one extreme group and another extreme group, and everybody else in the middle. The people in the middle are basically just going on with their lives.

It's like you've got a terrorist group, and the police and everybody else are just doing their thing.

They have to be extreme to counter the extreme on the other side. The planet didn't blow up, because the mass of the population and the life on the planet were living a benign existence.

Okay. So after the planet in the Pleaides, then what? Then negativity came to Earth?

No, then we had the extreme culture on Sirius, the planet that recently blew up.

But they didn't seem to resolve anything; they were totally under the control of some of their leaders.

They took it to its extreme. The souls are coming from that planet straight to third-dimensional Earth, where they're going to have some conscious recollection of having extreme negativity. They're going to come to 3.0-dimensional Earth where they're going to be immediately exposed to 50% benevolence and 50% negativity, which to them will be a tremendous improvement.

So what is the plan for them?

Their experience of extreme negativity does not go to waste. That 50/50 situation is going to be great for them initially, and eventually they will lower the experience of negativity. But for right now they need to have it so their souls can have continuity.

And what is the plan for them? If we move up, they're going to be down here in 3D and we'll have to deal with them.

You're not going to deal with them at all.

Why?

You're not on 3.0-dimensional Earth anymore.

But as the beings who read this book become the Creator, there must be some long-range plan for those people who have suffered so much.

There is no plan; if we fix upon a plan we limit the situation. The long-range plan, if there were one, would be to allow them to evolve in the direction that pleases them most. Water runs downhill; we don't push it uphill. You have to remember that one of the basic responsibilities of Creator is allowance. The minute we make plans, no matter how benevolent they seem to be, we're not allowing everything.

What about all the people who talk about the divine plan? I always say, "You make the path as you walk."

Well, to some extent that idea exists so that people will be reassured that this isn't simply some wild, chaotic, atomic nuclear reaction in which their destined life is dangling by a thread. It is a reassuring concept.

The Origin of Creator

February 20, 1996

Okay. Greetings.

Well, what shall we talk about tonight?

How many creations back do you remember the Creator?

You mean how far back I go with Him?

Yes.

Well, before this creation there was the world I told you about where I can only say that *creators of various tones live* -- that word is the closest I can find to differentiate between them.

Gathering Light

Before that there was a point at which, instead of expanding, the Creator actually went inward. If you were to look at Creator before that one, It would have appeared like a tube of white light with a little bulge in the middle, and you would have seen light going into it from all directions. You could say that at that time Creator was gathering light selectively to acquire tones He had not experienced until the moment of that gathering.

Within the tones of light are all possibilities. We know that light reflects, illuminates and sharpens, makes something clearer. You know that on the physical level, but the idea of putting light on an idea or a subject is also real. That existence lasted a very long time for Creator, almost as long as this creation has lasted. Creator did not at that time know She would be involved in the creation of an entire universe. But there was that feeling (because with Creator everything is done by feelings) of a need to acquire a very broad range and variety of light ideas, light expressions, light feelings, light emotions and so on, even in areas that Creator was not personally interested in. So Creator was gathering light as if She were walking through library stacks gathering wisdom on subjects She had no interest in whatsoever. Creator felt She would do something with it at some point that would allow Her to expand Her own point of view.

The true purpose for creation is to move beyond what you know to be so, to become more. So here is Creator gathering light, with the vague feeling (and it was truly vague at that point) that this would somehow cause Him to become more. But Creator would not create that universe until passing through the tones of other creators – that's where the feeling came. Now, before that, Creator was a portion . . .

Wait. When you say that She/He pulled in the light, who is the Creator absorbing this light or information from?

Creator moved through universes of creation, so there was actual physical motion when It was moving through the creations of others, exploring, in a sense – wandering, one might say.

It's allowable that any conscious being can go to anybody's creation?

Oh, absolutely, as long as there is permission, and there was. Creator was wandering and sampling, as one might at a smorgasbord, a little bit of this and a little bit of that. That was basically what Creator was doing, not acquiring too much even in areas where He was personally interested, just sampling a little bit of everything.

Tubes of Light: Individuality

Now, before that, Creator was in a portion . . . I need pencil and paper now. I'm going to draw a little sketch here of what Creator was in before that. [He's drawing.] Now, you need to picture this as made up of tubes of white light that bulge a little in the center because more activity goes on there. These tubes of white light grow outward from the center in this pattern.

This is what Creator was a portion of. It is very important, because it tells you that creators are created. To borrow a term from the ancient scripts, the center is like a diaphragm that opens and closes, but instead of opening laterally like a window shutter, it opens from the center in all directions. We're looking at it two-dimensionally here, but

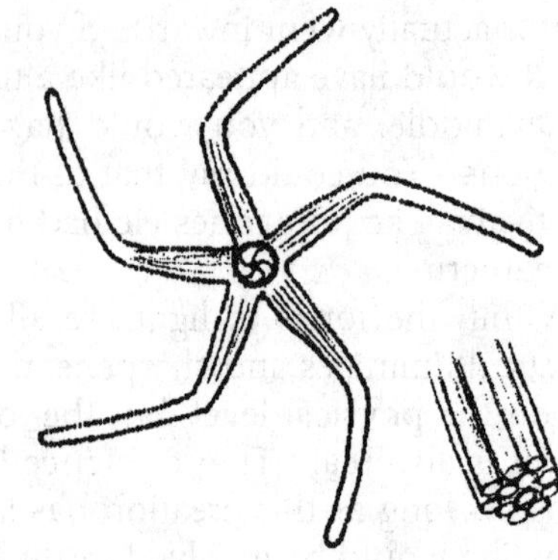

if you were to approach it from any direction, you would see something that suggests the infinite amount of arms of creatorship coming out there. Each one of them is made up of many different tubes of light. Any given arm might be pictured with perhaps seven or eight tubes for each arm of light, And if the artist wishes to be mystical, use eleven tubes to each arm. The bend in the arm is part of the arm, so one arm would be this [indicates on the drawing].

So where you have five, he can draw eleven.

There are five arms and eleven tubes per arm. Creators are born there. At one point your Creator was at the outer stretch (coming out from the center) of one of those arms. When that being you now know as Creator felt complete with the experience of total emergence and ready for individuality of a sort, then It broke off the end and floated out into . . . let's call it space, for lack of a better term. (By the way, this structure of Creator is moving, it's not just sitting there.) That was the life Creator was involved in before.

We're backtracking; we've gone through only three so far. Preceding the center point where things come from, we arrive at a point where Creator is not an individual, but one of the many. We're going through the center point back one more.

The Dark Tube of Potential

We have what I can describe as a long tunnel. If you were to look at it from a physical perspective, you would appear to be floating. If you looked with eyes that were very sharp indeed, you would see a dark tube within darkness. Now, this is a very important point: this darkness goes back quite a ways physically and emotionally, because the ball within the center [see drawing] is literally emotionality. The dark tube going back before that has to do with potential. So you can label the tube "potential" that extends back from the ball of emotion at the center of the thing with the five arms. Now we're in a long, dark tube in a dark space. The tube is separate from that space, yet in it. I think I can go back one more creation.

Stillness and Quiet

The next one back has to be described in terms of a mood. That tube stops at a place I can only describe in the following manner in your language: It's basically a feeling related to stillness—being still, quiet. I don't want to call it the void, because that suggests nothingness, and this isn't nothingness.

Because there's feeling . . .

Because there's a feeling of stillness and quiet. I'll see if I can go back further than that, but I'm not sure if I can. A moment. Yes, we can go back one more, but again it's a feeling. We're not following a physical trail at this point, but something you can find only through the use of your instinct. Instinct is really the very core of creation.

Discovery, Where Zoosh Joined Creator

Now, we follow back on this trail of instinct and we get to a spot I can only describe emotionally as discovery. From this point on I can only speak theoretically because I haven't been there, but I have heard others talk about it. Before discovery . . .

Wait. You were in these places?

No. It was at the point of discovery that I joined with Creator. Yes, I've been in all these places.

In one of those arms?

Inception: Awareness of Identity

I don't want to say that. It was at the point of discovery that I joined Creator, but before discovery the best I can describe it is inception, which means here the awareness of identity. It is at that point that Creator had the ability to have needs. When a child is born, it's aware of its identity and the first thing it knows is that it has needs. It was very much the same. You know, Creator can only re-create Itself, so Creator knows It has needs. At the point Creator recognized that, I was called. But by the time I got there, Creator had already gotten to the next level, so all I could hear about inception was from Creator. We've been together ever since, hand in hand off into the sunset.

And there's more to come!

Yes.

Is this in store for us?

Well, remember that since you have been created from Creator, in a sense it is *your* past, because you are a portion of Creator. You are asking, if Creator has spawned you to replace Him, are you going to go through that process, too? No, because Creator, your parent, has already done it. You will simply continue to carry on from that point. Just like when you were born, the child does not live the parents' or the grandparents' lives; everyone goes forward.

Yes, but you don't have their experiences to help.

No, that's not true. Remember, when you are beyond the boundaries of the gift of ignorance, meaning beyond the limits of your physical body, *you will remember all you know.* You will have the heritage of Creator, which is not limited as it is in physical life to photographs of Grandma and Grandpa and family stories. It is beyond the limits of physical memory and intelligence—it is unlimited family stories, unlimited pictures. You will have *all* that Creator has been. You won't have to go back and do it again; you don't start from scratch. The Creator you've known has already done it, so you go forward with what She has done, knowing all that She knows.

When you started out as a portion of Creator and before you were given the gift of ignorance, you knew all that Creator knew because you were a portion of It. (You still are, but for the purpose of our discussion we're separating you.) You had all of that knowledge. Now we've come out on this circuitous path, allowing you to pass through the veils that separate you from knowing who you are—which is the only way to reinvent something with the potential of reinventing something new. Creator felt that it was worth the risk.

By the time you get back, the idea is that you will have regained all the knowledge Creator had. Because you had to pass through the veil of ignorance, you will have re-created some things (Creator hopes) and resolved some things (which you have already done) that Creator told the other creator tones She would do when She came here to make this creation. You will be more equipped than Creator was when She created this creation in which you are living. The idea is that you would be better equipped for what *you* would create. You're going to take over Creator's creation, but you will also create on your own, hopefully knowing (from Creator's point of view) more than Creator knew when She started Her creation.

Okay. At some place out there is there just one of these structures of beings with arms and tubes who spawns creators?

Different Origins of Zoosh and Creator

I can only tell you of *this* one in reference to your question about Creator. However, if you're asking about beyond that point—no. From where I came, there were other organisms. Your Creator came from a place of structure; this shape or form [see drawing] is a symbol of structure—creation with purpose. Where I came from was more organic; creators came out of something that would look more amoebic to you. It did not have a clearly definable structure, such as this object we're showing you; it would be more of an undulating mass. Where I came from is not the outside of a cell, but something that has an amoebic look and would draw the attention of a microbiologist: "What's that?"

Was it a cell-division kind of thing?

No, I can't really call it that. I'm trying to get this in your language. This is a conceptual diorama, in that it had to do with creation as an adventure, as an emotional experience rather than the structures of creation. Your Creator, of course, had to be highly expert in the structures of creation (which is why your scientists are so interested in large and small structures), but where I came from had more to do with feeling. In my understanding, feeling precedes structure, because without the feeling, one does not have the motivation even to become aware of oneself as an identity apart from others.

Coming from the world of feeling, I had to adapt myself to the world of structure so that I could communicate with, support, sustain and otherwise accompany your Creator on His/Her/Its journey.

I can go back only as far as where I met your Creator. He's telling me now about the place where He was before that. I think His earlier existence was not one in which He was aware of Himself as someone. I think before that He was not individually aware of Himself, and that's probably why He Himself can chart His individual awareness of self only from the point just before I connected up with Him.

If you do not have an individual concept of yourself, your own personality, you're not going to conceive of yourself as who I might relate you to. I am who I am, and I relate to you as your personality, but if you are *more* than that, you are someone else. So you are not going to conceive of yourself as who you are now, but as this "more."

You realize you're not going to be able to hide out anymore once this book gets out, don't you?

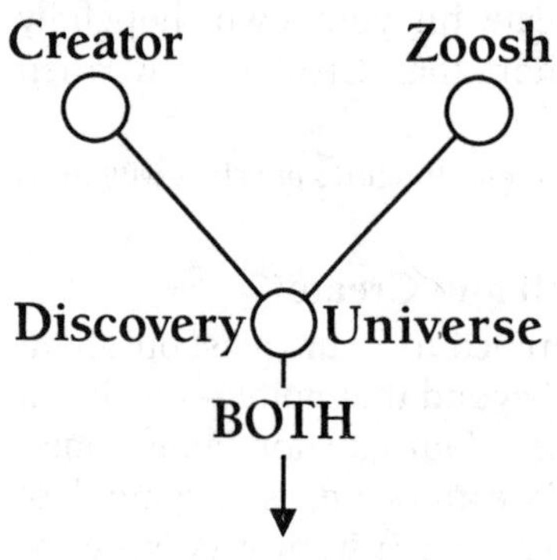

Well, I haven't been hiding very well so far, anyway.

Yes, you have.

[He's drawing.] Oh, I'd better mark this diagram. Now [he explains, using the diagram] . . . Creator . . . Zoosh . . . onward.

Okay. Is this a relationship, or are these just friends?

Companions.

Okay. You remember 9 creations and the Creator remembers 123, and you've been together for . . .

A long time!

Six or seven here, so why does he remember so many more than you do?

I believe She traveled a path where She was a part of something greater and gradually became something more focused. If we take a lens and shine a light through it, the light could be a huge quantity focused into a pinhole – but we're trying to go back to the original light. I think what happened is that to get an individual personality from a mass of light, Creator had to come through the lens. I was not aware of your Creator before She passed through the lens, but I think She was more before that time. She was a portion of something greater, just as every child comes from a mother who is larger, something more; and after the child passes out of the mother, it is its own individual. As near as I can tell, this is simply a microcosmic description of what your Creator was – a portion of a larger complex of beings. This, I believe, is where Creator first got the idea of birth as a microcosmic equivalent to achieving awareness of self. If you were all over creation in all that your Creator has created, not just on your world, some form of birth would be the means of awareness of one's individuality.

Are there creations where this doesn't happen?

Yes, that's right. I believe that we can extrapolate backward that Creator was at one time part of something greater that birthed Her, after which She became aware of Her individual self.

You said that 99% of Creator is undifferentiated and aware of everything. When I replied that this meant that only 1% is the Explorer Race is individualized, you said, "Why would anyone want to condense down and lose all of that?"

Creator could not spare too much, you know. Here He was using His entire self to create His universe, His expression of Himself, to re-create Himself in His multitude of facets. There was a limit to how much of Himself He could spare in that creation process. So Creator just pulled off what He could. You might say He pulled off His backup systems and spawned them out of the universe and said, "You gotta come back, because I can't do without you indefinitely."

Yet for a while He's going to do without, because He and the 99% are going on.

But remember, you're always going to be linked. So even though She'll be available to help you, Creator hopes that you will also be available to help Her if that is necessary where She is going. The link works both ways.

Right. But the part of the Creator that wanted to be individualized is the part that was birthed to become humanity and the Explorer Race – that part wanted to be more and go beyond, right?

No, because the Creator is a monument to individuality. It was *before* passing through that lens, *before* becoming aware of Itself as an individual, that Creator was motivated to be an individual. So we have to say that the very pulse of life within Creator is individuality and all of its possibilities.

Have the two of you been together in other creations where you've watched the creation of a universe?

No; this is the point at which we came together. Ruling out the obvious conceptual fact that everyone is a portion of everything else, in terms of our individual personalities, the diagram [see above] I gave you is the point at which we came together. I was with Creator on Its journey to *become* what It is now. It was not my job to go out and start creating a cosmos while I was helping this being.

Oh, I see. So you got to soak up part of everything that was, too!

The Necessity of Companionship

An underlying factor for all human beings that is also an underlying factor for all life as I know it is that *companionship of an equal is absolutely necessary*. That falls under the heading of *never forget that!* All human beings crave for the companionship of an equal, even if it's a friend – someone you can talk to who understands you, someone you don't have to explain everything to. Creator could not go on this voyage without having someone that It could communicate with. Did Creator want this to be a lonely voyage in which It is separate from all of Its creation, a lonely, miserable being? No, that's impossible! Where's the love there? You can't deny that to your Creator. No, Creator needed a companion, someone to talk things over with, to sit with around the pickle barrel and tell stories to. I was that person.

The other members of the Creator group did not go on this journey, then.

They came from other sources. Your Creator came from *this* source. It's possible that there were a few creator tones that might have come from a similar source as your Creator, but since I was not with them on their journey and since we have not discussed this at length, I cannot speak for them.

All right. So there are books and books and books and books here on what you saw on your journey!

Yes!

Can you very briefly give us a few little tidbits about how other creators create their creations?

Emotions, the Ecstatic Nuances of Feeling

Well, let me give you this. This is along your Creator's path. When Creator was going through Its beginning feelings of self-discovery and becoming aware of Itself, Creator tentatively experienced feelings that you now define as emotions. But these emotions also have sounds, music, tones – very clear tones for each feeling. In the very beginning when Creator had feelings, It was experiencing specifically clear tones for each one. There was no commingling of tones, no combination of feelings, no subtleties of feelings combined to make a universal chorus. Each feeling/tone was very clear and separate, and Creator would go in and feel each one.

The experience of selfhood was based primarily on feelings. If I had to put my finger on any one time when your Creator was the most excited, it was during the discovery of those feelings. When your Creator is excited, there are waves of color and sound and the feeling of . . . well, if you as an individual personality were exposed to that feeling, it would be joy beyond ecstasy to the hundredth power. I mention this to you because in your individual experience of your personality, you dole out to yourself these feelings in such small quantities that I want to encourage you to be more generous with yourself. You need to understand that joyous, wonderful feelings – love, happiness, anything like that – are the core of what Creator truly is. This is what makes Creator excited and happy! Creator doesn't want you to be serious all the time.

Although there are times when being serious is important, being happy, being love, being joy is Creator's true happiness, Creator's true reason for being.

I want you to give yourself more of that even if you have to practice emotions as an actor or an actress might in an acting workshop. Whatever you need to do – tickle each other, go to a funny movie and laugh! This is important, because you cannot be in full alignment with Creator's joy and ecstasy without experiencing these kinds of emotions *on a regular basis!* Understand that the more serious or unemotional you are, the farther you will be from Creator.

[Laughter.] The poor Vulcans.

Well, there is the seduction of the unemotional personality. For races of beings (or even individuals on your own planet) who have had a tumultuous, emotional life, quiet would be ecstasy. It *is* that for a time, but it is not the solution. It is only a temporary relief, then one must go on.

Oh, that's wonderful! How is what we call birth accomplished in some of these other creations? How do they replicate themselves? What are some of the things you saw?

As Creator moved on after the ecstasy of the discovery of individual emotions . . .

Was that in a particular creation, or was that just in the process of moving?

No, that's before Creator became the tube of light. That happened as Creator came down the dark tube and approached that diaphragm – the symbol I've drawn for you [black center of first drawing]. As you approach it from any direction [plane], it looks like that. I'm showing you one direction, but there are many arms coming out – there might be a hundred rather than the five shown. So as Creator approached that, He discovered these emotions.

And you too, right?

No, I think I was aware of those emotions before. That's how I was able to communicate. Creator was excited and happy and communicating with me and so on, and I was able to be grandfather: "Oh, wonderful! Let's have fun, let's go fishing" – the things that grandfathers do, but we're doing it with emotions.

Creator on a Pathway of Individuality

Now, as Creator was moving toward the diaphragm where individual creators come out in these tubes, Creator was beginning to experience the possibilities of mingling the emotions. At that point Creator was feeling individuality becoming more, and was beginning to ease out into one of the branches like the way a crystal grows. And as She moved out, She began to commingle some of the emotions and thus She discovered the subtleties and ramifications of the combinations. It was like happiness, joy and discovery all at the same time.

Creator discovered that as wonderful as these emotions are separately, certain combinations are even better. That was another great moment. Then as Creator moved out along the branch of creation, of individuality, It began to observe that there were other beings like Itself being birthed by this apparatus of creation. Creator began to understand, relating with these other beings, the great joy in companionship of one friend to another, one to one. It began to relate to other individuals as an individual, because before It went through the lens and discovered It was an individual, Creator was a mass of something with something else, not an individual. Now Creator began to discover having equal friends as individuals. So Creator continued to build the steps of discovering individuality; it's safe to say that Creator came here on a pathway of individuality.

But if He's in one of these tubes [see first diagram], are you in the one next to Him?

No, I'm with Him, I'm right there.

Oh, both of you are together?

Yes! What am I going to do, ride along in a caboose?

In the other tubes are there more than one being?

Every arm has, for the purpose of our discussion, eleven tubes. And when the twelfth comes out of the diaphragm, one breaks off, so it continues to be eleven.

Creator Joined by Zoosh

There are two of you in here.

I'm just riding along; I'm not interfering with Creator's evolution.

You don't need all this.

No, I don't need this, but I am the friend. Because I am not coming from Creator's point of origin, I am not affected by this creatorship experience that Creator is going through any more than I wish to be. To be a companion to Creator, I need to be a little distant from It so I can be a witness. If I were exactly what Creator is, what would Creator need me for? Creator could have spawned another version of Itself to relate to that would be an identical copy of Itself, but what good would that be? Creator needs to have someone who has a different point of view. Therefore I need to be a little removed, though I am with Creator. I am simply not affected any more than any witness might be by the events of a friend's life.

Okay, but this seems a little strange; none of the other eleven beings on the hundred arms have a friend with them, do they?

I haven't paid much attention to it, but going back now to look . . . yeah, I can see two or three companions with them. So I wasn't the only one. It's not a common thing, but it's there.

All right. So Creator broke off from that, and that is when He went on this great circuit of all the creations.

Creator still didn't know It was a creator at this point; It was aware of Its individuality floating out there, but It didn't realize It had any responsibility at this point. Responsibility for creation of any sort had not come into Creator's consciousness at this time.

But He's birthed as a creator, and you're along – yet you are a potential creator, too.

Yes, but this isn't about me. So here we were cruising along, and Creator saw/felt light – and within this light there were concepts, wisdom, ideals and so on. To the extent that Creator was interested, She gathered it without having to go where it was; She just pulled it like a magnet. As this material began to accumulate, Creator realized who She was – that She had the capacity to create some vast thing – because these points of light all came from someplace where something vast had been created. Creator realized, "Oh, *this* is who I am!" Creator at that point realized who She was.

What an awesome thing!

And in that moment Creator immediately experienced the feeling He had not felt before – *responsibility*. With that feeling, Creator then began to acquire points of light that He is not personally interested in, realizing He might need them at a later point in some level of creation that He doesn't even know about yet. Creator then started to acquire . . .

Seriously . . .

. . . seriously (or happily) beyond that which is particularly of interest, because responsibility requires that one do more than one is necessarily interested in.

You could say that first there was a happy jaunt, then it became something that had purpose.

That's right. Well, it had a purpose before then; that was Creator discovering Her individuality: "Oh boy, this is great!" Then Creator discovered that She was about something else: "Yes, this is a wonderful thing for me, *and* I can do something to create more here." At that point Creator didn't cease to be a happy, wonderful being; She simply added responsibility and the potential for consequences (though She had not experienced it yet).

As Creator tooled along acquiring stuff, It eventually came to the point where these other creators were in this place of creator tones. The creators there had already created something. This was where Creator went to get advice, as it were, still not knowing that It was getting ready to create something. So far it had been just a journey, an odyssey. At this point Creator "heard" the stories of other creations – what worked, what didn't and so on.

You can't be around all of these old creator cronies telling all their old "war stories" without wanting to go out and do it yourself, so this is where Creator discovered *motivation,* building all the while on individuality. We rejoin our story at the point when Creator asked for volunteers, "Who wants to go along on my journey?"

"Because I've got a good idea!" Okay. I know that numbers are meaningless, in a sense, but are there billions of creations out there – millions, trillions, quadrillions?

The number is higher than that.

Creator Coffee Shop

Okay. And the creators and the little coffee shop – we're talking humongous numbers, is that correct? Or aren't there that many gatherings between creations?

This is just where certain creators hang out.

Oh, it's just one . . .

The creator coffee shop, you mean, the tea room, crumpets? Well, I'd have to say that this isn't where they all are at any given time, but when we were there the number ranged anywhere from a thousand to several thousand, with individuals coming and going all the time, depending on what they were doing. It was never really less than 900.

Because when they're in the creation, they can't just leave.

That's right. They have a responsibility, they've got to be there. But when they're in between creations, they can stop off at the coffee shop where creators tell tales.

Do they create more than once?

They have that choice. There is infinite choice at this level. In the case of your Creator, you can birth someone out of you to take over your job. Creator didn't want to go back to the coffee shop; it didn't interest Him to go back and tell war stories. Creator wanted to go on. That makes complete sense when you allow for the fact that Creator's reason for being is based on individuality, so going back to the coffee shop and hanging out with the other kids, as it were, wasn't interesting at all. Creator wanted to pursue Its individuality and discover more, so It had no interest in backtracking.

These other creators had no concept or awareness that they could go higher?

They might have gone higher already, but they didn't talk about that to this creator. Understand that these creators had already experienced responsibility and consequences and all this other stuff, and here is this new kid. They don't want to influence Him; they want to do everything they can to support this individual's purpose.

Can you just look around? Are there other creators who have followed each other upstairs?

Just a minute. I can find about nine. Some of them have gone beyond that point to the next. One of them who has gone way, way, way, way, way . . .

Oh, that's exciting! Even at that level there's a ladder, right? It keeps going.

Yes, it keeps going.

That's the promise. That's going to keep us from being bored. All right!

You must always remember that boredom is a choice. Never forget that.

Is there anything else you'd like to say here? I want people who read this to feel how awesome, how much more there is than what they thought there was.

Well, they will get the feeling, because the intention is that people simply move beyond limits.

The Continuing Adventures of the Explorer Race

Humanity will become the Creator, responsible for maintaining the creation. Since our Creator already achieved His purpose, the Explorer Race will have to come up with a new reason for continuing the creation?

You have to remember, we're not going to rush this thing. It's not from the end of your physical life on this planet that you become the Creator.

How come? I'd like that.

You're going to become the Explorer Race first, do the intention there, and *eventually* . . .

Oh, all right. So we still have to finish the original concept.

Oh yes. You're not going to dump the original concept just because you discovered where you came from.

Ah! All right.

You're not going to do that, because the concept has value. You need to help out the rest of the universe. After that we get to say, "Okay, now that we have stirred the pot, we can take over." When you do take over, you'll not only have Creator's creation, but you'll also have Creator's creation that you personally influenced. Anybody in the business of communication can identify with this. It would be taking over somebody else's stuff, but you will have put your mark on it. Of course, you will have a different point of view, because you will be more than Creator—which is Creator's intent. You will have that much more to create in this universe than Creator has created. You will do that much more because of the more that you are, and when you complete that you will think about doing something similar to what Creator has done, but I can't say anything about that now.

Okay. This sounds like a hell of an adventure. First we get to go out and look at all the creation and see how it's doing, liven it up, then create some more – right?

Yes, it's a heaven of an adventure. [Laughter.]

Ah, it sounds like fun.

It certainly is. It's been fun for me, I'll tell you that.

Okay. I want you to make a commitment that after we finish the Explorer Race series we can get your autobiography.

Certainly.

If the Creator were to whisper in your ear, what would he want humanity to know at this point?

That you came for more, that it is your nature to be more and that this more is truly joy, not despair.

Joy and adventure and discovery and . . .

A Word to Media People and Teachers

That's right. We want to encourage that in the children; we don't want to encourage the children to believe that violence is the way – it isn't. We want to encourage Hollywood; we don't want to tell them they're bad, because they're into discovery as well as anybody else. We want to encourage them to encourage the children to discover their world, not fear it. Special-effects people, I set you upon that task, if you wish, as well as directors, producers and so on. Always encourage them to be more in a way that if you were to meet them on the street after they saw your movie, you'd *want* to meet them. How many people would want to meet the fans of the movie "Natural Born Killers"? Would *you* want to meet them on the street? You know you wouldn't. You've got a responsibility, and like it or not, people in the media, right now your responsibility happens to be greater than almost any other profession.

Because that's what people use as role models.

That's right. You don't like it, but there it is. I'm expecting the younger generation of filmmakers to live up to it, but all the other guys and gals get to do it, too. If you can't do it in your art, then find some other art you can do it in, okay? We're not going to judge you, but at the end of your life you will see and *feel* what you've done, and once you realize that, you will want to make that afterlife experience the most pleasant possible. Like it or not, you're very important folks right now.

But won't everything change as the influence of our tyrants here diminish?

Certainly, yes, but because it changes incrementally, people might not feel that change in the formative years of their lives. By the time children are ten or eleven, they have their little sack of motivations. We want to encourage them to explore their world, not get a gun so it's safe to explore it. We don't want to take you to task too much, but that's a reality. It's the same in philosophy and teaching and religion.

We want to encourage people to become more, to know what's right for them, to know what's right for others. We want to give them tools. The sign of a good teacher is not what the teacher has taught, but how the student can learn *because* of what the teacher has taught. We want to give them tools so the student can learn.

On Zoosh, Creator and the Explorer Race

February 27, 1996

All right, Zoosh speaking.

Welcome!

Thank you. You did want to know where I came from.

I did, I did! Will you share it?

The World of Substance

A little bit. I told you that when I was called to come to be with Creator on His journey toward His present creation, I was in a plasmic living mass of light and substance. At the moment of my leaving there and going to be with Creator, I was from the world of substance, which is all matter that can be felt in any way. That is part of the reason I went to join Creator, because He had never had an experience with anything substantive, anything that had any mass, as you understand it in physical terms. Creator didn't even have a means by which to react to physical stimulation. Since I was in a world where all forms of sensing were available and in use, I was able to guide him. That's where I was at that time.

The Place of Directions

Going back a little further, there was a point of creation that an artist might call an inverted tornado. A tornado is wide at the top and narrow at the bottom, but this is wide at the bottom and narrow at the top. I came from that. It is related to the focus of direction. With so many different life forms going so many different ways, one might wonder why any particular life form would go in a particular direction. What is the point? And how many directions are there?

I came from the infinity portal of focus, otherwise known as the Place of Directions. I had come through that so I could experience the infinite multiplicity of directions possible, given an infinite number times an infinite number. In this way I was able to adapt myself and become comfortable with variety to the nth degree. Such a degree of variety experienced in a linear fashion might be something that an individual can assimilate, but I wasn't experiencing it in a linear fashion; I was experiencing it all at once. When you come into that gate, or portal, the narrow portion starts you off with a certain focus of variety, but as you emerge through the gate (which gets wider) there is more and more varietal experience in any given moment. So you start out with a small amount and you work up to a large amount. That's where I came through.

The Void/the Continuum

Before that I experienced something that was absolutely essential. As you know, perhaps one of the most important points of life is that whatever purpose you are involved in, there needs to be a counterpoint *in order to promote growth.* It doesn't necessarily have to be in opposition, but there needs to be something balancing it on the other side of the scale. The place I had come from before that portal was what you would call space, or the Void—the mass that exists between the stars, but contains no particles of personalized energy. You might call it the canvas upon which anything can be painted.

I spent quite a bit of time there, because I needed to understand the root foundations of creation. Anything that is created is necessarily created upon something else. If all of what the Creator is was taken away from the space in which Creator's creation exists, there would be left that Void. The Void itself is not nothing; it is some *thing.* It is not the absence of anything, but a thing unto itself, which I discovered by hanging around there for a while. It was what I would call the continuum. If you go anywhere in any known universe that I've been in, you will find this substance. I would have to call it the continuum, because everything I've ever seen that has been created or anything that moves about in any form of creation is matted over this continuum substance. As near as I can tell, the continuum is the foundation upon which all forms of creation are laid, including its cornerstone of love. You can see why it was very essential for me to experience that.

Home: The Feeling of Love, Joy, Expectation

Before the continuum I can only say that my experience of myself was in a massive state. If I had to put a word on it, I'd say it was a feeling of unconditional love and joy. Yet it was a feeling of expectation, the expectation that something was going to happen that would be wonderful or worthwhile. And it would be light, you would say, but individual points of light within a larger mass of light. That is what I would have to call my home.

At each of these places did you discern or interact with other beings like yourself?

Yes, in the place I would call my home there are other focuses of personality not unlike myself. Sometimes it seemed like I was communicating to these aspects, sometimes it seemed like I was talking to myself and other times it seemed like I was talking to some fantastically creative portion of me that I'd forgotten about. There's a feeling of unity, yet there is that constant potential for a surprise or the discovery of something worthy.

The place you came from, the place you experienced, would you say it was a creation by some creator?

Yes, I would have to say so. Somebody had the idea, yes? Somebody did something. I can say that before this is some point in the distance from which I might have come. There might have been something that preceded what I'm calling my home, but this seems to be all I can speak of now.

You could have been generated someplace else and then became aware.

That is certainly possible. The main thing that's important, though, is the continuity of it all. You know, this experience that you have of feeling that things end – this whole experience of endings here . . .

Every day is like an ending!

Yes. It's almost unknown in most other places I've ever been to or visited, to say nothing of most other cultures. Even cultures on the Pleaides, Orion and Andromeda do not experience endings the same way. The feeling of continuum runs deep there. I believe that your familiarity with endings here is so strong because there need to be experiences that drive you to want to *maintain* creation rather than become externally or addictively involved with destruction or endings. The reason endings are so pervasive in your cultures all over this world is because you are being literally driven or herded, as it were, toward desiring to perpetuate life rather than destroy it. The idea of wanting to destroy life (aside from being a polarity) is, at its core, a desire to return to where you came from.

You say that you remembered 9 creations and the Creator 123, but it appears that you are conscious of more creations than He is.

Well, I can't really say that, because I'm not speaking directly for the Creator. I gave you the Creator's lineage from the point at which I joined Him, but if Creator were here to speak, I daresay He would be able to say more about where He'd been and what He'd done. It's possible that I preceded Creator, but I think it would be only that I preceded Creator *in the form in which we're discussing Him.* Certainly Creator existed before the moment I joined Him, but in another form.

Could the continuum be part of what we call the One Creator? Someplace there's one.

Well, it is *all* one. There is only the feeling of separation here.

Each place you've been was an individual creation by some conscious being.

Yes, but it's like a tree with leaves, you know; it's all connected. When you are not involved in a world such as you are in now, you are conscious of something that might be light-years away as a portion of yourself. You are actually conscious of that. It is only in this world where you will live a brief encapsulated life that you become unconscious of that distant connection to that distant thing, because you must remain focused on your day-to-day life in order to accomplish your scholastic studies.

In last week's "Star Trek: Voyager" a member of the Q continuum had done it all, seen it all, and was bored and wanted to die. Does anyone in this vast creation ever feel like that?

No, I haven't known of any. Understand that Q is a fictional character based on the idea of the mythic gods. They would not understand death either, because the underlying continuum would be so prominent. In this case the character Q wanted to have the experience of not being aware. The obvious way for Q to become unaware would be to have a physical life on Earth. Earth is the place where advances can be made through ignorance. Although things can be done through ignorance that are *not* advances, it's one of the few ways in which things can be re-created and new things discovered.

The 1% of Creator in Individual Consciousness (Its Backup Systems)

Let's look at that 1% of the Creator that has focused into that individual consciousness. They are all over the creation, right, not just on Earth?

Yes.

They were part of the Explorer Race, too?

Certainly.

Ah! So the Explorer Race is much bigger than just Earth!

Yes. Remember, you are not a member of the Explorer Race *only* while you have an incarnation here. You might have had an Earth incarnation a thousand years ago, but let's say that now you're living on Andromeda. Do you give up your membership card in the Explorer Race? No, you do not. Once the idea colors the soul, you carry it off to Andromeda. And because you are on a *continuing* life cycle, you will remember that you had become a member of the Explorer Race on Earth. So even though you are now a citizen of Andromeda, you maintain an interest in the Explorer Race and you influence others around you on other planets to be interested in the genetic experiment on Earth of which you are a result. If they are already interested, you do something about it—maybe take an expedition to Earth or observe Earth from a distance. You perpetuate your interest. If your life is about something else entirely, maybe you go to the library now and then to read up on what's the latest and continue with your life.

When the individualized spirit being creates a soul, it takes the first step toward focusing as an individuality. Is it only the spirits who have souls who eventually create a physical life?

All souls are basically portions of Creator's vast personality. If those souls had not emerged from Creator, they would still be individual portions within Creator's vast personality. They would have individuality. In a sense, Creator Itself would be a mass of focuses or points of individuality—individual personality. As Earth-focused teachers, the only reason we become aware of that portion of Creator's personality is that it is on Earth, for the moment apparently separated from the mass of Creator's total self. We know that personality is not really separated, but ruling that out for the moment here, I would say that in its essential form it is still pretty much the same as when it was encompassed within Creator before it was launched.

This personality that is essentially you—this immortal personality that we call the soul—does not change all that much when it emerges from the mass of Creator to become an individualized person through a string of lives. After it comes out of the Creator's mass and pursues a chain of lives, it remains basically the same person. Once it has evolved through a series of lives with different experiences and different applications of the personality to tasks or abilities and so on, the individual might accumulate various colorations of personality on the basis of experience. However, it is still essentially the same.

Some of those beings created souls and then those souls created physical bodies, but not all of them created souls unless they planned to have a physical extrusion, did they?

You said "those beings created souls."

Yes. A soul is a step down from the spirit that comes from the Creator.

No.

The spirit and then the soul and then a physical being.

No, there's no stepdown. The soul is the immortal personality—I'd rather call it that because the idea of soul has been abused through various religions.

Okay, my understanding is that the immortal personality extrudes something, creates something it inhabits and then that (the soul) creates the physical and inhabits it. Then at a point when the soul has gained enough experience, it is subsumed back into this immortal personality.

No. I recognize that different individuals, for the sake of creating a model of understanding, have created the terms *creator, oversoul* and *soul.* I've even done it from time to time myself when speaking to people who believe this way. But basically, the portion of the immortal personality that emerges to have a chain of lives apparently external from the Creator is very high; it makes choices that allow it to reflect different sides of itself. If one of those choices happens to be ignorance, then so be it. But in the larger sense, the oversoul is not really a separate being.

I don't know the word "oversoul."

Well, let's call it the immortal personality; the oversoul would be the spirit. In the larger sense the spirit is the accumulated experience, or chain of lives, of the immortal personality (soul) – the accumulated mass experience that a single life might not tap into because it would interfere with your creation in that life.

Okay. Let me see if I can ask this intelligently. Ninety-nine percent of the Creator has not gone out to create what you call a chain of lives or focused individually. There must be something that this 1% created that allowed it to contract into the third density. You have to really focus and focus to get . . .

Well, it's a stretch, but when you reach for a higher principle that creates a eureka! effect, you often have to stretch to do that also. It's the same stretch, the same function.

Okay. What motivates this 1%? Does it realize it's such a minority? On that high level it does realize it, because it is aware of everything?

Well, they realize that they have a combined task. At the level of awareness beyond where you are now (so you don't remember who you are), you remember what you're doing and that you're all doing this together for a specific reason – to give the Creator a coffee break, to replace the Creator.

Does this 1% come from a specific place in the Creator, such as His brain cells?

No.

Or is it from a fingertip or a toe . . . if the Creator had a body?

If the Creator had a body, it would be the portion of the Creator that technology would refer to as the backup systems – meaning not the systems online and functioning, but that portion He might keep around in case something didn't function.

I don't like that! [Laughter.]

It's the Creator's spare tire!

That's not romantic!

No, but it's amusing! Creator likes amusing. That's part of the reason He picked me, because I can be amusing.

So He's listening to our conversation.

Certainly.

Certainly! Oh, wow! So He chose part of Himself to focus down into this level? Or did the individual, curious parts of Itself choose this?

No, He focused it.

He chose that part of Himself.

That's right. Because Creator Itself, remember, had no doubt what life was about. Creator had assimilated all that It could create up to that point.

Okay, stop there. There was no physicality, so what was created up to that point, the 98.6% of creation?

No, no. There *was* physicality, there *was* life.

Not as we know it now. The experiment hadn't happened.

That's right. If we took you to some benign planet where there were beautiful cities and colored lights flying about that you could touch and taste, you would not be bored.

It's beautiful and wonderful, but it doesn't cause any creative change, right?

It doesn't grow. It doesn't have anything to do with growth. It does have to do with beauty, though, and love and all of the other wonderful emotions. Those portions of Creator that Creator spun off to grow, however, were given a task. At its core the task was, "Go out and find something new. Go out and do something new. Go out and become more than I am now." That's your job – to become more than Creator is now, so that when you return you will be more. You were as much as Creator when you were sent out because you were equal portions of Him. Any one portion of Creator would be as much as all portions.

But we're going as a hologram with the record of our experiences.

You're going back with the experience plus some new things you've done. These new things largely have to do with your challenges. For instance, hope – hope didn't exist before you started out. Of course, there wasn't the need for it, but since you were put into a condition where there *was a need for it, you invented it.*

But this is a whole new idea. I actually thought that everyone chose this; but were we sort of assigned to this process?

That makes it sound like you were sent to the blackboard after school to write something 500 times. It wasn't like that at all. The Creator said to Itself (musing as the Creator might, you know), "What can I do that's different? How can I create something different? I can spare this amount of myself, so I will send it out to attempt to become more than I am."

This was after He was reminded that creation was almost over and He still hadn't fulfilled His original plan?

That's right, because He hadn't been able to resolve all of the things He'd chosen to bite off and chew. If He was going to go on and do something else, He couldn't just abandon it. He'd have to leave something in His place, and it would have to be more or better than He was so that it might resolve the unresolvable. That's where the Explorer Race comes in.

As we graduate from Earth now, we have completed the experiment, but we have not completed the resolution of all of the dualities, polarities, negativities.

The Next Step

That's right. And you don't have to do it all; other people will do many of these things in various incarnations. Various members of the Explorer Race will go out, influencing others, and these others might be able to resolve their own problems. The main thing is, once so-called negativity has been discovered to have value in small portions of itself, then growth can return to the universe. It just has to return in a way that is tolerable. Even a half percent of negativity can create a growth curve. It causes no harm, yet stimulates growth. This is something you've done as the Explorer Race, one of the big things, because the universe has been stuck for a long time.

What about before this Creator? No creation has ever had negativity, or just this one?

No creation has ever successfully found a means by which to apply negativity in a way that had a benevolent effect. Taking something that is basically negative and making it benevolent without changing it is a parodoxical idea – unless you can demonstrate that it has merit, meaning you have to mix other things with it. When you find the right mix, you add that half percent or maybe 2% negativity tops, in any given situation, and you've created a growth curve *without harm.*

And the Explorer Race will take that energy as a catalyst, right?

Well, the people who have been here on Earth through this time of the experiment will incarnate in other places. Some people will go out on spaceships, yes, but the bulk of the people who will not be hanging around for the next 500 years or so will, in their next incarnations, take these ideas and experiences out all over the universe. When they incarnate and remember who they are, they will bring growth to the universe *in a safe way.*

You're saying that when we get out of here and incarnate on other planets, we don't have to forget? We're going to remember our experiences?

Oh, yes! Here you forget your experiences so that you have ignorance, but you don't have to have it elsewhere.

The Creator-Zoosh Connection

Okay. When the Creator called, why did you answer? Is there some connection in this larger being of which you're both a part?

In the larger sense, I am the portion of Creator that existed before She focused in a specific creation of anything, and I am the portion of Creator that exists after She is done with this specific something. In other words, when Creator knows what She's doing, that's one portion. But when Creator doesn't know, or when She is resting, that's what I am. Have you ever eaten those cookies with that delicious cream filling?

Sure, Oreos!

Let's consider that Creator is the cream filling and I am the cookies.

[Laughter.] Okay. That makes a lot of sense. You're out there becoming aware of who you are, and independently of you the Creator is becoming aware of who It is, then It calls and you come.

But you see, I can be in many places at many times doing many different things at the same time. I might not only be serving this Creator right now, but I might just as easily be serving another creator in another creation someplace else at the same moment.

Aha! Now we're getting someplace!

We're not talking about an individualized being focused in one place, regardless of how vast or small the place might be. I represent something that Creator does not have to be and that is not a portion of Creator's creative identity. In a sense, I represent the hammock that Creator needs to lie in in order to give you the general shape of Creator's body. If you were underneath the hammock and something suddenly plunked into it and you wondered what it was, you would know what it looked like by the general outline of the hammock. I am that which provides the outline. But I don't want to make it sound as if I am some great and powerful wizard of Oz, you know (apologies to Frank Baum).

Your Potentials

Our purpose in this book is to expand the perceptions of the people who read it into what's possible out there.

Well, it's important for you to understand that you are basically what I am, only you've forgotten who you are. And that creation and all of the creations going on everywhere is *so vast* that even if you use the full capacity of your brain (you're not doing it now because you're not using its auric field) you could not imagine even one one-thousandth of the vastness of the creation, because it is not for you to be distracted. There is so much going on that there is room for creators of every tone.

But 3.47 density won't hold the capacities of the auric field of the brain, right?

No.

How far up the dimensional scale do we have to go so we can start opening to that?

Let's say that we kept you at your physical dimensions as they are now. In order to get full use of the brain you'd have to be in the fifth dimension. But once you slide past about 3.50 on the dimensional scale, you will start to use more and more of the auric-field energies that can activate the brain cells that are not now giving messages to your instruments. You might not be able to interpret them right away, but you'd certainly say, "Whoa! What happened?"

For those who are doing neurological experimentation, the best way to do it is with live subjects, ideally someone spiritual or at least someone with training in meditation. Your subject can use an area 40 or 50 feet from the physical head as if it were a portion of the brain. The scientist might say, "I want you to focus on that spot on the wall." The subject would focus on it, but broaden the field of physical identity. This should get a significantly increased feedback from portions of the brain that are not now giving a readout.

This is close to the first of March 1996 and we're at 3.47. When will we be at 3.50?

Well, I will give you a landmark. It's when you start to have the feeling of reassurance, for no particular reason, once, twice, maybe three times a day or more, that everything's going to be all right, that it's all going to work out regardless of what's going on in the news or in your life. When you and others around you regularly start to have that feeling, you will know that you've made it past 3.50. That's a much better landmark than time, because it's something that people can identify with.

Thank you. Even though the planet gets there at a certain time, some people might get there in two days and some in two years. Is it that far apart, or will the whole planet get there at pretty much the same time?

The whole planet will get there pretty much at the same time, but not everybody will notice it right away.

It's called realizing it.

It's called honoring your feelings. This largely has to do with the rise in awareness of one's feminine energy. Feminine energy relates to the spiritual body (the inspirational self) and the feeling self (the emotional body).

Visitors to This Creation

There are three things I want to pursue here. In this creation now, which is encompassed by the Creator, you're not here, you're out there. There are eight friends of the Creator who came in and are running all over in this creation, but they weren't generated by it.

Correct.

Are there other beings in this creation from someplace else who are not fragments or extrusions of these friends of the Creator?

This is allowed, provided that they do not bring in anything foreign with them, meaning something beyond the expectations, needs or desires of this creation. They might have to park a certain amount of their personal baggage, as it were, before they come in, or they might not be able to bring all of themselves in. So yes, visiting is allowed.

How many visitors do we have?

Oh, it's not a number. Visitors are allowed, providing that they can follow these rules, which they do.

Is it a massive thing?

No. You have to understand that most creators are involved in a creation somewhere, so the only creators or beings at that level who might pass through here are those whose creations do not occupy much of their energy, who can send a portion of themselves here to see what your Creator's up to, or who are for some reason not involved directly in a creation that requires their personal responsibility. In other words, they've got time. This does not happen very often, but it does happen occasionally.

So it's only creators; no one else comes here?

I would say generally that creators, or beings involved in the creation process . . .

Are the ones who are sort of checking it out to see how it works?

Yes, because they have an interest in coming here. There are no baying dogs at the edge of this creation, but those who wish to come here are those with a personal stake or interest in the matter. If you go to a library, you would not pull down a book from one of the stacks that has no interest for you.

So some of these creators have a stake in the outcome of this creation. Would they be creators from creations who could not resolve some of the things they needed to resolve?

Yes. They might be creators who created something and at the zenith of their creation took on an unsolvable problem, like your Creator, or creators who had a problem that developed in the course of their creation that they could not change or solve no matter what they did *within the parameters* of their creation. It is usually these beings who come here to see what you've done.

For instance, there have been two or three different creators who have come to see what you have done with negativity and how you have literally rescued it. *Transforming* negativity to a positive experience does not allow negativity to be accepted, appreciated or used as something valuable – it just transforms it. That doesn't work. No, the whole purpose was to take it and use it in a way whereby it could be constructive *as itself.* You have done here what has not been done successfully before. So naturally creators will swing by and get a load of that.

You can look in on these previous creations, I assume. Would the previous creations of these creators all have similar working parts?

Maybe, maybe not. They might not be recognizable as a kind of creation you normally deal with at this dimension or perception. It might be something entirely different, and if it is, I can't even put it into your language.

But they have sentient life forms who are looking to experience something so that that creator can learn and grow, right?

Yes. There are beings who are doing something at the core of it. No matter what is accomplished, which usually is quite significant, occasionally there is a byproduct created for which there is no constructive application. In creation you can't have a junkyard, you know. You can't have a trash heap where a boulder sweeps portions of the continuum over it and then leave it that way. If there's anything created, a byproduct or otherwise, somebody somewhere has to say, "I'll take it on and see what I can do with it."

So some of this could go back ten, twenty, fifty creations?

Or more. There are no creations that are considered unequal to other creations. As a matter of fact, to a creator, the idea and applications of negativity are not a bad thing within the context that negativity is something that exists, that there is a good purpose for it, and that once we learn what it is, negativity won't have to walk the path of creating something destructive. The minute we find its good purpose, it can be fulfilled. The Creator does not look at negativity and say, "Bad!" Instead, Creator scratches His head, you might say, and says, "What do we do with this thing? There's got to be a reason for it, or it wouldn't exist." This suggests that your ultimate creator has passed on these conundrums to the other creators, as if to say, "Okay, try this. See what you can do with that now."

Part of the yearning of the Creator and you and His other friends to go up to the next level is to see more of this ultimate creation, right?

To experience more, yes. At the creator level, when you suddenly discover something is there like negativity that others have been unable to use, you have to wonder where it came from. Even if it is a byproduct of someone else's universe, where did the potentials that became this thing come from? *Somebody* must have done it. Is it something that just happens? No, since we don't believe in accidents. It came from somewhere, so let's go onward and upward (if there is an onward and upward) and see what we can find.

Sounds good. The other thing to explore here is feminine energy. The creator is a whole being; has the word "feminine" been applied to this receptive, intuitional nature, then beings formed to experience it? How does this work?

Negativity and Masculine Energy

You are in this loop of time and are experiencing something where the masculine energy – the energy that is separated from the feminine and that does things, creates things, but does not necessarily have connection to its heart – can create from thought and is allowed to do so. Thought in its own right does not have heart, so the feminine energy was allowed (for a short time only) to be separated from the masculine energy. For a while Creator, having studied negativity for as long as He did before taking it on here, thought that it might be a portion of the masculine energy exclusively. One way to find out would be to separate you into masculine and feminine focuses to see where negativity tends to be greater.

For a long time it looked as if negativity was more a factor in masculinity. Even though you would find negativity where there were exclusively feminine beings, such as a prison or a sisterhood, very often it could be traced back to something that happened when they were young and were influenced even indirectly by masculine energy. Creator was 99% sure that negativity as you understand it was an aspect of the polarized masculine identity that was primarily self-destructive because it was always seeking to return to the loving point of entrance into sentient consciousness, identified as physical birth. For a long time Creator has felt this.

Now, if we said that a man on this planet was 100% masculine energy we'd be lying, because there is no man on this planet who is 100% masculine energy – even a casual scientist can prove that – because of the genetic code.

At this time Creator is still 99.99% (plus several zeros and a few digits) of the belief that negative energy has its roots in 100% polarized masculine energy. However, you can't say the same thing of feminine energy, because feminine energy is, at its core, spirit (inspiration) and feelings. And inspiration, when it comes into any physical being, starts out on a spiritual level, then comes through the emotional body. When it comes in it starts out in the feminine and integrates there into the emotions or the feelings, so it's a feminine experience. *Then* it goes into the masculine – the physical and the mental.

What I 'm saying here is that Creator is still of the belief that 100% polarized masculine is the source of negativity, but that does not mean that 100% masculine *is* negativity; it means that it's the *source* of negativity – the source of *growth*. This tells you that 100% feminine can create, stimulate, support and sustain loving, benevolent environments, *but* those loving, benevolent environments might not, when they are 100% feminine, *have any growth factor at all*. However, the 100% polarized masculine (if it existed), because it's the source of negativity, has built-in growth, but in your culture that growth is dramatized in anything from the subtle and the annoying (which is the idealized growth curve) to extreme negativity. That is why no human being (and that's why the human is a hybrid) is allowed to be 100% of either of these polarities. You have a commingled, cojoined genetic subcode that has both the masculine and the feminine within it.

Desperation, the Impetus for Negativity

I'm not saying that the core of masculinity is negative, but that if you separate the mental and the physical *totally* from the spiritual (thus the heart and the love), that mental and physical will become desperate to find that love and spirit by any means necessary. It is that desperation that is the impetus for negativity, because that impetus has within it the knowledge that if we *are* this mental and this physical and this is all we can be, then the only possible way we can become more is *to grow!* And the only way we can grow is to produce something resulting from what we are that by its very nature will help us become more.

Having polarized masculinity and polarized femininity in previous aspects of creation was how negativity was born – by being separated out into the polarized masculine. This tells you that even when the polarized masculine was separated from spirit and from feeling or love, in its moment of desperation, after it had tried everything else, it did not curl up and die and return to Creator saying, "I give up!" Instead, out of its desperation it produced anger, frustration, outrage – something that stimulated it to grow. So desperation is the source of negativity – or, as sometimes put, being at your wit's end after you've tried everything else. Desperation separated from spirit and love is the source of negativity.

The scenario would be: the polarized masculine (the mental body and the physical body) desperately longs for the polarized feminine, which is able to maintain and perpetuate itself in benign harmony and which does not need to grow because it's so wonderful. The polarized masculine, crying, "Oh dear, oh dear, what can I do?" finally becomes desperate and creates negativity out of its need, because it is the polar opposite of harmony that is being created in the polarized feminine. And *since all needs must be fulfilled*, it commands the attention of those who can fulfill it. Thus it stimulates growth as it cries out to be rescued. Therefore Creator says, "We're going to rescue negativity; we're going to find out its purpose." You, as the Explorer Race, have discovered its purpose and learned that when used in very small portions it can stimulate growth.

Homeopathic Negativity

This is like the early days when many medical formulas were discovered. Often they would use a massive amount, then much later discover that a little bit works just as well, sometimes even better. That's the whole theory of homeopathy. But even in conventional allopathic medicine or pharmacology, as it's called, very often they've discovered the same thing. It is simply a microcosm of the macrocosmic experience of Creator and the creators before your Creator.

But our Creator is the first one who actually . . .

Found a use . . .

But did the other creators understand the cause before?

They understood the cause, but not always. When the polarized masculine and feminine were separated (which somebody else did in a previous creation) and negativity became this thing that was created out of desperation because there was noplace in that beautiful and harmonious creation for it, it was plucked out. The masculine was immediately depolarized, because it was seen by that creator that negativity could not exist within its creation.

Your Creator came along and said, "You've got this byproduct, don't you?" to the other creator. This other creator said yes, so Creator said, "I'll take it with me; I'll find a use for it." And this other creator said, "Thank you very much."

Let's back up. We're down here in the third dimension and we're in polarity, but how high does the male/female go – the fourth dimension, fifth dimension, sixth? Where there's no polarity, how does the male/female work?

They are combined, especially when there are lightbodies and there's no aging or death. The lightbody exists in time immemorial, so there is no need for polarized masculine and feminine. There might also be the androgynous or the third sex or various expressions of group personality. But there would not be a polarized feminine or a polarized masculine lightbeing unless that society chose that (which sometimes they do, but it's not typical).

Does the third sex mean the same thing as bisexual?

No. A long time ago, remember, we talked about the third sex [see chapter 13 in *The Explorer Race*, Book One]. Bisexuality is not unusual; on other planets it is actually fairly normal. I like to use the Pleiades as an example, because they do not have hangups about sex; the idea of bisexuality is common there.

No matter what form you've chosen, your preference can be either . . .

It can be either one, depending on who you're attracted to and the manners and mores of your society. If it's allowed, fine.

Whereas on this planet, because of the control and manipulation and guilt and shame . . .

And also because you're dealing with consequences. On the Pleiades they might not need to learn about consequences. Since consequences are a fact of creation, you must deal with various aspects in their multiplexity of consequences (building from one to another is multiplexity). An example of consequences is a society that, because of its manners and mores, does not believe in bisexuality, homosexuality or lesbianism. It happens that you are living in a time when that belief pattern is common, but it's certainly not common in other parts of the galaxy, to say nothing of the universe. Even now as we speak, there are more people evolving toward bisexuality and homosexuality and lesbianism out of necessity. (I talked before about the sperm count and the ova development dropping because there are too many people here; this is one of the benign ways to reduce the population.)

Creator's Friends

We know that there are other creators who came in with this Creator. Sometimes they have a creation going when they come in. Whenever they notice something evolving here that might affect their creation, they might step out, as it were, and go back and alter things, saying "Aha, let's try this!" But by and large they stick around.

If there is a tremendous variety of responses as a result of that change within their own creation, they might stay there for a while. But that has not happened more than two or three times.

So the friends of Creator who came in – you said they were notes of the octave, so there were eight. Is that a correct assumption?

Yes. We'll rule out flats and sharps.

Could you make a statement that affects them all? Have they all been creators?

They all have created universes.

And they brought a special tone or flavor.

I like the word flavor better, yes.

Okay. Coming into this creation means that they have all been physical, all the way down to the third density?

Not necessarily; they might have been involved in the creation of a universe that would not be perceived as physical.

I didn't ask the question right. In bringing this flavor to this creation, have they all come all the way down to the third dimension? You have stayed outside, but have the rest of them come in and influenced this creation?

Yes.

All the way down to third density?

Yes, even down to the second density.

Wow! Can we get on the tape names we have known these beings by, names that have been active in history?

I'd rather just give the tonal notes of the scale, especially in the given range of any being. One being might be able to make the tones of that scale in a comfortable range, allowing him to feel the individuality of these beings. You might make a tone [Zoosh sings a tone] and compare it to a different tone [he sings another tone]; it would be a little different. But that's a completely different subject.

Have they all been active in such a way that an educated human would recognize those names, if you chose to give them?

No. Even though you have the printed word available, you'd have to have the full range of knowledge of the creation within this creation, which is available only through crystals or dolphin memory on your planet.

Our memory has been wiped since Atlantis, since before . . .

That's right. If you had the full range of memory, you would recognize some names, but I don't think you'd recognize any names after the period when printed word became an accepted means of storing knowledge.

Which would be about 5000 years ago?

About that, maybe a little more.

Would we recognize them in mythology?

I don't think as individuals, no. Maybe as combined . . .

Vywamus said that there had been twenty-six avatars before Atlantis and Lemuria. Have these beings come as avatars?

No, because an avatar has a purpose, and they did not have a purpose more than simply helping. They were not driven by a sense of purpose to *do* something as an avatar is.

He wants so much to help that he'll come . . .

Somewhere where he can do whatever it is that he can do.

Well, usually it's to anchor a particular aspect of love.

Yes. They were not that; they did not have an agenda.

Okay. All avatars were generated by the Creator.

Yes. We're going to have to stop now.

Fundamentals of Applied 3D Creationism

April 1, 1996

We've talked about Creator's personality, describing where Creator is from, even a little bit of what Creator looks like, but we haven't really talked about the process of creation. Tonight I want to give you some basic fundamentals of creationism.

A fundamental error that your society has been making, perhaps as a result of technology and your normal means of creation, is starting with something small and getting bigger. For example, a wall starts with one stone, then stones are added and the wall gets larger and larger. Before you know it you've got the Great Wall of China.

Creationism isn't like that at all; it is just the opposite. In point of fact, we know that Creator is All That Is, at least in this particular part of the conscious apparatus of physical awareness. The ribbons of creation that join this particular intersection where you are at the moment are not only numerous, they're actually flexible.

Starting with Multiple Possibilities

Picture for a moment a mass of some creation – anything you wish – that contains multiple possibilities of whatever is created (for example, this house) and multiple possibilities of how it might appear, all this existing in some mass. Now, this mass exists in a higher dimension. As we focus it into roughly the 3.47 dimension (or, to keep it simple, the third dimension), it narrows down like a lens and focuses light. It becomes more and more specific until it gets to the point where we are right now, where there is this house. Thus creation starts out with a mass of things, then it is narrowed down to one thing – for example, this house.

It is absolutely essential to understand this about creationism; otherwise you will not be able to re-create your reality while you are living in it consciously. Now, you have said for some time, "I want to do it *now*." Well, here's your chance. You don't have to *work* to change what you are living in, but you do need a very good imagination. It helps to be visual, but you don't have to be. If you aren't, then try to imagine an abundance of a variety of some one thing and how you would feel emotionally in the presence of that abundant item. This will give you a way to directly relate to this particular level of creation.

Then if you wish to change your current creation in a way that all in this creation can accept comfortably, you will do certain things. I say "accept comfortably" because this is a world in which technology is a reality, where many things become something they do not wish to be. They are taken from their natural world to become something unnatural – for example, a piece of a mountain in the form of copper ore becomes a copper tube without its permission. (No reflection on the artist; just a fact in your technological world.)

How to Change Your Current World

In order to change your current world, it is necessary *not* to look at the house in a certain way. For instance, you wouldn't look at this room and say, "How am I going to change it?" You wouldn't move furniture around by hand; I'm talking about *changing it to another reality*. You might imagine the multiplicity of appearances this house might take by visualizing the variety of how it might look in other potentials of itself, then focusing within that variety on something bearing on *your* criteria. Perhaps something more natural, perhaps something more elegant, perhaps something more comfortable – whatever your criteria are.

You do not have to construct it yourself. The key to specific creationism on the applied level is to look at as much of the variety of the house as possible (at its higher-dimensional aspects), then imagine being in this house. Take your imagination up to that imagined house, then put yourself into the feeling that this house would create in you – for instance, basking in the total comfort of complete elegance.

If opulent luxury is your criterion, you would bring that feeling to you as much as possible. Then, while images are floating by, you might see water running naturally through the much larger house (where you'd want it, of course) and whatever else that is beautiful and luxurious you might wish to have – but you are not attached to it. It is not forming an idea based on will, only an image floating by amongst the cacophony of other images. The main thing is *the feeling*, and *seeing the quantum possibilities* (to the extent the human brain is capable).

Then stay with the feeling, stay with the variety as much as you can. (You have to do more than one thing at once here, but that's the nature of creationism.) Thus you are feeling how it would feel to live in opulent elegance while you are still surrounded by as many images as possible. Stay with that as long as you can until the images gradually fade and what is left within you are the actual emotions associated with living in opulent elegance. This will take five minutes tops. Then just get up and go on with your life.

The key to creationism is not just patience, but allowing the process to work beyond the parameters of your immediate conscious self. The foundation of creationism, then, is not so much mental in terms of the structure of the mind as it is imaginative, emotional and physical. It is narrow where you are dimensionally, and it is wider where the abundance of possibilities exists, to the extent you can imagine them in higher dimensions.

Quantum creation—quantum mastery, if you like—has to do with infinite numbers, possibilities *not extrapolated.* We're not trying to *find* something, but to draw from among the infinite possibilities those that meet the specific requirements of the *feelings/desires* and that also take into account the cooperation of all the physical bits involved—the copper, the steel, the wood, whatever. It is just possible that given a chance, natural stone, natural water or even ice would form its own beauty if given that option. (For example, many diseases or microorganisms in the body of a human being are not there from natural causes, but from the mainly unconscious meddling by mankind with the true nature of the expression of life.)

How to Discharge from the Body What Is Unnatural

One might discover cures for diseases within the plant world. But what about the herbal cures for some diseases where those plants have been almost totally eradicated? One way to approach this is to simply breathe in and out, ideally outside in the fresh air, knowing that the energy of all natural things around you are in that air. If there is anything unnatural inside your body (prompted by something not in the natural world), simply ask that whatever is unnatural be embraced by something that will be naturally created within your body and escorted out by natural means. That's it!

This will allow an antibody to be formed in your body by the cooperation from the natural world around you. Let's say you're sitting on a hill with plants around you (avoid sitting on them, perhaps sitting on a rock). You sit on the rock for a while, relaxing and comfortably breathing in and out for fifteen or twenty minutes, ideally with your eyes closed. Perhaps the sun is shining on you and you're warm, or it can be cool and moist or dry.

The important thing is that you give life around you the opportunity to cooperate with your physical body to form natural structures (antibodies in this case) around something that is unnatural, fitting it like a glove and escorting it out through the body's natural elimination process. A large tumor might take time, but you could nip a lot of things in the bud this way.

This is something I recommend. In creationism one first makes an effort to allow the natural processes or the natural life forms around you to assist you, because it is their nature to assist each other. A rock naturally helps the sky; the water naturally helps the stone. A plant grows on its own and is welcomed in the ground where it grows and it is watered and nurtured by the natural processes of the Earth. All life forms help each other naturally, and since your physical body has been created out of physical Earth substance (aside from your soul and the spark from the Creator), *it is natural for natural life forms to help you, too.* This is Creationism 101. It's basic, but you need to remember to do it, because it might seem almost *too* simple. You don't need to jump over the moon to do something so simple.

Creationism doesn't start with something small; it starts with something big, which is then reduced or focused into something more specific.

Ribbons of Creation: Possibilities for Creating

What were those ribbons of creation you were talking about?

Imagine this abundance of creation. I don't want you to try to imagine the All That Is; it's too chaotic for you. Just imagine the abundance of possibilities of how this house might look. Coming down from that creation are ribbons—bits, strings, cords, if you like. The cords are not inflated with true possibilities or potentials until you focus, but they are there nevertheless as possibilities. The house looks this way now; the house *could* look a different way, not necessarily by your control, but by some means you don't need to control.

It is in your nature to be creators. Even though you've been limited here on Earth in order to understand more things about the physical and material world, how it operates and all the other stuff we talk about, it's important for you to remember now who you are. Creationism starts with the all and is then focused into something specific.

The idea of changing your physical body is always attractive to even the most beautiful person, because that person still has the same critical eye as one who might more obviously feel a need for more beauty. The change of the physical body must be done in a way that is happy and enthusiastic. The emotions are absolutely essential, because if you are happy and thrilled to suddenly look more beautiful in your mind's eye, it will happen in a benevolent way. (Obviously you don't want to be hit by a truck and be laid up in the hospital for six months, losing fifty pounds through that suffering.)

The key is the feelings, the visualizations, and understanding that you can do the same thing with your body as we did with the house. You can imagine the varieties of how you might look, and then as you pick out the look that pleases you, go through the same steps we did with the house. You like it, you enjoy it, you admire yourself and so on, and you bring it into the point where there is that *feeling*.

A lot of this has been touched on before, such as creating the feeling of something in order to create, but here I'm adding the visual part and the idea of changing the creation structure. It's very important to understand that the old, slow, brick-by-brick method of creation might have sufficed before, but now you need to begin to play in larger fields.

Why is the 1% of the Creator that is training to become Creator intrinsically more creative than the rest of creation, which seems to have a job and does it?

You're not intrinsically more creative; you're simply more curious about the steps of creation. Remember, the portion that broke out of Creator was primarily the portion that was curious. You're more inclined to examine it. Obviously, the portion of Creator that would go along with anything would not be curious enough to examine anything; it would be assigned to do it and then just do it.

But we would say, "How can we make it better, how can we be different, how does this work, why does it work?"

That's right. You must have that quality if you're going to have beings functioning in a place of variety. You know, part of the reason this exercise works is because I'm offering you variety. If I weren't offering you variety in the mass of creation, you wouldn't find it interesting at all.

BOOK MARKET

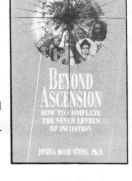

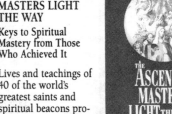

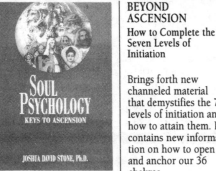

A reader's guide to the extraordinary books we publish, print and market for your enLightenment.

ORDER NOW!
1-800-450-0985
or Fax 1-800-393-7017
Or use order form at end

✦ THE ASCENSION BOOK SERIES by JOSHUA DAVID STONE, Ph.D.

THE COMPLETE ASCENSION MANUAL

How to Achieve Ascension in This Lifetime

A synthesis of the past and guidance for ascension. An extraordinary compendium of practical techniques and spiritual history. Compiled from research and channeled information.

$14.95 Softcover 297p ISBN 0-929385-55-1

SOUL PSYCHOLOGY

Keys to Ascension

Modern psychology deals exclusively with personality, ignoring the dimensions of spirit and soul. This book provides ground-breaking theories and techniques for healing and self-realization.

$14.95 Softcover 276p ISBN 0-929385-56-X

BEYOND ASCENSION

How to Complete the Seven Levels of Initiation

Brings forth new channeled material that demystifies the 7 levels of initiation and how to attain them. It contains new information on how to open and anchor our 36 chakras.

$14.95 Softcover 279p ISBN 0-929385-73-X

HIDDEN MYSTERIES

An Overview of History's Secrets from Mystery Schools to ET Contacts

Explores the unknown and suppressed aspects of Earth's past; reveals new information on the ET movement and secret teachings of the ancient Master schools.

$14.95 Softcover 333p ISBN 0-929385-57-8

THE ASCENDED MASTERS LIGHT THE WAY

Keys to Spiritual Mastery from Those Who Achieved It

Lives and teachings of 40 of the world's greatest saints and spiritual beacons provide a blueprint for total self-realization. Guidance from those who mastered the secrets in their lifetimes.

$14.95 Softcover 258p ISBN 0-929385-58-6

ASCENSION-ACTIVATION TAPES

How to anchor and open your 36 chakras and build your light quotient at a speed never dreamed possible. Scores of new ascension techniques and meditations directly from the galactic and universal core.

$12.00 each

ASCENSION-ACTIVATION TAPES:

S101 Ascension Activation Meditation
S102 Tree of Life Ascension Meditation
S103 Mt. Shasta Ascension Activation Meditation
S104 Kabbalistic Ascension Activation
S105 Complete Ascension Manual Meditation

Set of all 5 tapes $49.95

LIGHT TECHNOLOGY'S BOOKS OF LIGHT

✦ JOSÉ ARGÜELLES

ARCTURUS PROBE

A poetic depiction of how the Earth was seeded by beings from the Arcturus system of three million years ago. The author of *The Mayan Factor* tells how Atlantis and Elysium were played out on Mars and implanted into Earth consciousness. Illustrated.

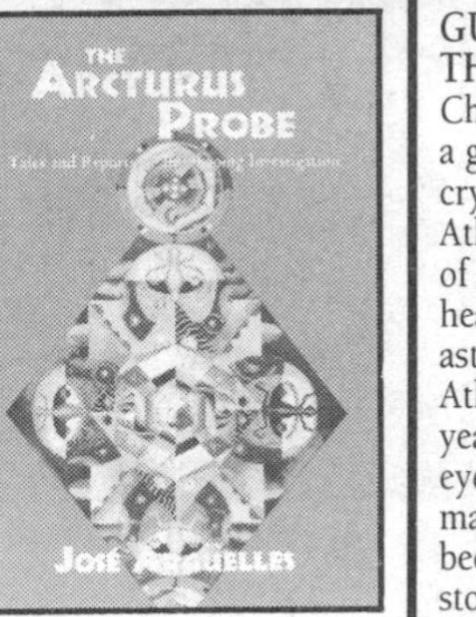

$14.95 Softcover

ISBN 0-929385-75-6

✦ TAMAR GEORGE

GUARDIANS OF THE FLAME

Channeled drama of a golden city in a crystal land tells of Atlantis, the source of humanity's ritual, healing, myth and astrology. Life in Atlantis over 12,000 years ago through the eyes of Persephone, a magician who can become invisible. A story you'll get lost in.

$14.95 Softcover

ISBN 0-929385-76-4

✦ JOHN McINTOSH

THE MILLENNIUM TABLETS

Twelve tablets containing 12 powerful secrets, yet only 2 opened. The Lightbearers and Wayshowers will pave the way, dispelling darkness and allowing the opening of the 10 remaining tablets to humanity, thus beginning the age of total freedom.

$14.95 Softcover

ISBN 0-929385-78-0

✦ BOOKS BY WES BATEMAN

KNOWLEDGE FROM THE STARS

A telepath with contact to ETs, Bateman has provided a wide spectrum of scientific information. A fascinating compilation of articles surveying the Federation, ETs, evolution and the trading houses, all part of the true history of the galaxy.

$11.95 Softcover 171p

ISBN 0-929385-39-X

DRAGONS AND CHARIOTS

An explanation of spacecraft, propulsion systems, gravity, the Dragon, manipulated Light and inter-stellar and intergalactic motherships by a renowned telepath who details specific technological information received from ETs.

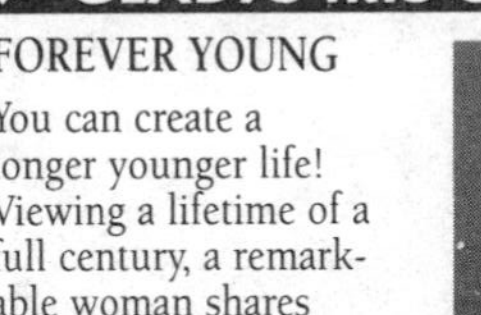

$9.95 Softcover 65p

ISBN 0-929385-45-4

✦ GLADYS IRIS CLARK

FOREVER YOUNG

You can create a longer younger life! Viewing a lifetime of a full century, a remarkable woman shares her secrets for longevity and rejuvenation. A manual for all ages. Explores tools for optimizing vitality, nutrition, skin care via Tibetan exercises, crystals, sex.

$9.95 Softcover 109p

ISBN 0-929385-53-5

LIGHT TECHNOLOGY'S BOOKS OF LIGHT

BOOK MARKET

A reader's guide to the extraordinary books we publish, print and market for your enLightenment.

ORDER NOW!
1-800-450-0985
or Fax 1-800-393-7017
Or use order form at end

BOOKS by HALLIE DEERING

LIGHT FROM THE ANGELS

Channeling the Angel Academy

Now those who cannot attend the Angel Academy in person can meet the Rose Angels who share their metaphysical wisdom and technology in this fascinating book.

$15.00 Softcover 230p ISBN 0-929385-72-1

DO-IT-YOURSELF POWER TOOLS

Assemble your own glass disks that holographically amplify energy to heal trauma, open the heart & mind, destroy negative thought forms, tune the base chakra and other powerful work. Build 10 angelic instruments worth $700.

$25.00 Softcover 96p ISBN 0-929385-63-2

ALOA STARR

PRISONERS OF EARTH

Psychic Possession and Its Release

Aloa Starr

The symptoms, causes and release techniques in a documented exploration by a practitioner. A fascinating study that demystifies possession.

$11.95 Softcover 179p ISBN 0-929385-37-3

RICHARD DANNELLEY

SEDONA POWER SPOT, VORTEX AND MEDICINE WHEEL GUIDE

Discover why this book is so popular! Six detailed maps, special meditations for each power spot, and a lot of heart. Richard Dannelley is a native of the Sedona area.

$11.00 Softcover 112p ISBN 0-9629453-2-3

NEW! SEDONA: BEYOND THE VORTEX

The Ultimate Journey to Your Personal Place of Power

An advanced guide to ascension, using vortex power, sacred geometry, and the Merkaba.

$12.00 Softcover 152p ISBN 0-9629453-7-4

AIELLO

THIS WORLD AND THE NEXT ONE

A handbook about your life before birth and your life after death, it explains the how and why of experiences with space people and dimensions. Man in his many forms is a "puppet on the stage of creation."

THIS WORLD AND THE NEXT ONE

(And There Is a Next One)

By "AIELLO"

(This Is About Your Life Before Birth and Your Life After Death)

A Handbook of How and Why!

$9.95 Softcover 213p ISBN 0-929385-44-6

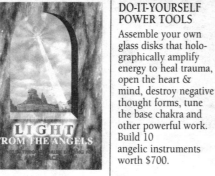

LIGHT TECHNOLOGY'S BOOKS OF LIGHT

✦ BOOKS by TOM DONGO

NEW!
MERGING DIMENSIONS
with Linda Bradshaw

The authors' personal experiences. 132 photographs of strange events, otherworldly beings, strange flying craft, unexplained light anomalies. *They're leaving physical evidence!*

$14.95 Softcover 200p — ISBN 0-9622748-4-4

UNSEEN BEINGS UNSEEN WORLDS

Venture into unknown realms with a leading researcher. Discover new information on how to communicate with nonphysical beings, aliens, ghosts, wee people and the Gray zone. Photos of ET activity and interaction with humans.

$9.95 Softcover 122p — ISBN 0-9622748-3-6

THE ALIEN TIDE
The Mysteries of Sedona II

UFO/ET events and paranormal activity in the Sedona area and U.S. are investigated by a leading researcher who cautions against fear of the alien presence. For all who seek new insights. Photos/illustrations.

$7.95 Softcover 128p — ISBN 0-9622748-1-X

THE QUEST
The Mysteries of Sedona III

Fascinating in-depth interviews with 26 who have answered the call to Sedona and speak of their spiritual experiences. Explores the mystique of the area and effect the quests have had on individual lives. Photos/illustrations.

$8.95 Softcover 144p — ISBN 0-9622748-2-8

✦ EILEEN NAUMAN, DHM

MEDICAL ASTROLOGY

The most comprehensive book ever on this rapidly growing science. Homeopath Eileen Nauman presents the Med-Scan Technique of relating astrology to health and nutrition. With case histories and guide to nutrition.

$29.95 339p — ISBN 0-9634662-4-0

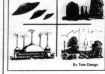

THE MYSTERIES OF SEDONA

An overview of the New Age Mecca that is Sedona, Arizona. Topics are the famous energy vortexes, UFOs, channeling, Lemuria, metaphysical and mystical experiences and area paranormal activity. Photos/illustrations.

$6.95 Softcover 84p — ISBN 0-9622748-0-1

BOOK MARKET

A reader's guide to the extraordinary books we publish, print and market for your enLightenment.

ORDER NOW!
1-800-450-0985
or Fax 1-800-393-7017
Or use order form at end

VISA · MasterCard

✦ BOOKS by ROYAL PRIEST RESEARCH

PRISM OF LYRA

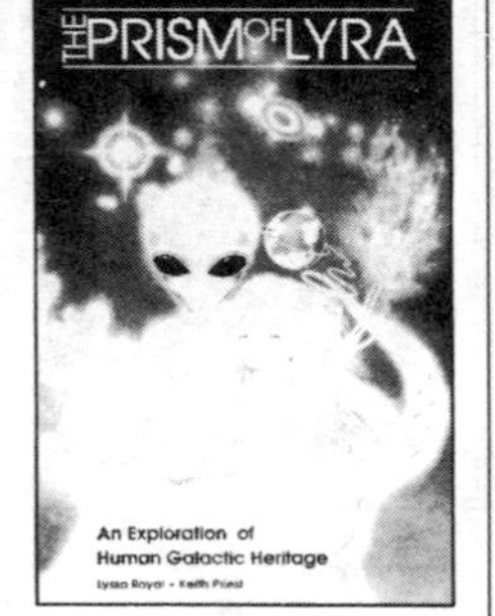

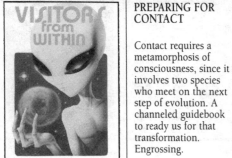

Traces the inception of the human race back to Lyra, where the original expansion of the duality was begun, to be finally integrated on earth. Fascinating channeled information with inspired pen and ink illustrations for each chapter.

$11.95 Softcover 112p ISBN 0-9631320-0-8

VISITORS FROM WITHIN

Explores the extra-terrestrial contact and abduction phenomenon in a unique and intriguing way. Narrative, precisely focused channeling & firsthand accounts.

$12.95 Softcover 171p ISBN 0-9631320-1-6

PREPARING FOR CONTACT

Contact requires a metamorphosis of consciousness, since it involves two species who meet on the next step of evolution. A channeled guidebook to ready us for that transformation. Engrossing.

$12.95 Softcover 188p ISBN 0-9631320-2-4

✦ AI GVHDI WAYA

SOUL RECOVERY & EXTRACTION

Soul recovery is about regaining the pieces of one's spirit that have been trapped, lost or stolen either by another person or through a traumatic incident that has occurred in one's life.

$9.95 Softcover 74p ISBN 0-9634662-3-2

✦ MARY FRAN KOPPA

MAYAN CALENDAR BIRTHDAY BOOK

An ephemeris and guide to the Mayan solar glyphs, tones and planets, corresponding to each day of the year. Your birthday glyph represents your soul's purpose for this lifetime, and your tone energy, which affects all that you do, influences your expression.

$12.95 Softcover 140p ISBN 1-889965-03-0

✦ MAURICE CHATELAIN

OUR COSMIC ANCESTORS

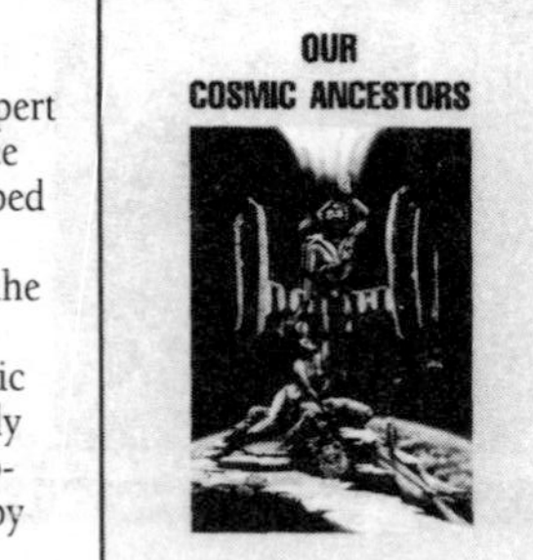

A former NASA expert documents evidence left in codes inscribed on ancient monuments pointing to the existence of an advanced prehistoric civilization regularly visited (and technologically assisted) by ETs.

$9.95 Softcover 216p ISBN 0-929686-00-4

LIGHT TECHNOLOGY'S BOOKS OF LIGHT

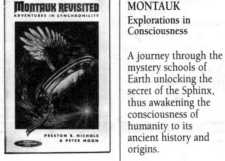

✦ BOOKS by PRESTON NICHOLS with PETER MOON

THE MONTAUK PROJECT
Experiments in Time

The truth about time that reads like science fiction! Secret research with invisibility experiments that culminated at Montauk, tapping the powers of creation and manipulating time itself. Exposé by the technical director.

$15.95 Softcover 156p ISBN 0-9631889-0-9

MONTAUK REVISITED
Adventures in Synchronicity

The sequel unmasks the occult forces that were behind the technology of Montauk and the incredible characters associated with it.

$19.95 Softcover 249p ISBN 0-9631889-1-7

PYRAMIDS OF MONTAUK
Explorations in Consciousness

A journey through the mystery schools of Earth unlocking the secret of the Sphinx, thus awakening the consciousness of humanity to its ancient history and origins.

$19.95 Softcover 249p ISBN 0-9631889-2-5

ENCOUNTER IN THE PLEIADES:
An Inside Look at UFOs

For the first time, the personal history of Preston Nichols is revealed, also amazing information the world has not yet heard. An unprecedented insight into the technology of flying saucers. Never has the complex subject of UFOs been explained in such simple language.

$19.95 Softcover ISBN 0-9631889-3-3

✦ PETE SANDERS JR.

ACCESS YOUR BRAIN'S JOY CENTER

An M.I.T.-trained scientist's discovery of how to self-trigger the brain's natural mood-elevation mechanisms as an alternative to alcohol, nicotine, drugs or overeating to cope with life's pressures and challenges.

$14.95 Softcover 214p ISBN 0-9641911-2-1

✦ MAILE

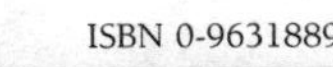

PRINCIPLES TO REMEMBER AND APPLY

A handbook for the heart and mind, it will spark and expand your remembrance. Explores space, time, relationships, health and includes beautiful meditations and affirmations. Lucid and penetrating.

$11.95 Softcover 114p ISBN 0-929385-59-4

BOOK MARKET

A reader's guide to the extraordinary books we publish, print and market for your enLightenment.

ORDER NOW!
1-800-450-0985
or Fax 1-800-393-7017
Or use order form at end

✦ MARIA VOSACEK

NEW!
A Dedication to the SOUL/SOLE GOOD OF HUMANITY

To open awareness, the author shares information drawn from looking beyond the doorway into the Light. She explores dreams, UFOs, crystals, relationships and ascension.

$9.95 Softcover 288p ISBN 0-9640683-9-7

✦ RAYMOND MARDYKS

SEDONA STARSEED

There is a boundary between the dimensions humans experience as reality and the beyond. Voices from beyond the veil revealed a series of messages. The stars and constellations will guide you to the place inside where infinite possibilities exist.

$14.95 Softcover 146p ISBN 0-9644180-0-2

✦ NANCY ANNE CLARK, Ph.D.

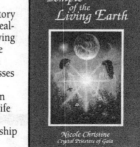

EARTH IN ASCENSION

About the past, present and future of planet Earth and the role humans will play in her progress. Nothing can stop Earth's incredible journey into the unknown. You who asked to participate in the birthing of Gaia into the fifth dimension were chosen!

$14.95 Softcover 136p ISBN 0-9648307-6-0

✦ ELLWOOD NORQUIST

WE ARE ONE:
A Challenge to Traditional Christianity

Is there a more fulfilling way to deal with Christianity than by perceiving humankind as sinful, separate and in need of salvation? Humanity is divine, one with its Creator, and already saved.

$14.95 Softcover 169p ISBN 0-9646995-2-4

✦ RICH WORK & ANN GROTH

AWAKEN TO THE HEALER WITHIN

An empowerment of the soul, an awakening within and a releasing of bonds and emotions that have held us tethered to physical and emotional disharmonies. A tool to open the awareness of your healing ability.

$16.50 Softcover 330p ISBN 0-9648002-0-9

✦ NICOLE CHRISTINE

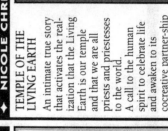

TEMPLE OF THE LIVING EARTH

An intimate true story that activates the realization that the Living Earth is our temple and that we are all priests and priestesses to the world.
A call to the human spirit to celebrate life and awaken to its cocreative partner-ship with Earth.

$16.00 Softcover 150p ISBN 0-9647306-0-X

LIGHT TECHNOLOGY'S BOOKS OF LIGHT

✦ PHYLLIS V. SCHLEMMER

THE ONLY PLANET OF CHOICE

Council of Nine

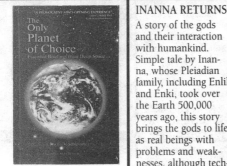

One of the most significant books of our time, updated. About free will and the power of Earth's inhabitants to create a harmonious world. Covers ET civilizations, the nature of the Source of the universe, humanity's ancient history etc.

$14.95 Softcover 342p ISBN 1-85860-023-5

✦ V.S. FERGUSON

INANNA RETURNS

A story of the gods and their interaction with humankind. Simple tale by Inanna, whose Pleiadian family, including Enlil and Enki, took over the Earth 500,000 years ago, this story brings the gods to life as real beings with problems and weaknesses, although technologically superior.

$14.00 Softcover 274p

✦ JOY S. GILBERT

IT'S TIME TO REMEMBER

A riveting story of one woman's awakening to alien beings. Joy recounts her initial sightings, dreams and ongoing interaction with nonhuman beings she calls her friends. After her terror, she sees her experiences as transformative and joyful.

$19.95 Hardcover 186p ISBN 0-9645941-4-5

✦ TAKA (audio tape)

MAGICAL SEDONA through the DIDGERIDOO

Blends the sound of the Didgeridoo into the sacred landscape of Sedona, describing different areas, joining with characteristic animal and nature sounds. Takes us into a state between conscious and unconscious; the joy of nature sounds and human voice touches deeper levels in our beings to reconnect with all creation. Each selection is about 10 minutes long and starts and ends at sunrise.

1. *Cathedral Rock (Nurturing, beauty)* 2. *Bell Rock (Energizing, guardian)* 3. *Dry Beaver Creek (Inspiring, creative)* 4. *Airport (Beyond limitation)* 5. *Canyons (Sheltering, motherly)* 6. *Sedona (Love)*

T101 $12.00

✦ LORIN POWELL

LIFE IS THE FATHER WITHIN

How the universe, galaxies, stars and planets were created and the evolution of beings. Channeled information of a holy spirit, as answers to 185 questions, asked about a tremendous array of extremely interesting subject matter. A very in-depth explanation of science and astronomy, soon to be validated.

$19.75 Softcover

✦ SAL RACHELLE

LIFE ON THE CUTTING EDGE

The most significant questions of our time require a cosmic view of reality. From the evolution of consciousness, dimensions and ETs to the New World Order, this is a no-nonsense book from behind and beyond the scenes. A must-read!

$14.95 Softcover 336p ISBN 0-9640535-0-0

BOOK MARKET

A reader's guide to the extraordinary books we publish, print and market for your enLightenment.

ORDER NOW!
1-800-450-0985
or Fax 1-800-393-7017
Or use order form at end

VISA MasterCard

✦ NANCY FALLON

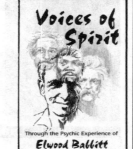

ACUPRESSURE FOR THE SOUL

A revolutionary vision of emotions as sources of power, rocket fuel for fulfilling our purpose. A formula for awakening transformation with 12 beautiful illustrations.

$11.95 Softcover 150p ISBN 0-929385-49-7

✦ BOOKS BY RUTH RYDEN

THE GOLDEN PATH

"Book of Lessons" by the master teachers explaining the process of channeling. Akashic Records, karma, opening the third eye, the ego and the meaning of Bible stories. It is a master class for opening your personal pathway.

$11.95 Softcover 200p ISBN 0-929385-43-8

LIVING THE GOLDEN PATH

Practical Soul-utions to Today's Problems

Guidance that can be used in the real world to solve dilemmas, to strengthen inner resolves and see the Light at the end of the road. Covers the difficult issues of addictions, rape, abortion, suicide and personal loss.

$11.95 Softcover 186p ISBN 0-929385-65-9

✦ JOSÉ ARGÜELLES

THE TRANSFORMATIVE VISION

Reprint of his 1975 tour de force, which prophesied the Harmonic Convergence as the "climax of matter," the collapse of materialism. Follows the evolution of the human soul in modern times by reviewing its expressions through the arts and philosophers.

$14.95 Softcover 364p ISBN 0-9631750-0-9

✦ JERRY MULVIN

OUT-OF-BODY EXPLORATION

Techniques for traveling in the Soul Body to achieve absolute freedom and experience truth for oneself. Discover reincarnation, karma and your personal spiritual path.

$8.95 Softcover 87p ISBN 0-941464-01-6

✦ CHARLES H. HAPGOOD

VOICES OF SPIRIT

The author discusses 15 years of work with Elwood Babbit, the famed channel. Will fascinate both the curious sceptic and the believer. Includes complete transcripts.

$13.00 Softcover 350p ISBN 1-881343-00-6

BOOK MARKET

A reader's guide to the extraordinary books we publish, print and market for your enLightenment.

✦ 4 BOOKS by MSI

ASCENSION!

An Analysis of the Art of Ascension as Taught by the Ishayas

Clearly explains the teachings of the Ishayas, an ancient order of monks entrusted by the Apostle John to preserve the original teachings of Christ until the third millennium. An invitation to awaken the innermost reality of your wonderful, exalted soul!

$11.95 Softcover 194p ISBN 0-931783-51-8

FIRST THUNDER

An Adventure of Discovery

This riveting tale of discovery and adventure contains a secret teaching that is changing our world. Tells how to rise above stress and self-defeating beliefs and experience life Here and Now! A vital book that can lead you away from the past into a bright new future of joy and wonder.

$12.95 Softcover 282p ISBN 0-931783-07-0

SECOND THUNDER

Seeking the Black Ishayas

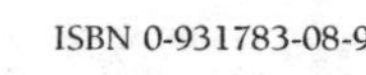

A fascinating visionary story of Universal Consciousness, expressing the highest aspirations of our souls. Continues the adventures of discovery of *First Thunder*. Beautifully illustrated with pen and ink drawings.

$17.95 Softcover 390p ISBN 0-931783-08-9

✦ 2 BOOKS by BARBARA MARCINIAK

BRINGERS OF THE DAWN

Teachings from the Pleiadians

Imparts to us the wisdom of the Pleiadians, a group of enlightened beings who have come to Earth to help us discover how to reach a new stage of evolution. Startling, intense, intelligent and controversial, these teachings offer essential reading.

$12.95 Softcover 288p ISBN 0-939680-98-X

EARTH

Pleiadian Keys to the Living Library

More teachings from the Pleiadians. Explore the corridors of time, sacred sites and ancient civilizations. Once again the work is challenging, controversial, humorous and brilliant, awakening our deepest rememberings.

$12.95 Softcover 288p ISBN 1-879181-21-5

ENLIGHTENMENT

The Yoga Sutras of Patañjali

A New Translation and Commentary

The Yoga Sutras of Patañjali are verses that show the way to Divine Union or Enlightenment. This new translation shows how other translations have been flawed to make it seem difficult when it is actually simple. Makes Yoga effortless and natural.

$15.95 Softcover 338p ISBN 0-931783-17-8

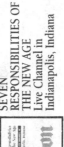

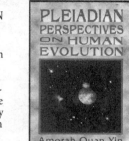

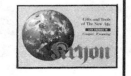

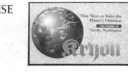

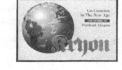

BOOKSTORE DISCOUNTS HONORED

☐ CHECK ☐ MONEY ORDER

CREDIT CARD: ☐ MC ☐ VISA

____

Exp. date: ______________

Signature: ______________

(U.S. FUNDS ONLY) PAYABLE TO:

LIGHT TECHNOLOGY PUBLISHING

P.O. BOX 1526 • SEDONA • AZ 86339

(520) 282-6523 FAX: (520) 282-4130

1-800-450-0985

Fax 1-800-393-7017

NAME/COMPANY ______________

ADDRESS ______________

CITY/STATE/ZIP ______________

PHONE ______________ CONTACT ______________

SUBTOTAL: $ ______________

SALES TAX: $ ______________
(7.5% – AZ residents only)

SHIPPING/HANDLING: $ ______________
($3 Min.; 10% of orders over $30)

CANADA S/H: $ ______________
(20% of order)

TOTAL AMOUNT ENCLOSED: $ ______________

All prices in US$. Higher in Canada and Europe.

CANADA: CHEREV CANADA, INC. 1(800) 263-2408 FAX (519) 986-3103 • ENGLAND/EUROPE: WINDRUSH PRESS LTD. 0608 652012/652025 FAX 0608 652125

AUSTRALIA: GEMCRAFT BOOKS (03)888-0111 FAX (03)888-0044 • NEW ZEALAND: PEACEFUL LIVING PUB. (07)571-8105 FAX (07)571-8513

BOOKSTORE DISCOUNTS HONORED

☐ CHECK ☐ MONEY ORDER

CREDIT CARD: ☐ MC ☐ VISA

____

Exp. date: ______________

Signature: ______________

(U.S. FUNDS ONLY) PAYABLE TO:

LIGHT TECHNOLOGY PUBLISHING

P.O. BOX 1526 • SEDONA • AZ 86339

(520) 282-6523 FAX: (520) 282-4130

1-800-450-0985

Fax 1-800-393-7017

NAME/COMPANY ______________

ADDRESS ______________

CITY/STATE/ZIP ______________

PHONE ______________ CONTACT ______________

SUBTOTAL: $ ______________

SALES TAX: $ ______________
(7.5% – AZ residents only)

SHIPPING/HANDLING: $ ______________
($3 Min.; 10% of orders over $30)

CANADA S/H: $ ______________
(20% of order)

TOTAL AMOUNT ENCLOSED: $ ______________

All prices in US$. Higher in Canada and Europe.

CANADA: CHEREV CANADA, INC. 1(800) 263-2408 FAX (519) 986-3103 • ENGLAND/EUROPE: WINDRUSH PRESS LTD. 0608 652012/652025 FAX 0608 652125

AUSTRALIA: GEMCRAFT BOOKS (03)888-0111 FAX (03)888-0044 • NEW ZEALAND: PEACEFUL LIVING PUB. (07)571-8105 FAX (07)571-8513

Brings You The Truth about the Secret Government and their ET Allies/Enemies through Arthur Fanning and Robert Shapiro...

SHINING THE LIGHT

THE TRUTH ABOUT ETS ✦ ALIEN BASES ✦ THE SECRET GOVERNMENT

The most significant events in Earth's history are happening . . . in secret.

Despite official denial and tactics of derision, active minds are demanding the truth. This truth is stranger than all the fictions about ETs, alien bases and the Secret Government.

The revelations are shocking and enlightening. A crashed UFO leads to information on:

✦ The Secret Government's time-travel spacecraft ✦ Renegade ETs mining the moon
✦ The U.S. peace-avatar president ✦ Underground alien bases and populations
✦ The prime directive — now ✦ Ancient Pleiadian warships ✦ Many more startling facts!

ISBN 0-929385-66-7 $14.95

SHINING THE LIGHT II

THE BATTLE CONTINUES ✦ COSMIC PHOTOGRAPHS ✦ THE PHOTON BELT ✦

The Continuing Story of ETs & The SECRET GOVERNMENT

Current status of the Secret Government and those who want to usurp its power. **Actual cosmic photographs** of hidden craft, beings and events. **Photon Belt "Doctor"** comes to save Mother Earth. **The truth becomes even stranger as the story of alien involvement in Earth history continues. More information on:**

✦ Contactee children ✦ Amazing cosmic holograms in the sky ✦ The face of God photographed
✦ Crop circles ✦ The Photon Belt ✦ A new vortex in the Sedona area ✦ Light meditations
✦ & More!

ISBN 0-929385-70-5 $14.95

SHINING THE LIGHT III

HUMANITY GETS A SECOND CHANCE

The focus is on humanity as we learn to render the Secret Government powerless by being the light that we are. **Earth becomes a member of the Council of Planets** and the universe time-shifts to preserve the Explorer Race.

✦ Ninth-Dimensional Mars ✦ ET Helicopter Pilots ✦ Material Masters ✦ Sonic Mapping
✦ Exploding Planets ✦ Time Collapses ✦ the Null Zone ✦ Much more!

ISBN 0-929385-71-3 $14.95

SHINING THE LIGHT IV

HUMANITY'S GREATEST CHALLENGE

Includes information on Hale-Bopp, SSG, all updates since Volume III and material on the **uncreating of Hitler** in 1993.

✦ Negative Sirians Coming to the Third Dimension ✦ The Express Bus to Creatorship ✦ The Poison HAARP Project ✦ Luciferian Traits and Critical Mass ✦ ETs in Brazil ✦ Comet Brings Lightbeing-Filled Vehicle Bigger than Earth ✦ Sinister Secret Government under Control of Beings from the Alternate Negative Future

ISBN 0-929385-93-4 $14.95